Executive Editor, Geosciences: Christian Botting
Director, Courseware Portfolio Management: Beth Wilbur
Courseware Managing Producer: Mike Early
Content Producer: Becca Groves, SPi-Global
Courseware Portfolio Management Specialist:
Jonathan Cheney
Courseware Director, Content Development:
Ginnie Simione Jutson
Courseware Editorial Assistant: Sherry Wang
Senior Rich Media Content Producer: Tim Hainley
Rich Media Content Producer: Ziki Dekel

Full-Service Vendor: Patty Donovan, SPi-Global
Design Manager: Mark Ong, Side By Side Studios
Interior & Cover Designer: Tamara Newnam
Rights & Permissions Management: Matt Perry, Cenveo Publishing
Services
Photo Research: Cenveo Publishing Services
Senior Procurement Specialist: Stacey Weinberger
Product Marketing Manager: Alysun Burns
Executive Field Marketing Manager: Mary Salzman
Marketing Assistant: Kelsey Nieman

Cover Photo Credit: Grant Ordelheide
Part Opening Photo Credits: PART 1 Rob Cicchetti/123RF, PART 2 Alberto Loyo/123RF, PART 3 Solarseven/123RF, PART 4 Vadim Sadovski/123RF, PART 5 Edward Lam/123RF

Library of Congress Cataloging-in-Publication Data

Names: Tarbuck, Edward J, author. | Lutgens, Frederick K, author. |
Tasa, Dennis, illustrator.
Title: Applications and investigations in earth science / Edward J. Tarbuck,
Frederick K Lutgens ; illustrated by Dennis Tasa.
Description: Ninth edition | Hoboken, NJ : Pearson Education, [2019]
Identifiers: LCCN 2017053149| ISBN 9780134746241 | ISBN 0134746244
Subjects: LCSH: Earth sciences—Laboratory manuals. | Earth sciences—Problems, exercises, etc.
Classification: LCC QE44 .T37 2019 | DDC 550.78—dc23 LC record
available at https://lccn.loc.gov/2017053149

About Our Sustainability Initiatives

Pearson recognizes the environmental challenges facing this planet, and acknowledges our responsibility in making a difference. This book has been carefully crafted to minimize environmental impact. The binding, cover, and paper come from facilities that minimize waste, energy consumption, and the use of harmful chemicals. Pearson closes the loop by recycling every out-of-date text returned to our warehouse.

Along with developing and exploring digital solutions to our market's needs, Pearson has a strong commitment to achieving carbon neutrality. As of 2009, Pearson became the first carbon- and climate-neutral publishing company. Since then, Pearson remains strongly committed to measuring, reducing, and offsetting our carbon footprint.

The future holds great promise for reducing our impact on Earth's environment, and Pearson is proud to be leading the way. We strive to publish the best books with the most up-to-date and accurate content, and to do so in ways that minimize our impact on Earth.

To learn more about our initiatives, please visit: https://www.pearson.com/corporate/sustainability.html

Contents

Preface

As with previous editions, *Applications and Investigations in Earth Science* is intended to be a supplemental tool for achieving an understanding of the basic principles of geology, oceanography, meteorology, and astronomy. While enrolled in what may be their first, and possibly only, Earth Science course, students will benefit from putting the material *presented* in the classroom to *work* in the laboratory. Learning becomes more significant when accomplished by discovery.

Ninth Edition Features

One of the goals for this ninth edition was to minimize the need for lengthy presentations at the start of each lab session. This allows more time for student involvement in lab activities, as well as more time for instructors to interact with students individually or in small groups. Here is a list of important features associated with this new edition:

- **Revised organization.** Previous users of this lab manual will see that the order of the exercises in Part I Geology has been changed. It now reflects the topic order associated with *Earth Science 15th edition*, the textbook most often used in conjunction with this lab manual. Of course, because exercises are largely self contained, they may be assigned in a different order.

- **New lab exercise on *Volcanism and Volcanic Hazards*.** At the urging of several reviewers and past users of this lab manual, a new exercise has been added. *Exercise 5: Volcanism and Volcanic Hazards* examines the nature of volcanic eruptions and the formation and characteristics of different types of volcanoes. The new exercise also explores some of the geologic hazards associated with volcanoes.

- **Pre Lab Videos.** Each lab is accompanied by a Pre Lab Video, prepared and narrated by Professor Callan Bentley. Each lesson examines and explains the key ideas explored in the exercise, thereby largely eliminating the need for a pre-lab lecture by the instructor.

- **A design and layout that promotes user flexibility.** Each exercise is divided into sections that include background material and one or more related activities for students to complete. This layout makes it easier for instructors to customize each exercise to fit the allotted lab period and their individual teaching preferences. The design also effectively ties figures and tables to the associated activities.

- **Mastering™ Geology.** Mastering Geology is an online homework, tutorial, and assessment program designed to work with this lab manual to engage students and improve results. Interactive, self-paced activities provide individualized coaching to help students stay on track. With a wide range of activities available, students can actively learn, understand, and retain even the most difficult Earth Science concepts. Materials in Mastering Geology include Pre-Lab Videos, Geoscience Animations, Mobile Field Trips, "Project Condor" Quadcopter videos, GIS-inspired MapMaster 2.0 interactive maps, *In the News* articles, Key Term Study Tools, and an optional Pearson eText.

- **SmartFigures—art that teaches.** Inside most exercises are *SmartFigures*. Students may use a mobile device to scan the Quick Response (QR) code next to a SmartFigure to view enhanced, dynamic art. Each 2- to 4-minute feature is a mini-lesson that examines the concepts illustrated by the figure. Several new SmartFigures have been added to this new edition. In addition to the Tutorials that appeared in the previous edition, there are now several SmartFigure Animations and SmartFigure Videos. SmartFigures is truly *art that teaches*.

- **Exercises that are largely self-contained.** Significant effort has been put into making the exercises less reliant on traditional text material and/or direct faculty instruction. In some cases, additional background material is provided within the exercise. Questions that rely heavily on outside material have been modified or replaced. We are confident that this approach makes exercises more useful and meaningful for students as well as instructors.

- **Inquiry-based lab experiences.** Whenever possible, the exercises provide hands-on learning. We also endeavor to engage students in gathering and analyzing scientific data to improve their critical reasoning skills.

- **Content and illustrations revised to improve clarity.** Our many years in the classroom have made us keenly aware of the frustration that students and instructors face when instructions, illustrations, and questions are unclear. Likewise, we recognize that instructors are genuinely interested in making learning experiences meaningful for their students. With those ideas in mind, the exercises were reviewed not only by Earth Science faculty but also by a support team with educational backgrounds other than Earth Science—a reflection, essentially, of the majority of the students who utilize this manual.

We sincerely hope that this ninth edition enhances the planning and implementation of instructional goals of all faculty—those who have used our materials for many years as well as those who bring fresh ideas and perspectives to the classrooms of the twenty-first century.

Acknowledgments

Writing a laboratory manual requires the talents and cooperation of many people. It is truly a team effort, and we authors are fortunate to be part of an extraordinary team at Pearson Education. In addition to being great people to work with, all are committed to producing the best textbooks possible. Special thanks to our Executive Editor at Pearson Education, Christian Botting. We appreciate his enthusiasm, hard work, and quest for excellence. We also want to acknowledge our conscientious Content Producer, Becca Groves for the skills she exhibited in keeping this project on track.

As always, we want to acknowledge the production team, led by Patty Donovan at SPi-Global, who turned our manuscript into a finished product. The team included copy editor Kitty Wilson, compositor SPi Global, and proofreader Linda Duarte. These talented people are true professionals, with whom we are very fortunate to be associated.

We owe special thanks to a number of other people who were critical to this project:

- Working with Dennis Tasa, who creates the manual's outstanding illustrations, is always enjoyable and rewarding. We value his amazing artistic talent, imagination, and extraordinary patience with extensive revisions. Dennis and his excellent staff have definitely strengthened an already outstanding art program.

- We value the support of Teresa Tarbuck of Vincennes University, whose editorial assistance greatly enhanced this ninth edition. She helped make the exercises more current, readable, and engaging.

- Callan Bentley has been an important contributor to this edition of *Applications and Investigations*. Callan is a professor of geology at Northern Virginia Community College in Annandale, where he has been honored many times as an outstanding teacher. He is a frequent contributor to *Earth* magazine and author of the popular geology blog *Mountain Beltway*. Callan was responsible for preparing the Pre Lab Videos as well as many of SmartFigures that appear throughout this manual.

Appreciation also goes to our colleagues who prepared in-depth reviews of prior editions and the current edition. Their critical comments and thoughtful input helped guide and strengthen our efforts. Special thanks to:

Glenn Blaylock, Laredo Community College
Nahid Brown, Northeastern Illinois University
Brett Burkett, Collin County Community College
James Cunliffe, Nashville State Community College
Dora Devery, Alvin Community College
Carol Edson, Las Positas College
Ethan Goddard, St. Petersburg College
Roberta Hicks, Memorial University of Newfoundland
Jane MacGibbon, University of North Florida
Remo Masiello, Tidewater Community College
Mark Peebles, St. Petersburg College
Colleen Petosa, Tarrant County College
Melissa Ranhofer, Furman University
Jeffery Richardson, Columbus State Community College
James Sachinelli, Atlantic Cape Community College
Brian Scheidt, Mineral Area College
Jana Svec, Moraine Valley Community College
Krista Syrup, Moraine Valley Community College

Last, but certainly not least, we gratefully acknowledge the support and encouragement of our wives, Joanne Bannon and Nancy Lutgens. Preparation of *Applications and Investigations in Earth Science*, ninth edition, would have been far more difficult without their assistance, patience, and understanding.

Ed Tarbuck
Fred Lutgens

The Study of Minerals

LEARNING OBJECTIVES

Each statement represents an important learning objective that relates to one or more sections of this lab. After you complete this exercise you should be able to:

- List the main characteristics that an Earth material must possess to be considered a mineral.
- Describe the physical properties commonly used to identify minerals.
- Identify minerals using a mineral identification key.
- Identify the most common rock-forming minerals and list the uses of several economic minerals.

MATERIALS

set of mineral specimens
streak plate
dilute hydrochloric acid
contact goniometer

hand lens
magnet
glass plate

PRE-LAB VIDEO https://goo.gl/8Wkho7

 Prepare for lab! Prior to attending your laboratory session, view the pre-lab video. Each video provides valuable background that will contribute to your understanding and success in lab.

INTRODUCTION

For a student learning about our planet, identifying minerals using relatively simple techniques is an important skill. Knowledge of common minerals and their properties is basic to an understanding of rocks. This exercise introduces the physical and chemical properties of minerals and how these properties are used to identify common minerals.

Geology

PART 1

1.1 Minerals: Building Blocks of Rock

■ **List the main characteristics that an Earth material must possess to be considered a mineral.**

Earth's continental and oceanic crust is home to a wide variety of useful and essential rocks and minerals. Many of them have economic value. In addition, all the processes that geologists study are in some way dependent on the properties of these basic Earth materials. Events such as volcanic eruptions, mountain building, weathering and erosion, and earthquakes involve rocks and minerals. Consequently, a basic knowledge of Earth materials is essential to understanding all geologic phenomena.

What Is a Mineral?

We begin our discussion of Earth materials with an overview of minerals—the building blocks of rocks. Geologists define **mineral** as *any naturally occurring inorganic solid that possesses an orderly crystalline structure and a definite chemical composition that allows for some variation.* Thus, Earth materials that are classified as minerals exhibit the following characteristics:

- **Naturally occurring** Minerals form by natural geologic processes. Synthetic materials—that is, those produced in a laboratory or by human intervention—are not considered minerals.
- **Generally inorganic** Inorganic crystalline solids, such as ordinary table salt (halite), that are found naturally in the ground are considered minerals. (Organic compounds, on the other hand, are generally not. Sugar, a crystalline solid that comes from sugarcane or sugar beets, is a common example of such an organic compound.)
- **Solid substance** Only solid crystalline substances are considered minerals. Ice (frozen water) fits this criterion and is considered a mineral, whereas liquid water and water vapor do not.
- **Orderly crystalline structure** Minerals are crystalline substances, which means their atoms (ions) are arranged in an orderly, repetitive manner. This orderly packing of atoms is reflected in regularly shaped objects called *crystals*. Some naturally occurring solids, such as volcanic glass (obsidian), lack a repetitive atomic structure and are not considered minerals.
- **Definite chemical composition that allows for some variation** Minerals are chemical compounds having compositions that can be expressed by a chemical formula. For example, the common mineral quartz has the formula SiO_2, which indicates that quartz consists of silicon (Si) and oxygen (O) atoms, in a ratio of one to two. This proportion of silicon to oxygen is true for any sample of pure quartz, regardless of its origin, size, or when it formed. However, the compositions of some minerals vary *within specific, well-defined limits.* This occurs because certain elements can substitute for others of similar size without changing the mineral's internal structure.

What Is a Rock?

Most minerals occur as components of rocks. Simply, a **rock** *is any solid mass of mineral, or mineral-like matter (such as volcanic glass), that occurs naturally as part of our planet.* Most rocks, like the sample of granite shown in **Figure 1.1**, occur as aggregates of several different minerals. The term *aggregate* implies that the minerals are joined in such a way that their individual properties are retained. Note that the different minerals that make up granite can be easily identified. However, some rocks are composed almost entirely of one mineral. A common example is the sedimentary rock *limestone*, which occurs as an impure mass of the mineral calcite.

Physical Properties of Minerals

Minerals have definite crystalline structures and chemical compositions that give them unique sets of physical and chemical properties shared by all specimens of that mineral, regardless of when or where they form. For example, if you compare two samples of the mineral quartz, they will be equally hard and equally dense, and they will break in a similar

Granite
(Rock)

Quartz
(Mineral)

Hornblende
(Mineral)

Feldspar
(Mineral)

VIDEO
https://goo.gl/dLJ6f7

manner. However, the physical properties of individual samples may vary within specific limits due to ionic substitutions, inclusions of foreign elements (impurities), and defects in the crystalline structure.

Some mineral properties, called **diagnostic properties**, are particularly useful in identifying an unknown mineral. The mineral halite, for example, has a salty taste. Because so few minerals share this property, a salty taste is considered a diagnostic property of halite. Other properties of certain minerals vary among different specimens of the same mineral. These properties are referred to as **ambiguous properties**.

Next, we examine the most common physical properties used to identify minerals, which include luster, color, streak, crystal shape (or habit), hardness, cleavage, fracture, and specific gravity. We will then look at some special properties that are useful in the identification of a few specific minerals.

ACTIVITY 1.1

Minerals: Building Blocks of Rock

1. List the five characteristics an Earth material must have in order to be considered a mineral.

a. _____   b. _____

c. _____   d. _____

e. _____

2. Use the geologic definition of a mineral to determine which of the items listed in **Figure 1.2** are minerals and which are not. For each item listed, check either Yes or No and explain your choice.

Mineral	Yes	No	Explanations
Rain water			
Quartz			
Coal			
Silver			
Wood			
Synthetic diamonds			
Halite			

◀ **Figure 1.2** Which of these materials are minerals?

continued

Activity 1.1 continued

3. **Figure 1.3** provides images of some rocks and minerals. Which of these appear to be rocks, and which are most likely minerals? (Identify the samples by letter.)

Rocks: _____   Minerals: _____

A.

B.

C.

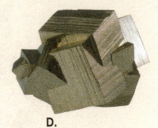

D.

▲ **Figure 1.3** Rock or mineral?

4. The five samples shown in **Figure 1.4** are all specimens of the mineral fluorite. Is color a diagnostic or ambiguous property of fluorite? Explain.

_____

_____

▶ **SmartFigure 1.4** Color variations exhibited by the mineral fluorite. (Photo by Dennis Tasa)

VIDEO
https://goo.gl/qbLhc7

1.2   Luster

■ **Describe the physical properties commonly used to identify minerals.**

The appearance of light reflected from the surface of a mineral is known as **luster**. Minerals that have the appearance of metals, regardless of color, are said to have a **metallic luster**. Some metallic minerals, such as native copper and galena, develop a dull coating or tarnish when exposed to the atmosphere. Because they are not as shiny as samples with freshly broken surfaces, these samples exhibit a *submetallic luster*.

Most minerals have a **nonmetallic luster** and are described using various adjectives, such as *vitreous* (*glassy*), *dull* or *earthy* (a dull appearance like soil), or *pearly* (such as a pearl or the inside of a clamshell). Still others exhibit lusters that are *silky* (like silk or satin cloth) or *greasy* (as though coated in oil).

ACTIVITY 1.2

Luster

1. Examine the luster of each mineral in **Figure 1.5**. Place the letter A, B, C, D, or E in the space provided that corresponds to the luster exhibited. Letters may be used more than once. **A.** Metallic luster, **B.** Nonmetallic luster—glassy, **C.** Nonmetallic luster—dull, **D.** Nonmetallic luster—silky, **E.** Nonmetallic luster—greasy.

▲ **Figure 1.5** Various types of mineral luster.

2. Examine the mineral specimens provided by your instructor and separate them into two groups—metallic and nonmetallic.

Metallic: _____

Nonmetallic:_____

1.3 ## Color and Streak

■ **Describe the physical properties commonly used to identify minerals.**

Color is generally the most conspicuous mineral characteristic. However, because color is often variable, it usually is not a diagnostic property of most minerals. There are exceptions. For example, the mineral sulfur is usually bright yellow.

The *color* of a mineral in powdered form, called **streak**, is often useful in identification. A mineral's streak is obtained by rubbing it across a *streak plate* (a piece of unglazed porcelain) and observing the color of the mark it leaves. Although the color of a particular mineral may vary from sample to sample, the streak is usually consistent.

Streak can also help distinguish between minerals with metallic luster and those with nonmetallic luster. Minerals with a metallic luster generally have a dense, dark streak (**Figure 1.6**), whereas minerals with a nonmetallic luster typically have a light-colored streak.

Not all minerals produce a streak when rubbed across a streak plate. For example, the mineral quartz is harder than a streak plate and, therefore, produces no streak using this method.

▲ **SmartFigure 1.6** Using streak to help identify a mineral.

VIDEO
https://goo.gl/GzAZMk

ACTIVITY 1.3
Color and Streak

1. Based on the samples of quartz in **Figure 1.7**, explain why color is not a diagnostic property of this mineral.

   _____

   _____

   _____

▶ **Figure 1.7** Color variations
exhibited by the mineral quartz.

2. **Figure 1.8** shows two specimens of the mineral hematite and their corresponding streaks. For both samples, describe the color of the specimen and the streak.

A.

B.

◀ **Figure 1.8** Hematite, an ore of
iron, is found in both nonmetallic and
metallic forms.

	COLOR OF SPECIMEN	STREAK
Specimen A:	_____	_____
Specimen B:	_____	_____

3. Select three of the mineral specimens provided by your instructor. Do they exhibit a streak? If so, is the streak the same color as the mineral specimen?

	COLOR OF SPECIMEN	STREAK
Specimen A:	_____	_____
Specimen B:	_____	_____
Specimen C:	_____	_____

4. To some observers, the mineral shown in **Figure 1.9** exhibits a metallic luster, while others describe its luster as nonmetallic. Based on the streak of this sample, how would you describe its luster?

   Luster: _____

Light yellow streak

▶ **Figure 1.9** Using streak to assist
in describing luster.

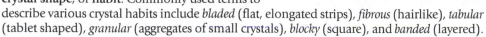

1.4   Crystal Shape, or Habit

■ **Describe the physical properties commonly used to identify minerals.**

Recall that all minerals are crystalline, and when they form in unrestricted environments, they develop **crystals** that exhibit geometric shapes. For example, well-developed quartz crystals are hexagonal, with pyramid-shaped ends, and garnet crystals are 12-sided (**Figure 1.10**). In addition, some crystals tend to grow and form characteristic shapes or patterns called **crystal shape**, or **habit**. Commonly used terms to describe various crystal habits include *bladed* (flat, elongated strips), *fibrous* (hairlike), *tabular* (tablet shaped), *granular* (aggregates of small crystals), *blocky* (square), and *banded* (layered).

Although crystal shape, or habit, is a diagnostic property for some specimens, many of the mineral samples you will encounter consist of crystals that are too tiny to be seen with the unaided eye or are intergrown such that their shapes cannot be determined.

A. Quartz

B. Garnet

▲ **Figure 1.10** Characteristic crystal forms of **A.** quartz and **B.** garnet.

ACTIVITY 1.4
Crystal Shape, or Habit

1. Select one of the following terms to describe the crystal shape, or habit, of each specimen shown in **Figure 1.11**: cubic crystals, hexagonal crystals, fibrous habit, banded habit, blocky habit, bladed habit, tabular habit.

   Specimen A: _____

   Specimen B: _____

   Specimen C: _____

   Specimen D: _____

2. Use a *contact goniometer*, illustrated in **Figure 1.12**, to measure the angle between adjacent faces on the quartz crystals on display in the lab.

   a. Are the angles about the same for each quartz specimen, or do they vary from one sample to another?

   _____

   _____

   b. Write a generalization that describes how the angle between crystal faces relates to the size and/or shape of the sample.

   _____

   _____

   _____

Specimen A

Specimen B

Specimen C

Specimen D

▲ **Figure 1.11** Crystal shapes and habits.

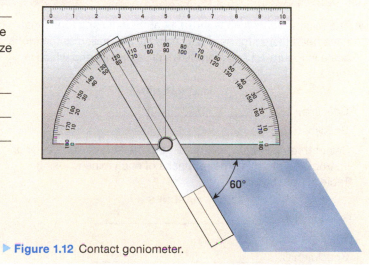

▶ **Figure 1.12** Contact goniometer.

1.5 Hardness

■ Describe the physical properties commonly used to identify minerals.

▼ **SmartFigure 1.13** Hardness scales. **A.** Mohs scale of hardness, with the hardnesses of some common objects. **B.** Relationship between the Mohs relative hardness scale and an absolute hardness scale.

VIDEO
https://goo.gl/n4U6HL

One of the most useful diagnostic properties is **hardness**, a measure of the resistance of a mineral to abrasion or scratching. This property is determined by rubbing a mineral of unknown hardness against one of known hardness or vice versa. A numerical value of hardness can be obtained by using the **Mohs scale** of hardness, which consists of 10 minerals arranged in order from 1 (softest) to 10 (hardest), as shown in **Figure 1.13**. It should be noted that the Mohs scale is a relative ranking; it does not imply that mineral number 2, gypsum, is twice as hard as mineral 1, talc.

In the laboratory, common objects are often used to determine the hardness of a mineral. These objects include a human fingernail, which has a hardness of about 2.5, a copper penny (3.5), and a piece of glass (5.5). The mineral gypsum, which has a hardness of 2, can be easily scratched with a fingernail. On the other hand, the mineral calcite, which has a hardness of 3, will scratch a fingernail but will not scratch glass. Quartz, one of the hardest common minerals, will easily scratch glass. Diamonds, hardest of all, scratch anything, including other diamonds.

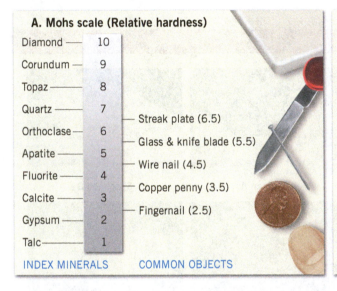

A. Mohs scale (Relative hardness)

INDEX MINERALS		COMMON OBJECTS
Diamond	10	
Corundum	9	
Topaz	8	
Quartz	7	
Orthoclase	6	Streak plate (6.5)
		Glass & knife blade (5.5)
Apatite	5	
		Wire nail (4.5)
Fluorite	4	
		Copper penny (3.5)
Calcite	3	
		Fingernail (2.5)
Gypsum	2	
Talc	1	

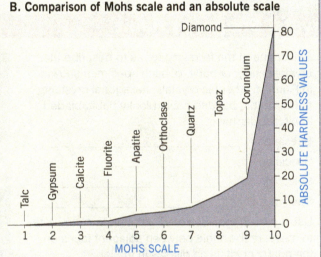

B. Comparison of Mohs scale and an absolute scale

ABSOLUTE HARDNESS VALUES — Diamond 80, Corundum, Topaz, Quartz, Orthoclase, Apatite, Fluorite, Calcite, Gypsum, Talc — MOHS SCALE 1 2 3 4 5 6 7 8 9 10

ACTIVITY 1.5

Hardness

1. The minerals shown in **Figure 1.14** are fluorite and topaz that have been tested for hardness using a wire nail. Use the Mohs scale in Figure 1.13 to identify which is fluorite and which is topaz.

   MINERAL NAME

   Specimen A: _____

   Specimen B: _____

2. Select three mineral specimens from the set provided by your instructor. Determine the hardness of each mineral, using Table 1.1 as a guide.

   HARDNESS

   Specimen A: _____

   Specimen B: _____

   Specimen C: _____

▲ **Figure 1.14** Hardness test.

Table 1.1 Hardness Guide

HARDNESS	DESCRIPTION
Less than 2.5	A mineral that can be scratched by your fingernail (hardness = 2.5)
2.5 to 5.5	A mineral that cannot be scratched by your fingernail (hardness = 2.5) and cannot scratch glass (hardness = 5.5)
Greater than 5.5	A mineral that scratches glass (hardness = 5.5)

1.6 Cleavage and Fracture

■ **Describe the physical properties commonly used to identify minerals.**

In the crystalline structure of many minerals, some chemical bonds are weaker than others. When minerals are stressed, they tend to break (cleave) along these planes of weak bonding, a property called **cleavage**. When broken, minerals that exhibit cleavage have smooth, flat surfaces, called **cleavage planes**, or **cleavage surfaces**.

Cleavage is described by first identifying the number of **directions of cleavage**, which is the number of different sets of cleavage planes that form on the surfaces of a mineral when it cleaves. Each cleavage surface of a mineral that has a different orientation is counted as a *different direction* of cleavage. However, when cleavage planes are parallel, they are counted only *once*, as one direction of cleavage.

Minerals may have one, two, three, four, or more directions of cleavage (**Figure 1.15**). For minerals with two or more directions of cleavage, you may also determine the **angle(s)** at which the directions of cleavage meet. The most common angles of cleavage are 60, 75, 90, and 120 degrees.

When minerals such as muscovite, calcite, halite, and fluorite are broken, they display cleavage surfaces that are easily detected. However, other minerals exhibit cleavage planes that consist of multiple offset surfaces that are not as obvious.

▶ **SmartFigure 1.15** Common cleavage directions of minerals. **A.** Basal cleavage produces flat sheets. **B.** This type of prismatic cleavage produces an elongated form with a rectangular cross section. **C.** This type of prismatic cleavage produces an elongated form with a parallelogram cross section. **D.** Cubic cleavage produces cubes or parts of cubes. **E.** Rhombic cleavage produces rhombohedrons. **F.** Octahedral cleavage produces octahedrons.

VIDEO
https://goo.gl/4w3QPn

A. Cleavage in one direction.
Example: Muscovite

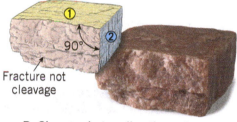

Fracture not cleavage

B. Cleavage in two directions at 90° angles. Example: Feldspar

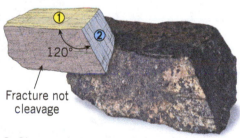

Fracture not cleavage

C. Cleavage in two directions not at 90° angles. Example: Hornblende

D. Cleavage in three directions at 90° angles. Example: Halite

E. Cleavage in three directions not at 90° angles. Example: Calcite

F. Cleavage in four directions.
Example: Fluorite

A reliable way to determine whether a specimen exhibits cleavage is to rotate it in bright light and look for flat surfaces that reflect light.

Do not confuse cleavage with crystal shape. When a mineral exhibits cleavage, it will break into pieces that all have the same geometry. By contrast, the smooth-sided quartz crystals shown in Figure 1.10A illustrate crystal shape rather than cleavage. If broken, quartz crystals fracture into shapes that do not resemble one another or the original crystals.

Minerals that do not exhibit cleavage when broken are said to **fracture** (**Figure 1.16**). Fractures are described using terms such as *irregular*, *splintery*, and *conchoidal* (smooth, curved surfaces resembling broken glass). Some minerals may cleave in one or two directions but fracture in another.

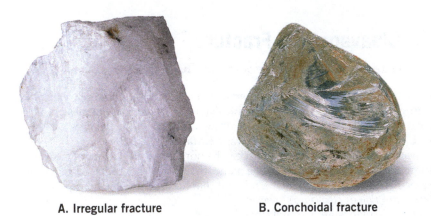

A. Irregular fracture   **B. Conchoidal fracture**

▲ **Figure 1.16** Minerals that do not exhibit cleavage are said to fracture.

ACTIVITY 1.6
Cleavage and Fracture

1. Describe the cleavage of the mineral shown in **Figure 1.17**.

_____

_____

_____

▲ **SmartFigure 1.17**
Identifying cleavage of muscovite.

ANIMATION
https://goo.gl/3ocvZJ

2. Refer to **Figure 1.18**, which shows a mineral that has several smooth, flat cleavage surfaces, to complete the following:

a. How many *cleavage surfaces* are present on the specimen?

_____

b. How many *directions of cleavage* are present on the specimen?

_____

c. Do the cleavage directions meet at 90-degree angles *or* angles other than 90 degrees?

_____

▲ **Figure 1.18** Identifying cleavage of calcite.

3. Select one mineral specimen supplied by your instructor that exhibits more than one direction of cleavage. How many directions of cleavage does it have? What are the angles of its cleavage?

Number of directions of cleavage: _____

Cleavage angles: _____ degrees

1.7   Specific Gravity

■ **Describe the physical properties commonly used to identify minerals.**

You are probably familiar with the term *density*, which is defined as mass per unit volume and is expressed in grams per cubic centimeter (g/cm^3). Mineralogists use a related measure called *specific gravity* to describe the density of minerals. **Specific gravity (SG)** is a number representing the ratio of a mineral's weight to the weight of an equal volume of water. Water has a specific gravity of 1.

Most common rock-forming minerals have a specific gravity between 2 and 3. For example, quartz has a specific gravity of 2.7. By contrast, some metallic minerals such as pyrite, native copper, and magnetite are more than twice as dense as quartz and thus are considered to have high specific gravity. Galena, an ore of lead, is even denser, with a specific gravity of about 7.5.

With a little practice, you can estimate the specific gravity of a mineral by hefting it in your hand. Ask yourself whether the mineral feels about as "heavy" as similar-sized rocks you have handled. If the answer is "yes," the specific gravity of the sample is likely between 2.5 and 3. (*Note:* Exercise 24, "The Metric System, Measurements, and Scientific Inquiry," contains a simple experiment involving determining the specific gravity of a solid.)

ACTIVITY 1.7
Specific Gravity

1. Heft each specimen supplied by your instructor. Using this technique, identify the minerals from this group that exhibit high specific gravity.

_____

2. Of those with a high specific gravity, did most of them have a metallic luster or a nonmetallic luster?

_____

1.8   Other Properties of Minerals

■ **Describe the physical properties commonly used to identify minerals.**

MAGNETISM   Magnetism is characteristic of minerals, such as magnetite, that have a high iron content and are attracted by a magnet. One variety of magnetite, called *lodestone*, is magnetic and will pick up small objects such as pins and paper clips (**Figure 1.19**).

▲ **Figure 1.19** Lodestone, a variety of magnetite, is a weak magnet and will attract iron objects.

TASTE   The mineral halite has a "salty" taste and is used for table salt.

ODOR   A few minerals have distinctive odors. For example, minerals that are compounds of sulfur smell like rotten eggs when rubbed vigorously on a streak plate.

FEEL   The mineral talc often feels "soapy," and the mineral graphite has a "greasy" feel.

STRIATIONS   Striations are closely spaced, fine lines on the crystal faces of some minerals. Certain plagioclase feldspar minerals exhibit striations on one cleavage surface (**Figure 1.20**).

▲ **Figure 1.20** These parallel lines, called *striations*, are a distinguishing characteristic of the plagioclase feldspars. Some other minerals also exhibit this characteristic.

CAUTION Do not taste any minerals or any other materials unless you know it is *absolutely* safe to do so.

CAUTION Hydrochloric acid can discolor, decompose, and disintegrate mineral and rock samples. Use the acid only after you have received specific instructions on its use from your instructor. Never taste minerals that have had acid placed on them.

▲ **SmartFigure 1.21** Calcite reacting to dilute hydrochloric acid. (Photo by Chip Clark/Fundamental Photographs)

VIDEO
https://goo.gl/UNQV3A

REACTION TO DILUTE HYDROCHLORIC ACID   A very small drop of dilute hydrochloric acid, when placed on the surface of certain minerals, will cause them to "fizz" (effervesce) as carbon dioxide is released (**Figure 1.21**). The acid test is used to identify the *carbonate minerals*, especially the mineral calcite ($CaCO_3$), the most common carbonate mineral.

TENACITY   The term **tenacity** describes a mineral's resistance to breaking or deforming. Some minerals, such as fluorite and halite, tend to be *brittle* and shatter into small pieces when struck. Other minerals, such as native copper, are *malleable*, or easily hammered into different shapes. Minerals, including gypsum and talc, that can be cut into thin shavings are described as *sectile*. Still others, notably the micas, are *elastic* and will bend and snap back to their original shape after the stress is released (Figure 1.22).

THE ABILITY TO TRANSMIT LIGHT   Minerals are able to transmit light to different degrees. A mineral is described as **opaque** when no light is transmitted; **translucent** when light, but not an image, is transmitted; and **transparent** when both light and an image are visible through the sample (see Figure 1.22).

▲ **Figure 1.22** Sheets of elastic minerals, like muscovite, can be bent but will snap back when the stress is released. Sheets of muscovite are transparent because they transmit both light and images. (Photo by Dennis Tasa)

ACTIVITY 1.8
Other Properties of Minerals

1. What do we mean when we refer to a mineral's *tenacity*? List three terms that describe tenacity.

   _____

   _____

2. Describe the simple chemical test that is useful in the identification of the mineral calcite.

   _____

   _____

3. Compare and contrast the terms *opaque*, *translucent*, and *transparent*. Select a mineral from the set supplied by your instructor that is an example of each.

   _____

   _____

1.9 Identification of Minerals

■ **Identify minerals using a mineral identification key.**

Now that you are acquainted with the physical properties of minerals, you are ready to identify the minerals supplied by your instructor. To complete this activity, you need the mineral data sheet in Figure 1.23 and the mineral identification key in Figure 1.24.

The mineral identification key divides minerals into three primary categories: (1) those with metallic luster, (2) those with nonmetallic luster that are dark colored, (3) and those with nonmetallic luster that are light colored. Hardness is used as a secondary identifying factor. As you complete the following activity, remember that the objective is to learn the *procedure* for identifying minerals through *observation* and *data collection* rather than simply to name the minerals.

MINERAL DATA SHEET #1

Sample number	Luster	Hardness	Color	Streak	Fracture or Cleavage (number of directions and angles)	Other Properties	Name	Economic Use or Rock-forming
								Refer to the section on "Mineral Groups" to complete this part

▲ **Figure 1.23** Mineral data sheet.

MINERAL DATA SHEET #2

Sample number	Luster	Hardness	Color	Streak	Fracture or Cleavage (number of directions and angles)	Other Properties	Name	Economic Use or Rock-forming

Refer to the section on "Mineral Groups" to complete this part

▲ **Figure 1.23** Mineral data sheet. (*continued*)

Identification of Minerals

1. Identify the specimens supplied by your instructor, using the following steps:

Step 1: Leaving enough space for each mineral, number a piece of paper up to the number of samples you've been assigned and place your specimens on the paper.

Step 2: Select a specimen and determine its physical properties by using the tools provided (glass plate, streak plate, magnet, etc.).

Step 3: List the properties of that specimen on the mineral data sheet (Figure 1.23).

Step 4: Use the mineral identification key (Figure 1.24) as a resource to identify the specimen.

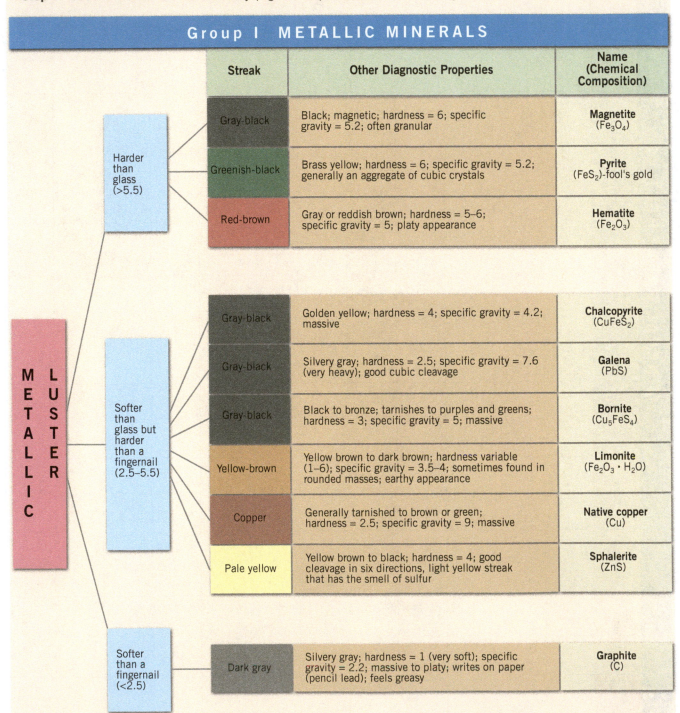

Group I METALLIC MINERALS

	Streak	Other Diagnostic Properties	Name (Chemical Composition)
Harder than glass (>5.5)	Gray-black	Black; magnetic; hardness = 6; specific gravity = 5.2; often granular	**Magnetite** (Fe_3O_4)
	Greenish-black	Brass yellow; hardness = 6; specific gravity = 5.2; generally an aggregate of cubic crystals	**Pyrite** (FeS_2)-fool's gold
	Red-brown	Gray or reddish brown; hardness = 5–6; specific gravity = 5; platy appearance	**Hematite** (Fe_2O_3)
Softer than glass but harder than a fingernail (2.5–5.5)	Gray-black	Golden yellow; hardness = 4; specific gravity = 4.2; massive	**Chalcopyrite** ($CuFeS_2$)
	Gray-black	Silvery gray; hardness = 2.5; specific gravity = 7.6 (very heavy); good cubic cleavage	**Galena** (PbS)
	Gray-black	Black to bronze; tarnishes to purples and greens; hardness = 3; specific gravity = 5; massive	**Bornite** (Cu_5FeS_4)
	Yellow-brown	Yellow brown to dark brown; hardness variable (1–6); specific gravity = 3.5–4; sometimes found in rounded masses; earthy appearance	**Limonite** ($Fe_2O_3 \cdot H_2O$)
	Copper	Generally tarnished to brown or green; hardness = 2.5; specific gravity = 9; massive	**Native copper** (Cu)
	Pale yellow	Yellow brown to black; hardness = 4; good cleavage in six directions, light yellow streak that has the smell of sulfur	**Sphalerite** (ZnS)
Softer than a fingernail (<2.5)	Dark gray	Silvery gray; hardness = 1 (very soft); specific gravity = 2.2; massive to platy; writes on paper (pencil lead); feels greasy	**Graphite** (C)

METALLIC LUSTER

▲ **Figure 1.24** Mineral identification key.

Group II NONMETALLIC MINERALS / DARK COLOR

Cleavage	Other Diagnostic Properties	Name (Chemical Composition)
Cleavage present	Greenish black to black; hardness = 5–6; specific gravity = 3.4; fair cleavage, two directions at nearly 90 degrees	**Augite** (Ca, Mg, Fe, Al silicate)
	Black to greenish black; hardness = 5–6; specific gravity = 3.2; fair cleavage, two directions at nearly 60 degrees and 120 degrees	**Hornblende** (Ca, Na, Mg, Fe, OH, Al silicate)
	White to dark gray; hardness = 6; specific gravity = 2.6; two directions of cleavage at nearly right angles; striations on some faces	**Plagioclase feldspar** $(Na, Ca, AlSi_3O_8)$
Cleavage poor or absent	Red to reddish brown; hardness = 6.5–7.5; conchoidal fracture; glassy luster	**Garnet** (Fe, Mg, Ca, Al silicate)
	Gray to brown; hardness = 9; specific gravity = 4; hexagonal crystals common	**Corundum** (Al_2O_3)
	Dark brown to black; hardness = 7; conchoidal fracture; glassy luster	**Smoky quartz** (SiO_2)
	Olive green; hardness = 6.5–7; small glassy grains	**Olivine** $(Mg, Fe)_2SiO_4$

Harder than glass (>5.5)

Cleavage	Other Diagnostic Properties	Name (Chemical Composition)
Cleavage present	Yellow brown to black; hardness = 4; good cleavage in six directions, light yellow streak that has the smell of sulfur	**Sphalerite** (ZnS)
	Dark brown to black; hardness = 2.5–3, excellent cleavage in one direction; elastic in thin sheets; black mica	**Biotite mica** (K, Mg, Fe, OH, Al silicate)
Cleavage absent	Generally tarnished to brown or green; hardness = 2.5; specific gravity = 9; massive	**Native copper** (Cu)

Softer than glass but harder than a fingernail (2.5–5.5)

Cleavage	Other Diagnostic Properties	Name (Chemical Composition)
Cleavage poor or absent	Reddish brown; hardness = 1–5; specific gravity = 4–5; red streak; earthy appearance	**Hematite** (Fe_2O_3)
	Yellow brown; hardness = 1–3; specific gravity = 3.5; earthy appearance; powders easily; not a true mineral	**Limonite** $(Fe_2O_3 \cdot H_2O)$

Softer than a fingernail (<2.5)

NONMETALLIC DARK COLOR

▲ **Figure 1.24** Mineral identification key. (*continued*)

continued

Activity 1.9 continued

Group III NONMETALLIC MINERALS / LIGHT COLOR

	Cleavage	Other Diagnostic Properties	Name (Chemical Composition)
Harder than glass (>5.5)	Cleavage present	Usually pink or white; hardness = 6; specific gravity = 2.6; two directions of cleavage at nearly right angles; lacks striations	**Potassium feldspar** ($KAlSi_3O_8$)
	Cleavage present	White to dark gray; hardness = 6; specific gravity = 2.6; two directions of cleavage at nearly right angles; striations on some faces	**Plagioclase feldspar** (Na, Ca, $AlSi_3O_8$)
	Cleavage absent	Any color; hardness = 7; specific gravity = 2.65; conchoidal fracture; glassy appearance; varieties: milky (white), rose (pink), smoky (gray), amethyst (violet)	**Quartz** (SiO_2)
Softer than glass but harder than a fingernail (2.5–5.5)	Cleavage present	White, yellowish to colorless; hardness = 3; three directions of cleavage at 75 degrees (rhombohedral); effervesces in HCl; often transparent	**Calcite** ($CaCO_3$)
	Cleavage present	White to colorless; hardness = 2.5; three directions of cleavage at 90 degrees (cubic); salty taste	**Halite** (NaCl)
	Cleavage present	Yellow, purple, green, colorless; hardness = 4; white streak; translucent to transparent; four directions of cleavage	**Fluorite** (CaF_2)
Softer than a fingernail (<2.5)	Cleavage present	Colorless; hardness = 2–2.5; transparent and elastic in thin sheets; excellent cleavage in one direction; light colored mica	**Muscovite mica** (K, OH, Al silicate)
	Cleavage present	White to transparent, hardness = 2; when in sheets; is flexible but not elastic; varieties: selenite (transparent, three directions of cleavage); satin spar (fibrous, silky luster); alabaster (aggregate of small crystals)	**Gypsum** ($CaSO_4 \cdot 2H_2O$)
	Cleavage poor or absent	White, pink, green; hardness = 1–2; soapy feel; pearly luster	**Talc** (Mg silicate)
	Cleavage poor or absent	Yellow; hardness = 1–2.5	**Sulfur** (S)
	Cleavage poor or absent	White; hardness = 2; smooth feel; earthy odor; when moistened, has typical clay texture	**Kaolinite** (Hydrous Al silicate)
	Cleavage poor or absent	Pale to dark reddish brown; hardness = 1–3; dull luster; earthy; often contains spherical particles; not a true mineral	**Bauxite** (Hydrous Al oxide)

NONMETALLIC LIGHT COLOR

▲ **Figure 1.24** Mineral identification key. (*continued*)

Repeat Steps 2 through 4 until you have identified all samples.

2. Read the following section, which examines some common rock-forming minerals and selected economic minerals. This will provide you with the information you need to complete the last column of the mineral data sheet (Figure 1.23).

1.10 Mineral Groups

■ **Identify the most common rock-forming minerals and list the uses of several economic minerals.**

More than 4000 minerals have been named, and several new ones are identified each year. Fortunately for students who are beginning to study minerals, no more than a few dozen are abundant! Collectively, these few make up most of the rocks of Earth's crust and, as such, are referred to as the **rock-forming minerals**.

Although less abundant, many other minerals are used extensively in the manufacture of products; they are called **economic minerals**. However, rock-forming minerals and economic minerals are not mutually exclusive groups. When found in large deposits, some rock-forming minerals are economically significant. For example, the mineral calcite has many uses, including the production of concrete.

Important Rock-Forming Minerals

FELDSPAR GROUP   Feldspar is the most abundant mineral group and is found in many igneous, sedimentary, and metamorphic rocks (**Figure 1.25**). One group of feldspar minerals contains potassium ions in its crystalline structure and is referred to as *potassium feldspar*. The other group, called *plagioclase feldspar*, contains calcium and/or sodium ions. All feldspar minerals have two directions of cleavage that meet at 90-degree angles and are relatively hard (6 on the Mohs scale). The only reliable way to physically distinguish the feldspars is to look for striations that are present on some cleavage surfaces of plagioclase feldspar (Figure 1.25D) but do not appear in potassium feldspar.

QUARTZ   *Quartz* is a major constituent of many igneous, sedimentary, and metamorphic rocks. Quartz is found in a wide variety of colors (caused by impurities), is

Potassium Feldspar

A. Potassium feldspar crystal (orthoclase)

B. Potassium feldspar showing cleavage (orthoclase)

Plagioclase Feldspar

C. Sodium-rich plagioclase feldspar (albite)

D. Plagioclase feldspar showing striations (labradorite)

▲ **Figure 1.25** Feldspar group. **A.** Characteristic crystal form of potassium feldspar. **B.** Like this sample, most pink feldspar belongs to the potassium feldspar subgroup. **C.** Most sodium-rich plagioclase feldspar is light colored and has a porcelain luster. **D.** Calcium-rich plagioclase feldspar tends to be gray, blue-gray, or black in color. Labradorite, the variety shown here, exhibits striations.

▶ **Figure 1.26** Quartz is one of the most common minerals and has many varieties. **A.** Smoky quartz is commonly found in coarse-grained igneous rocks. **B.** Rose quartz owes its color to small amounts of titanium. **C.** Milky quartz often occurs in veins that occasionally contain gold. **D.** Jasper is a variety of quartz composed of microscopically small crystals.

A. Smoky quartz   B. Rose quartz

C. Milky quartz   D. Jasper

quite hard (7 on the Mohs scale), and often exhibits conchoidal fracture when broken (**Figure 1.26**). Pure quartz is clear, and if allowed to grow without interference, it will develop hexagonal crystals with pyramid-shaped ends (see Figure 1.10A).

MICA   *Muscovite* and *biotite* are the two most abundant members of the mica family. Both have excellent cleavage in one direction and are relatively soft (2.5 to 3 on the Mohs scale) (**Figure 1.27**).

CLAY MINERALS   Most clay minerals originate as products of chemical weathering and make up much of the surface material we call soil. Clay minerals also account for nearly half of the volume of sedimentary rocks (**Figure 1.28**).

▶ **Figure 1.27** Two common micas: **A.** muscovite and **B.** biotite.

A. Muscovite   B. Biotite

▶ **Figure 1.28** Kaolinite, a common clay mineral. (Photo by Dennis Tasa)

Kaolinite

OLIVINE   Olivine is an important group of minerals that are major constituents of dark-colored igneous rocks and make up much of Earth's upper mantle. Olivine is black to olive green in color, has a glassy luster, and exhibits conchoidal fracture (**Figure 1.29**).

PYROXENE GROUP   The *pyroxenes* are a group of silicate minerals that are important components of dark-colored igneous rocks. The most common member, *augite*, is a black or greenish, opaque mineral with two directions of cleavage that meet at nearly 90-degree angles (**Figure 1.30A**).

AMPHIBOLE GROUP   *Hornblende* is the most common member of the amphibole group and is usually dark green to black in color (Figure 1.30B). Except for its cleavage angles, which are about 60 degrees and 120 degrees, it is very similar in appearance to augite. Found in igneous rocks, hornblende makes up the dark portion of otherwise light-colored rocks.

CALCITE   *Calcite*, a very abundant mineral, is the primary constituent in the sedimentary rock limestone and the metamorphic rock marble. A relatively soft mineral (3 on the Mohs scale), calcite has three directions of cleavage that meet at 75-degree angles (see Figure 1.18).

Economic Minerals

Many of the minerals selected for this exercise are metallic minerals that are mined to support our modern society. In addition, nonmetallic minerals such as fluorite, halite, and gypsum have economic value. Table 1.2 provides a list of some economic minerals and their industrial and commercial uses.

A. **B.**

▲ **Figure 1.29** Olivine. **A.** Solitary olivine crystal. **B.** Numerous olivine crystals in the igneous rock dunite. (Photos by Dennis Tasa)

A. Augite

B. Hornblende

▲ **Figure 1.30** These dark-colored silicate minerals are common constituents of igneous rocks: **A.** augite and **B.** hornblende. (Photos by Dennis Tasa)

Table 1.2 Economic Minerals

MINERAL	INDUSTRIAL AND COMMERCIAL USES
Calcite	Cement; soil conditioning
Chalcopyrite	Major ore of copper
Corundum	Gemstones; sandpaper
Diamond	Gemstones; drill bits
Fluorite	Used in steel manufacturing, toothpaste
Galena	Major ore of lead
Graphite	Pencil lead; lubricant
Gypsum	Wallboard; plaster
Halite	Table salt; road salt
Hematite	Ore of iron; pigment
Kaolinite	Ceramics; porcelain
Magnetite	Ore of iron
Muscovite	Insulator in electrical applications
Quartz	Primary ingredient in glass
Sphalerite	Major ore of zinc
Sulfur	Sulfa drugs; sulfuric acid
Sylvite	Potassium fertilizers
Talc	Paint; cosmetics

Mineral Groups

1. When feldspar minerals are found in igneous rocks, they tend to occur as elongated, rectangular crystals. By contrast, quartz (most commonly the smoky and milky varieties) usually occurs as irregular or rounded grains that have a glassy appearance. Which of the crystals (A, B, C, or D) in the igneous rocks shown in Figure 1.31 are feldspar crystals, and which are quartz?

   Feldspar: _____

   Quartz: _____

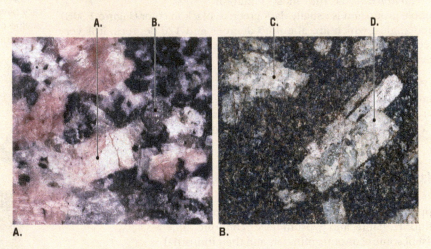

A.

B.

▲ Figure 1.31 Identifying crystals of feldspar and quartz in coarse-grained igneous rocks.

2. Complete the last column in the mineral data sheet (see Figure 1.23) by indicating "rock forming" or by listing the economic use of the samples used in this exercise (see Table 1.2). Which, if any, of the minerals you identified are *both* rock forming and economic?

   _____ _____

The Study of Minerals

Name _____   Course/Section _____

Date _____   Due Date _____

1. Name the physical property (hardness, color, streak, etc.) described by each of the following statements.

   DESCRIPTION   PHYSICAL PROPERTY

   Breaks along smooth planes: _____

   Scratches glass: _____

   Shines like a metal: _____

   Scratching produces a red powder: _____

   Looks like broken glass: _____

2. What term is used to describe the shape of a mineral that has three directions of cleavage that intersect at 90–degree angles?

   _____

3. Describe the cleavage of the minerals listed below. Include the number of directions and degrees of cleavage angles (if appropriate).

   MINERAL   CLEAVAGE

   Muscovite: _____

   Calcite: _____

   Halite: _____

   Feldspar: _____

4. What physical feature most distinguishes biotite mica from muscovite mica?

5. Name a mineral that exhibits the physical properties listed below. (Use the photos in this exercise, if needed.)

   PROPERTY   MINERAL

   One direction of cleavage: _____

   Striations: _____

   Multiple colors: _____

   Cubic cleavage: _____

   Nonmetallic, vitreous luster: _____

   Fracture: _____

   Metallic luster: _____

6. Figure 1.32 illustrates the common crystal form of the mineral fluorite and the characteristic shape of a cleaved sample of fluorite. Identify each specimen (A or B) next to its appropriate description below.

   Crystal form of fluorite: _____

   Cleavage specimen of fluorite: _____

A.

B.

▲ Figure 1.32 Comparing crystal shape and cleavage.

7. Refer to the photo in Figure 1.33 to complete the following.

   a. Describe the crystal form (habit) of this specimen.

      _____

   b. What term is applied to the lines on this sample?

      _____

   c. Based on what you can determine from this photo, use the mineral identification key (see Figure 1.24) to identify this mineral.

      _____

▲ Figure 1.33 Visually identifying mineral properties. (Photo by Dennis Tasa)

8. A photo of *agate*, a variety of quartz composed of microscopically small crystals, is provided in **Figure 1.34**. Based on this image, describe the habit of this sample.

_____

_____

▲ **Figure 1.34** Agate, a microcrystalline variety of quartz. (Photo by Dennis Tasa)

9. If a mineral can be scratched by a penny but not by a human fingernail, what is its hardness on the Mohs scale?

_____

10. What term is used to describe the tenacity of muscovite?

_____

11. Use the mineral identification key (see Figure 1.24) to identify a mineral that is nonmetallic, dark colored, harder than glass, lacking cleavage, and green in color.

_____

12. For each mineral listed below, list at least one diagnostic property that could be used to help distinguish this mineral from the others described in this exercise. (Refer to Figure 1.23 and Figure 1.24, if necessary.)

MINERAL NAME   DIAGNOSTIC PROPERTY

Halite: _____

Galena: _____

Magnetite: _____

Muscovite: _____

Hematite: _____

Fluorite: _____

Talc: _____

Graphite: _____

Calcite: _____

13. List the two most common rock-forming mineral groups.

_____

_____

14. Provide an economic use for each mineral listed below.

Galena: _____

Hematite: _____

Graphite: _____

Sphalerite: _____

Gypsum: _____

Calcite: _____

Rocks and the Rock Cycle

LEARNING OBJECTIVES

Each statement represents an important learning objective that relates to one or more sections of this lab. After you complete this exercise you should be able to:

- **Distinguish among igneous, sedimentary, and metamorphic rocks and the processes by which they are formed through the rock cycle.**
- **Describe the major igneous textures.**
- **List the dominant mineral(s) found in the four groups of igneous rocks: felsic (granitic), mafic (basaltic), intermediate (andesitic), and ultramafic.**
- **Use an identification key to identify igneous rocks based on their texture and mineral composition.**
- **Compare detrital and chemical sedimentary rocks and list the major constituents of sedimentary rocks.**
- **Use an identification key to identify sedimentary rocks based on their texture and mineral composition.**
- **List the environments in which sediments tend to be deposited.**
- **Use an identification key to identify metamorphic rocks based on their texture and mineral composition.**

MATERIALS

metric ruler	igneous rocks	hand lens
glass plate	sedimentary rocks	dilute hydrochloric acid
iron nail	metamorphic rocks	

PRE-LAB VIDEO https://goo.gl/QmmSoz

 Prepare for lab! Prior to attending your laboratory session, view the pre-lab video. Each video provides valuable background that will contribute to your understanding and success in lab.

INTRODUCTION

Why study rocks? You have already learned that some rocks and minerals have great economic value. In addition, all Earth processes depend in some way on the properties of these basic Earth materials. Events such as volcanic eruptions,

Geology

PART 1

mountain building, weathering, erosion, and even earthquakes involve rocks and minerals. Consequently, a basic knowledge of Earth materials is essential to understanding most geologic phenomena.

2.1   Rock Groups and the Rock Cycle

■ Distinguish among igneous, sedimentary, and metamorphic rocks and the processes by which they are formed through the rock cycle.

Most **rocks** are aggregates (mixtures) of mineral crystals or fragments (gravel, sand, and silt) of preexisting rocks. However, there are some important exceptions, including *obsidian*, a rock made of volcanic glass (a non-crystalline substance), and *coal*, a rock made of decayed plant material (an organic substance). Rocks are classified into three groups—igneous, sedimentary, and metamorphic—based on the processes by which they were formed.

Igneous, Sedimentary, and Metamorphic Rocks

Igneous rocks are the solidified products of once-molten material that was created by melting in the upper mantle or crust. Geologists call molten rock **magma** when it is found at depth and **lava** when it erupts at Earth's surface. The distinguishing feature of most igneous rocks is the interlocking arrangement of their mineral crystals that develops as the molten material cools and solidifies. *Intrusive* igneous rocks form beneath Earth's surface from magma, and *extrusive* igneous rocks form at the surface from lava.

Sedimentary rocks form at or near Earth's surface from the products of *weathering*. This material, called **sediment**, is transported by erosional agents (water, wind, or ice) as solid particles or ions in solution to their site of deposition. The process of **lithification** (meaning "to turn to stone") transforms the sediment into solid rock. Sedimentary rocks cover much of Earth's surface and may contain organic matter (oil, gas, and coal) and fossils. The layering that develops when sediment is deposited is the most recognizable feature of sedimentary rocks. These layers, called **strata**, or **beds**, usually accumulate in nearly horizontal sheets that can be as thin as a piece of paper or tens of meters thick (**Figure 2.1**).

Metamorphic rocks are produced from preexisting igneous, sedimentary, or other metamorphic rocks that have been subjected to conditions within Earth that are significantly different from those under which the parent rock originally formed. *Metamorphism*, the process that causes the transformation, generally occurs at depths where both temperatures and pressures are much higher than at Earth's surface.

The Rock Cycle

One very useful device for understanding rock groups and the geologic processes that transform one rock type into another is the **rock cycle** (**Figure 2.2**). We will begin with magma, molten rock that forms deep beneath Earth's surface. Over time, magma cools and solidifies. This process, called *crystallization*, may occur either beneath the surface or, following a volcanic eruption, at the surface. In either situation, the resulting rocks are called *igneous rocks*.

If igneous rocks are exposed at the surface, they will undergo *weathering*, in which the day-in and day-out influences of the atmosphere slowly disintegrate and decompose rocks. The materials that result are often moved downslope by gravity before being picked up and transported by any of a number of erosional agents, such as running water, glaciers, wind, or waves. Eventually these particles and dissolved substances, called *sediment*, are deposited. Although most sediment ultimately comes to rest in the ocean, other sites of deposition include river floodplains, desert basins, swamps, and sand dunes.

Next, the sediments undergo *lithification*, a term meaning "to turn to stone." Sediment is usually lithified into *sedimentary rock* when compacted by the weight of overlying layers or when cemented as percolating groundwater fills the pores with mineral matter.

▼ **Figure 2.1** Sedimentary layers, called strata, or beds in Canyonlands National Park, Utah. (Photo by SeBuKi/Alamy)

ROCK CYCLE

Viewed over long time spans, rocks are constantly forming, changing, and re-forming.

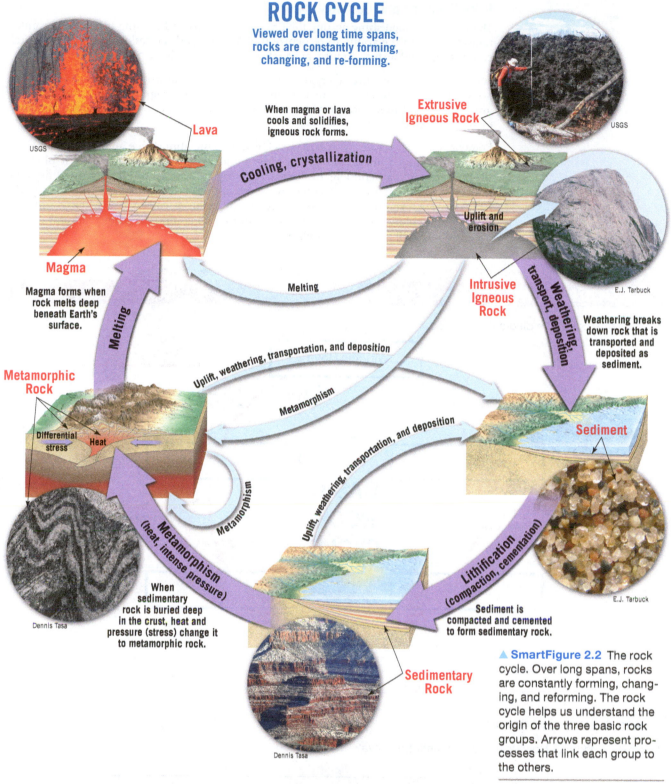

When magma or lava cools and solidifies, igneous rock forms.

Cooling, crystallization

Lava

USGS

Magma

Magma forms when rock melts deep beneath Earth's surface.

Melting

Melting

Metamorphic Rock

Differential stress   Heat

Dennis Tasa

Uplift, weathering, transportation, and deposition

Metamorphism

Metamorphism

Metamorphism (heat, intense pressure)

When sedimentary rock is buried deep in the crust, heat and pressure (stress) change it to metamorphic rock.

Extrusive Igneous Rock

USGS

Uplift and erosion

E.J. Tarbuck

Intrusive Igneous Rock

Weathering, transport, deposition

Weathering breaks down rock that is transported and deposited as sediment.

Sediment

E.J. Tarbuck

Uplift, weathering, transportation, and deposition

Lithification (compaction, cementation)

Sediment is compacted and cemented to form sedimentary rock.

Sedimentary Rock

Dennis Tasa

▲ **SmartFigure 2.2** The rock cycle. Over long spans, rocks are constantly forming, changing, and reforming. The rock cycle helps us understand the origin of the three basic rock groups. Arrows represent processes that link each group to the others.

 VIDEO https://goo.gl/y1Xs9G

If the resulting sedimentary rock is buried deep within Earth and involved in the dynamics of mountain building or intruded by a mass of magma, it will be subjected to great pressures and/or intense heat. The sedimentary rock will react to the changing environment and turn into the third rock type, *metamorphic rock*. When metamorphic rock is subjected to additional pressure changes or to still higher temperatures, it will melt, creating magma, which will eventually crystallize into igneous rock, starting the cycle all over again.

The paths described in the preceding section are not the only ones that are possible. To the contrary, other paths are just as likely to be followed. These alternatives are indicated by the light blue arrows in Figure 2.2.

Rock Textures and Compositions

The task of distinguishing among the three rock groups and naming individual samples relies heavily on the ability to recognize their *textures* and *compositions*. A rock's mineral composition and texture, in turn, are a reflection of the geologic processes that created it.

Texture   A rock's **texture** refers to the size, shape, and/or arrangement of its mineral grains. The shape and arrangement of mineral grains has a significant effect on a rock's appearance and therefore provides clues to whether a rock is igneous, sedimentary, or metamorphic.

Composition   The **composition** of a rock refers to the abundance and type of minerals it contains. Large mineral grains can often be identified by sight or by their physical properties, while a hand lens or microscope may be needed to identify small mineral grains. In addition, a few minerals can be identified based on a particular diagnostic property. For example, rocks composed of the mineral calcite fizz when a drop of dilute hydrochloric acid (HCl) is applied, and rocks composed of the mineral quartz are hard and will easily scratch glass.

ACTIVITY 2.1
Rock Groups and the Rock Cycle

1. Label the rock cycle diagram in **Figure 2.3**.

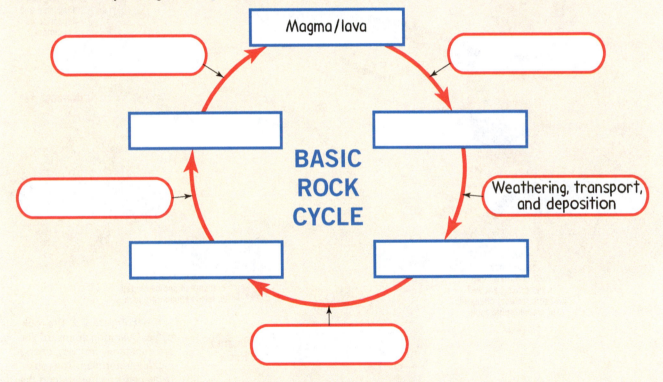

▲ **Figure 2.3** Rock cycle diagram.

2. What is magma? How does magma differ from lava?

_____

3. The terms *intrusive* and *extrusive* are used to describe which one of the three rock groups?

_____

4. In your own words, write a definition of *rock texture*.

_____

5. Name the process by which sediment is transformed into sedimentary rocks.

_____

6. What is the most characteristic feature of sedimentary rocks?

_____

7. What type of rock is generated when rocks that formed near the surface become deeply buried and exposed to intense pressures and/or heat?

_____

8. When determining rock composition, describe the test that can be used to determine whether it contains the mineral calcite.

_____

2.2 Igneous Rock Textures

■ **Describe the major igneous textures.**

Igneous rocks form when molten rock cools and crystallizes. The interlocking assemblage of silicate minerals that develop as the molten material cools gives most igneous rocks their distinctive crystalline appearance. The rate at which magma cools determines the size of the interlocking crystals in igneous rocks. The slower the cooling rate, the larger the mineral crystals.

Molten rock may solidify at Earth's surface, or it may solidify at depth. When molten rock solidifies *at the surface*, the resulting igneous rocks are classified as **extrusive rocks**, or **volcanic rocks** (after Vulcan, the Roman fire god). When magma solidifies at depth, it forms igneous rocks known as **intrusive rocks**, or **plutonic rocks** (after Pluto, the god of the underworld in classical mythology). Intrusive igneous rocks remain at depth unless portions of the crust are uplifted and the overlying rocks are stripped away by erosion.

A. Coarse-grained texture   **B. Fine-grained texture**

▲ **Figure 2.4** Coarse-grained and fine-grained igneous textures.

Coarse-Grained (Phaneritic) Texture

When a large mass of magma solidifies at depth, it cools slowly and forms *intrusive igneous rocks* that exhibit **coarse-grained texture** (**Figure 2.4A**). These rocks have intergrown crystals that are roughly equal in size and large enough that the individual minerals within can be identified with the unaided eye. A hand lens or binocular microscope can greatly assist in mineral identification.

Fine-Grained (Aphanitic) Texture

Igneous rocks that form when molten material cools rapidly at the surface, called extrusive igneous rock, or as small masses within the upper crust exhibit **fine-grained texture** (**Figure 2.4B**). Fine-grained igneous rocks are composed of individual crystals that are too small to be identified without strong magnification.

Porphyritic Texture

Porphyritic texture results when molten rock cools in two different environments. The resulting rock consists of larger crystals embedded in a matrix of smaller crystals (**Figure 2.5**). The larger crystals are termed *phenocrysts*, and the smaller, surrounding crystals are called *groundmass*, or *matrix*.

Glassy Texture

During explosive volcanic eruptions, molten rock is ejected into the atmosphere, where it is quenched (cooled to a solid state) very quickly. When the material solidifies before the atoms arrange themselves into an orderly crystalline structure, the rocks exhibit **glassy texture** that may resemble manufactured glass or fibers of spun glass (**Figure 2.6**).

Groundmass

Phenocryst

1 cm

Porphyritic texture

▲ **Figure 2.5** The large crystals in a porphyritic rock are called *phenocrysts*, and the matrix of smaller crystals is called *groundmass*.

Glassy texture

▲ **Figure 2.6** Rocks with a glassy texture are composed of unordered atoms and often resemble manufactured glass.

Vesicular texture

▲ **Figure 2.7** This extrusive igneous rock consists of glassy fibers and contains voids left by gas bubbles that escaped as lava solidified.

Fragmental (pyroclastic) texture

▲ **Figure 2.8** Fragmental texture results from the consolidation of fragments that may include ash, once molten blobs, or angular blocks that were ejected during an explosive volcanic eruption.

Vesicular Texture

Common features of some fine-grained and glassy extrusive igneous rocks are the voids left by gas bubbles that escape as lava solidifies. These somewhat spherical openings are called *vesicles*, and the rocks that contain them have **vesicular texture** (**Figure 2.7**).

Fragmental (Pyroclastic) Texture

Volcanoes sometimes blast fine ash, molten blobs, and/or angular blocks torn from the walls of the vent into the air during eruptions. Igneous rocks composed of these rock fragments have **fragmental (pyroclastic) texture** (**Figure 2.8**).

ACTIVITY 2.2
Igneous Textures

Use the close-up photos of igneous rocks in **Figure 2.9** to answer the following questions.

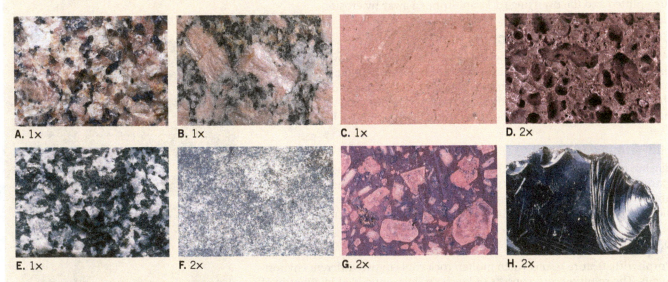

▲ **Figure 2.9** Identifying igneous textures.

1. Which sample(s) (A–H) exhibit(s) porphyritic texture?

_____

2. For the sample(s) you listed in Question 1, what terms are used to describe the larger crystals and the surrounding smaller crystals? Label the larger and smaller crystals on the photo(s).

Larger crystals: _____

Smaller crystals: _____

3. Which sample(s) in Figure 2.9 exhibit(s) coarse-grained texture?

_____

4. Complete the description of the environment in which *coarse-grained igneous rocks* form by choosing the appropriate terms: (1. at great depth *or* on the surface) (2. slowly *or* rapidly) (3. extrusive *or* intrusive).

When molten rock solidifies (1) _____, it cools (2) _____ and produces an (3) _____ igneous rock.

5. Which of the sample(s) in Figure 2.9 exhibit(s) fine-grained texture?

_____

6. Complete the description of the environment in which *fine-grained igneous rocks* form by choosing the appropriate terms: (1. at great depth *or* on/near the surface) (2. slowly *or* rapidly) (3. extrusive *or* intrusive).

When molten rock solidifies (1) _____, it cools (2) _____ and produces an (3) _____ igneous rock.

7. Which of the sample(s) in Figure 2.9 exhibit(s) vesicular texture?

_____

8. Which sample(s) in Figure 2.9 exhibit(s) glassy texture?

_____

9. Although Samples A and C in Figure 2.9 appear different, they have very similar mineral compositions. Briefly explain what accounts for their different appearances.

_____

2.3 Composition of Igneous Rocks

■ **List the dominant mineral(s) found in the four groups of igneous rocks: felsic (granitic), mafic (basaltic), intermediate (andesitic), and ultramafic.**

Despite their significant compositional diversity, igneous rocks (and the magmas from which they form) can be divided into four groups, based on the proportions of light and dark silicate minerals (**Figure 2.10**).

Felsic (or granitic) igneous rocks are composed mainly of the light-colored minerals quartz and potassium feldspar, with lesser amounts of plagioclase feldspar. Recall that feldspar crystals can be identified by their rectangular shapes, flat surfaces, and tendency to be pink, white, or dark gray in color. Quartz grains, on the other hand, are glassy and somewhat rounded and tend to be light gray. Dark-colored minerals account for no more than 15 percent of the minerals in rocks in this group.

Intermediate (or andesitic) rocks are mixtures of both light-colored minerals (mainly plagioclase feldspar) and dark-colored minerals (mainly amphibole). Dark minerals comprise between 15 percent and 45 percent of these rocks.

Mafic (or basaltic) rocks contain abundant dark-colored minerals (mainly pyroxene and olivine) that account for between 45 percent and 85 percent of their composition. Plagioclase feldspar makes up the bulk of the remainder.

Ultramafic rocks are composed almost entirely of the dark silicate minerals pyroxene and olivine and are seldom observed at Earth's surface. However, the ultramafic rock peridotite is a major constituent of Earth's upper mantle.

An important skill used in identifying igneous rocks is being able to classify them based on mineral composition—felsic, intermediate, mafic, or ultramafic. However, it is not possible to identify minerals in fine-grained igneous rocks without sophisticated equipment. Further, the dark silicate minerals are often difficult to differentiate—even in coarse-grained rocks. Consequently, geologists working in the field estimate mineralogy by using the **color index**, a value based on the percentage of dark silicate minerals, shown at the bottom of Figure 2.10. Rocks with a low color index (light in color) are felsic in composition, whereas those with a high color index (dark in color) are mafic or ultramafic.

It is important to note that although the dark silicate minerals are often black in color, the mineral pyroxene can be dark green, and olivine is often olive green in color. Therefore, rocks with a dark greenish color have a very high color index and are usually ultramafic in composition.

IGNEOUS ROCK IDENTIFICATION KEY

		MINERAL COMPOSITION			
		Felsic (Granitic)	**Intermediate** (Andesitic)	**Mafic** (Basaltic)	**Ultramafic**
Dominant Minerals		Quartz Potassium feldspar	Amphibole Plagioclase feldspar	Pyroxene Plagioclase feldspar	Olivine Pyroxene
Accessory Minerals		Plagioclase feldspar Amphibole Muscovite Biotite	Pyroxene Biotite	Amphibole Olivine	Plagioclase feldspar
TEXTURE	**Coarse-grained (Phaneritic)**	**Granite**	**Diorite**	**Gabbro**	**Peridotite**
	Fine-grained (Aphanitic)	**Rhyolite**	**Andesite**	**Basalt**	
	Porphyritic (two distinct grain sizes)	The name "porphyry" follows the rock name if it contains appreciable phenocrysts.			Uncommon
	Glassy	Obsidian (compact glass)			
	Vesicular (contains voids)	Pumice (frothy glass) (floats in water)		Scoria (Vesicular basalt)	
	Pyroclastic (fragmental)	Tuff (Most fragments less than 4 mm) Volcanic Breccia (Most fragments greater than 4 mm)			
Rock Color (based on % of dark minerals)		0% to 25%	25% to 45%	45% to 85%	85% to 100%

▲ **SmartFigure 2.10** Igneous rock identification key.

VIDEO
https://goo.gl/kCdkjt

ACTIVITY 2.3
Using Color Index to Determine Composition

Use the color index shown at the bottom of Figure 2.10 as well as the photos of Samples A–F in **Figure 2.11** to complete the following statements.

1. Sample(s) _____ has/have felsic (granitic) composition.

2. Sample(s) _____ has/have intermediate (andesitic) composition.

3. Sample(s) _____ has/have mafic (basaltic) or ultramafic composition.

▶ **Figure 2.11** Hand samples that illustrate igneous compositions.

A. Rhyolite   B. Gabbro   C. Andesite porphyry

D. Diorite   E. Peridotite   F. Granite

2.4 Identifying Igneous Rocks

■ **Use an identification key to identify igneous rocks based on their texture and mineral composition.**

Igneous rocks are identified by their texture and mineral composition. The texture of an igneous rock is mainly a result of its cooling history, whereas its mineral composition is largely a result of the chemical makeup of the parent magma. Because igneous rocks are classified on the basis of both mineral composition and texture, some rocks with similar mineral constituents but that exhibit different textures are given different names.

ACTIVITY 2.4
Identifying Igneous Rocks

Place each of the igneous rocks supplied by your instructor on a numbered piece of paper. Then complete the igneous rock chart in **Figure 2.12** and identify each specimen. To identify an igneous rock, follow these steps:

Step 1: Identify the texture of the rock.

Step 2: Determine the mineral composition of the rock by:
  a. Identifying the abundance of major minerals
     or
  b. Estimating the mineral composition using the color index

continued

Activity 2.4 continued

Step 3: Use the igneous rock identification key in Figure 2.10 to identify the rock.

IGNEOUS ROCK CHART

Sample Number	Texture	Color (light- intermediate- dark)	Dominant Minerals	Rock Name

▲ **Figure 2.12** Igneous rock chart.

2.5 Classifying Sedimentary Rocks

■ **Compare detrital and chemical sedimentary rocks and list the major constituents of sedimentary rocks.**

Weathering disintegrates and decomposes rock. The resulting products are transported and deposited or precipitated as sediment. Over long periods of time, compaction and cementation turn solid particles of sediment into sedimentary rock, whereas material dissolved in water may be precipitated as a crystalline rock.

Materials that accumulate as sediment have two principal sources. First, sediments may originate as solid particles from weathered rocks. These particles are called *detritus*, and the rocks they form are called *detrital sedimentary rocks*. The second major source of sediment is soluble material produced by chemical weathering.

Detrital sedimentary rocks consist of mineral grains or rock fragments derived from mechanical and chemical weathering that are transported and deposited as solid particles. Clay minerals are the most abundant solid products of chemical weathering. Quartz is abundant in detrital rocks because it is extremely durable. Geologists use particle size to distinguish among detrital sedimentary rocks.

Chemical and biochemical **sedimentary rocks** are products of mineral matter that were dissolved in water and later precipitated. Precipitation may occur as a result of processes such as evaporation or temperature change or as a result of life processes, such as those that result in the formation of shells. Sediment formed by life processes has a *biochemical* origin. Limestone, which is composed of calcite ($CaCO_3$), is the most common mineral in chemical sedimentary rocks and may originate either from chemical or biological processes.

Compositions of Sedimentary Rocks

Sedimentary rocks are classified as detrital or chemical, based on the nature of the sediment of which they are composed. Determining the mineral composition of sedimentary rocks is an important step in their identification. Most sedimentary rocks contain a high percentage of one of the following:

- **Clay minerals** Rocks composed of clay minerals are very fine grained and soft, and they can be scratched with an iron nail.
- **Quartz** Sedimentary rocks composed mainly of quartz usually consist of sand-size particles that are hard and can scratch glass.
- **Calcite** Rocks composed of calcite effervesce when a drop of dilute HCl is applied. They can also be easily scratched with an iron nail.
- **Evaporite minerals** These sedimentary rocks contain salts, usually halite or gypsum, that are deposited when saltwater evaporates. They are crystalline and can be scratched by an iron nail. Gypsum is soft enough to be scratched by a fingernail.
- **Altered plant fragments** Rocks composed of organic material are usually black in color, have a low density, and are easily broken.

Sedimentary Textures

Sedimentary rocks exhibit two basic textures: *clastic* and *nonclastic*. Rocks that display a **clastic texture** consist of discrete particles that are cemented or compacted together. Clastic textures are further divided based on particle size and shape.

Rocks that have a **nonclastic texture** are often crystalline and consist of minerals that form patterns of interlocking crystals. Although crystalline sedimentary rocks can be similar in appearance to some igneous rocks, they tend to consist of minerals—most often calcite, halite, or gypsum—that are easily distinguished from the silicate minerals found in igneous rocks.

ACTIVITY 2.5
Examining Sedimentary Compositions and Textures

Carefully examine the common sedimentary rocks shown in **Figure 2.13**. Use these photos and the preceding discussion to answer the following questions.

1. What characteristic can be used to distinguish Sample A (conglomerate) from Sample B (breccia)?

2. List the name(s) of the chemical sedimentary rocks that clearly exhibit(s) clastic texture.

3. Samples C, O, P, and Q are all primarily composed of quartz. What property of quartz would assist you in identifying these samples?

4. Samples G–L all contain calcite. What property of calcite would assist you in identifying these samples?

5. Compare and contrast Sample I (fossiliferous limestone) and Sample J (coquina).

continued

Activity 2.5 continued

CLASSIFICATION OF SEDIMENTARY ROCKS

Detrital Sedimentary Rocks

A. Conglomerate
(rounded fragments)

B. Breccia
(angular fragments)

C. Sandstone
(usually quartz)

D. Arkose
(feldspar, quartz)

E. Siltstone
(quartz, clay minerals)

F. Shale or Mudstone
(clay minerals)

Chemical and Biochemical Sedimentary Rocks

G. Crystalline limestone
(calcite)

H. Microcrystalline limestone
(calcite)

I. Fossiliferous limestone
(calcite)

J. Coquina
(calcite)

K. Chalk
(calcite)

L. Travertine
(calcite)

M. Rock salt
(halite)

N. Rock gypsum
(gypsum)

O. Chert
(microcrystalline quartz)

P. Flint
(microcrystalline quartz)

Q. Agate
(microcrystalline quartz)

R. Bituminous coal
(carbon)

▲ **Figure 2.13** Hand samples of common sedimentary rocks.

6. Sample G (crystalline limestone) in Figure 2.13 and Sample F (granite) in Figure 2.11 are both crystalline. How do these samples differ in appearance?

_____

7. What characteristic do Sample Q (agate) and Sample L (travertine) share?

_____

8. Which of these terms most accurately describes the texture of Sample N—*clastic* or *nonclastic*?

9. Sample H is a microcrystalline rock composed of calcite, and Sample O is a microcrystalline rock composed of quartz. What test(s) could be used to figure out which is which?

10. What mineral does Sample D contain that gives it a pinkish color?

2.6 Identifying Sedimentary Rocks

- **Use an identification key to identify sedimentary rocks based on their texture and mineral composition.**

The sedimentary rock identification key in **Figure 2.14** divides sedimentary rocks into two groups: detrital and chemical. The primary subdivisions for detrital rocks are based on grain size, whereas composition is the main criterion used to subdivide the chemical sedimentary rocks.

SEDIMENTARY ROCK IDENTIFICATION KEY

Detrital Sedimentary Rocks

Clastic Texture (particle size)	Distinctive Properties		Rock Name
Gravel Coarse (over 2 mm)	Rounded rock or mineral fragments, typically poorly sorted		**Conglomerate**
	Angular rock or mineral fragments, typically poorly sorted		**Breccia**
Sand Medium (1/16 to 2 mm)	Quartz grains, typically rounded, well sorted	**Sandstone**	**Quartz Sandstone**
	At least 25% feldspar, typically poorly sorted, angular fragments		**Arkose**
	Mixture of sand and mud, typically poorly sorted		**Graywacke**
Silt/mud Fine (1/16 to 1/256 mm)	Mostly silt-size quartz and clay, blocky, gritty	**Mudstone**	**Siltstone**
Clay/mud Very fine (less than 1/256 mm)	Mostly clay, splits into layers, may contain fossils		**Shale**
	Mostly clay, crumbles easily		**Claystone**

Chemical and Organic Sedimentary Rocks

Composition	Texture and Distinctive Properties		Rock Name
Calcite, $CaCO_3$ (effervesces in HCl)	Fine to coarse crystalline	**Limestone**	**Crystalline Limestone**
	No visible grains, may exhibit conchoidal fracture		**Micrite**
	Visible shells and shell fragments loosely cemented		**Coquina**
	Various size shells and shell fragments, well cemented		**Fossiliferous Limestone**
	Microscopic shells and clay, soft		**Chalk**
	Faint layering, may contain cavities or pores		**Travertine**
Quartz, SiO_2	Microcrystalline, may exhibit conchoidal fracture, will scratch glass		**Chert (light colored)**
			Flint (dark colored)
			Agate (banded)
Gypsum $CaSO_4 \bullet 2H_2O$	Fine to coarse crystalline, soft		**Rock Gypsum**
Halite, NaCl	Fine to coarse crystalline, tastes salty		**Rock Salt**
Altered plant fragments	Black brittle organic rock, may be layered		**Bituminous Coal**

▲ **Figure 2.14** Sedimentary rock identification key.

ACTIVITY 2.6 ///
Identifying Sedimentary Rocks

Place each of the sedimentary rocks supplied by your instructor on a numbered piece of paper. Then complete the sedimentary rock chart in Figure 2.15 for each rock. Use the sedimentary rock identification key in Figure 2.14 to identify each specimen.

SEDIMENTARY ROCK CHART

Specimen Number	Texture and Grain Size	Composition or Sediment Name	Detrital or Chemical (Biochemical)	Rock Name

▲ **Figure 2.15** Sedimentary rock chart.

2.7 Sedimentary Environments

■ **List the environments in which sediments tend to be deposited.**

Sedimentary rocks are extremely important in the study of Earth history. The texture and mineral composition of sedimentary rocks often suggest something about the location (environment) where the sediment accumulated. For example, rock salt forms in warm, shallow seas, where the loss of water by evaporation exceeds the annual rainfall. In addition, fossils and sedimentary structures provide important clues about a rock's history. Think of each sedimentary rock as representing a "place" on Earth where sediment was deposited and later lithified. Figure 2.16 illustrates and describes a few modern environments where sediment accumulates.

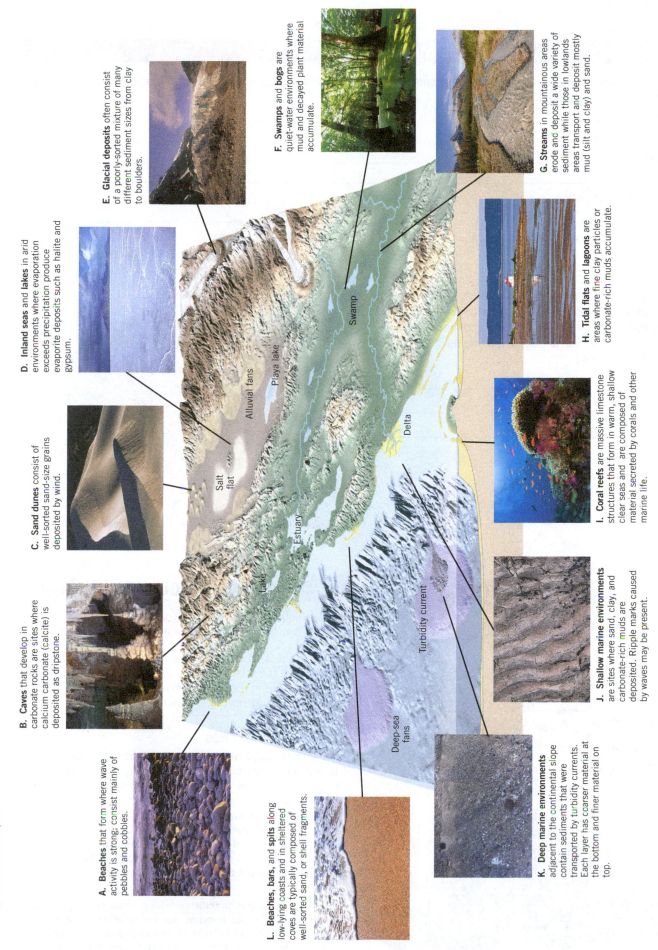

E. Glacial deposits often consist of a poorly-sorted mixture of many different sediment sizes from clay to boulders.

F. Swamps and **bogs** are quiet-water environments where mud and decayed plant material accumulate.

G. Streams in mountainous areas erode and deposit a wide variety of sediment while those in lowlands areas transport and deposit mostly mud (silt and clay) and sand.

D. Inland seas and **lakes** in arid environments where evaporation exceeds precipitation produce evaporite deposits such as halite and gypsum.

H. Tidal flats and **lagoons** are areas where fine clay particles or carbonate-rich muds accumulate.

C. Sand dunes consist of well-sorted sand-size grains deposited by wind.

I. Coral reefs are massive limestone structures that form in warm, shallow clear seas and are composed of material secreted by corals and other marine life.

B. Caves that develop in carbonate rocks are sites where calcium carbonate (calcite) is deposited as dripstone.

J. Shallow marine environments are sites where sand, clay, and carbonate-rich muds are deposited. Ripple marks caused by waves may be present.

A. Beaches that form where wave activity is strong; consist mainly of pebbles and cobbles.

K. Deep marine environments adjacent to the continental slope contain sediments that were transported by turbidity currents. Each layer has coarser material at the bottom and finer material on top.

L. Beaches, bars, and **spits** along low-lying coasts and in sheltered coves are typically composed of well-sorted sand, or shell fragments.

Alluvial fans

Playa lake

Swamp

Salt flat

Delta

Lake

Estuary

Turbidity current

Deep-sea fans

▲ **Figure 2.16 Sedimentary environments.** (Photos A and C by Michael Collier; photo B by Dennis Tasa; photo D by USGS; photo F by Mason Jar/Shutterstock; photo G by Alan Majchrowicz/AGE Fotostock; photo H by Barrett & MacKay/Glow Images; photo I by Radius/Superstock; photo J by E. J. Tarbuck; photo K by Marli Miller; photo L by Pixachi/Superstock)

ACTIVITY 2.7
Identifying Sedimentary Environments

1. Use Figure 2.16 to identify the environment(s) (A–L) where the sediment for the following sedimentary rocks could have been deposited.

   Rock gypsum: _____

   Conglomerate: _____

   Sandstone: _____

   Shale: _____

   Bituminous coal: _____

   Travertine: _____

2. Briefly describe the environment that is associated with the formation of coral reefs.

   _____

3. The rocks in Zion National Park, Utah, consist of layers of well-sorted quartz sandstone, shown in **Figure 2.17**. Describe the environment that existed when these sediments were deposited. (*Hint:* Examine the sedimentary environments shown in Figure 2.16.)

   _____

   _____

   _____

▶ **Figure 2.17** The orange and yellow cliffs of Utah's Zion National Park.

Dennis Tasa

FOLIATION

Before metamorphism
(Confining pressure)

After metamorphism
(Differential stress)

Metamorphism

◀ **Figure 2.18** Under directed pressure, linear or platy minerals, such as the micas, become reoriented or recrystallized so that their surfaces are aligned at right angles to the stress. The resulting planar orientation of mineral grains is called foliation.

2.8 Identifying Metamorphic Rocks

- **Use an identification key to identify metamorphic rocks based on their texture and mineral composition.**

Extensive areas of metamorphic rocks are exposed on every continent in the relatively flat regions known as *shields*. They are also located in the cores of mountains and buried beneath sedimentary rocks on the continents.

Metamorphic Textures

During metamorphism, new minerals often form and/or existing minerals grow larger as the intensity of metamorphism increases. Frequently, mineral crystals that are elongated (such as hornblende) or have a sheet structure (for example, the micas—biotite and muscovite) become oriented perpendicular to compressional forces. The resulting parallel, linear alignment of mineral crystals is called **foliation** (**Figure 2.18**). Foliation is associated with many metamorphic rocks and gives them a layered or banded appearance.

The mineral crystals in foliated metamorphic rocks are either elongated or have thin, platy shapes and are arranged in a parallel or layered manner. During metamorphism, increased heat and pressure can cause mineral crystals to *become larger* and *foliation to become more obvious*. The various types of foliation, shown in **Figure 2.19A–D**, are described below:

- **Slaty or rock cleavage** refers to closely spaced, flat surfaces along which rocks split into thin slabs when struck with a hammer (Figure 2.19A).
- **Phyllite texture** develops when minute mica crystals in slate begin to increase in size. Phyllite surfaces have a shiny, somewhat metallic sheen and often have a wavy surface (Figure 2.19B).
- **Schisosity** can be identified by a scaly layering of glittery, platy minerals (mainly micas) that are often found in association with deformed quartz and feldspar grains (Figure 2.19C).
- **Gneissic texture** forms during high-grade metamorphism when ion migration results in the segregation of light and dark minerals (Figure 2.19D).

Some metamorphic rocks *do not* exhibit foliated texture. These rocks, referred to as **nonfoliated**, typically develop in environments where deformation is minimal and the parent rocks are composed of one mineral, such as quartz or calcite.

Foliated Metamorphic Rocks

Slate is a very fine-grained foliated rock composed of minute mica flakes that are too small to be visible to the unaided eye (Figure 2.19A). A noteworthy characteristic of slate is its excellent rock cleavage, or tendency to break into flat slabs. Slate is usually generated by low-grade metamorphism of shale. Slate's color is variable; black slate contains organic material, red slate gets its color from iron oxide, and green slate is usually composed of chlorite, a greenish mica-like mineral.

Phyllite represents a degree of metamorphism between that of slate and that of schist. Its constituent platy minerals, mainly muscovite and chlorite, are larger than those in slate but not large enough to be readily identifiable with the unaided eye. Although phyllite appears similar to slate, it can be easily distinguished from slate by its glossy sheen and wavy surface (Figure 2.19B).

Schists are moderate to strongly foliated rocks formed by regional metamorphism (Figure 2.19C). They are platy and can be readily split into thin flakes or slabs. Many schists originate from shale parent rock. The term *schist* describes the *texture* of a rock, regardless of composition. For example, schists composed primarily of muscovite and biotite are called *mica schists*.

Gneiss (pronounced "nice") is the term applied to banded metamorphic rocks in which elongated and granular (as opposed to platy) minerals predominate (Figure 2.19D). The most common minerals in gneisses are quartz and feldspar, with lesser amounts of muscovite, biotite, and hornblende. Gneisses exhibit strong segregation of light and dark silicates, giving them a characteristic banded texture.

Nonfoliated Metamorphic Rocks

Nonfoliated metamorphic rocks consist of inter-grown crystals of various size and are most often identified by determining their mineral composition. The minerals that comprise them are most often quartz or calcite. Therefore, the hardness test and/or the acid test can be used.

Marble is a coarse, crystalline rock whose parent rock is limestone (Figure 2.19E). Marble is composed of large interlocking calcite crystals formed from the recrystallization of smaller grains in the parent rock. White marble is particularly prized as a stone from which to carve monuments and statues. Marble can also be colored—pink, gray, green, or even black—if the parent rocks from which it formed contain impurities that color the stone.

COMMON METAMORPHIC ROCKS

FOLIATED TEXTURES

A. Slate
(Slaty or rock cleavage)

B. Phyllite
(Phyllite texture)

C. Mica Schist
(Schisosity)

D. Gneiss
(Gneissic texture)

NONFOLIATED TEXTURES

E. Marble
(Coarse grained)

F. Quartzite
(Fine grained)

▲ **Figure 2.19** Metamorphic textures.

Quartzite is a very hard metamorphic rock most often formed from quartz sandstone (see Figure 2.19F). Under moderate- to high-grade metamorphism, the quartz grains in sandstone fuse. Pure quartzite is white, but iron oxide may produce reddish or pinkish stains, and dark minerals may impart a gray color.

Other Metamorphic Rocks

Numerous other metamorphic rocks exist, as shown in **Figure 2.20A–F**. During intermediate- to high-grade metamorphism, recrystallization of existing minerals often produces new minerals. These newly formed minerals, commonly referred to as *accessory minerals*, tend to form large crystals, called *porphyroblasts*. The large porphyroblast crystals are surrounded by smaller crystals of other minerals, such as muscovite and biotite. When naming a metamorphic rock that contains one or more easily recognizable accessory minerals, geologists add a prefix to the appropriate rock name. For example, Figure 2.20A shows a mica schist that contains large dark red garnet crystals embedded in a matrix of fine-grained micas; consequently, this rock is called a *garnet-mica schist*. The metamorphic rock gneiss also frequently contains accessory minerals, including garnet and staurolite—which are called *garnet gneiss* and *staurolite gneiss*, respectively.

Some metamorphic rocks exhibit stretched or deformed pebbles (Figure 2.20B). Another distinguishing characteristic of highly deformed metamorphic rocks is foliation that becomes contorted or folded, as shown in Figure 2.20C. Furthermore, some metamorphic rocks, such as *hornfels* (Figure 2.20D), form when rock surrounding a molten igneous body is baked (contact metamorphism). The metamorphic form of coal, called *anthracite*, is

OTHER METAMORPHIC ROCKS

A. Garnet mica schist (Porphyroblasts)

B. Metaconglomerate (Stretched pebbles)

C. Gneiss (Highly deformed metamorphic rock)

D. Hornfels (Contact metamorphism)

E. Anthrcite (Metamorphosed coal)

E Serpentinite (Hydrothermal metamorphism)

▲ **Figure 2.20** Other metamorphic rocks.

generated when deformation caused by mountain building metamorphoses bituminous coal (Figure 2.20E). *Serpentinite,* forms by the hydrothermal alteration of mafic rocks along the mid-oceanic ridge (Figure 2.20F).

Classifying Metamorphic Rocks

Metamorphic rocks are classified according to their texture and mineral composition, as shown in the metamorphic rock identification key in **Figure 2.21**. Note that the names of medium- and coarse-grained metamorphic rocks are often modified to include the mineral composition before the name (for example, *mica schist*). Use this figure to complete the following activity.

METAMORPHIC ROCK IDENTIFICATION KEY

Distinctive Properties	Grain Size	Parent Rock	Texture	Rock Name
Excellent rock cleavage, smooth dull surfaces	Very fine	Shale, or siltstone	Foliated	Slate
Breaks along wavy surfaces, glossy sheen	Fine	Shale, slate, or siltstone	 Foliated	Phyllite
Micas dominate, breaks along scaly foliation	Medium to Coarse	Shale, slate, phyllite, or siltstone	 Foliated	Schist
Compositional banding due to segregation of dark and light minerals	Medium to Coarse	Shale, schist, granite, or volcanic rocks	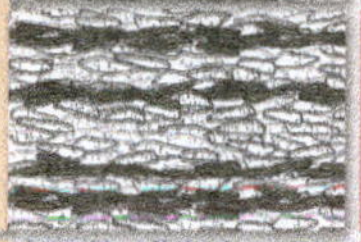 Foliated	Gneiss
Interlocking calcite or dolomite crystals nearly the same size, soft, reacts to HCl	Medium to coarse	Limestone, dolostone	Nonfoliated	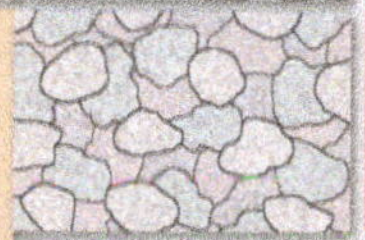 Marble
Fused quartz grains, massive, very hard	Medium to coarse	Quartz sandstone	 Nonfoliated	Quartzite
Round or stretched pebbles that have a preferred orientation	Coarse-grained	Quartz-rich conglomerate	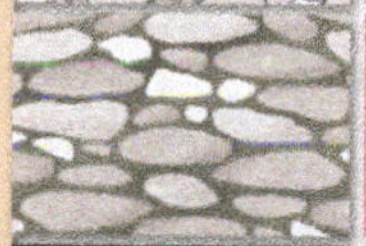 Nonfoliated	Metaconglomerate
Shiny black rock that may exhibit conchoidal fracture	Fine	Bituminous coal	Nonfoliated	Anthracite
Usually, dark massive rock with dull luster	Fine	Any rock type	Nonfoliated	Hornfels
Very fine grained, typically dull with a greenish color, may contain asbestos fibers	Fine	Mafic or ultramafic rocks	Nonfoliated	Serpentinite

▲ **Figure 2.21** Metamorphic rock identification key.

ACTIVITY 2.8 ///

Identifying Metamorphic Rocks

Place each of the metamorphic rocks supplied by your instructor on a numbered piece of paper. Then complete the metamorphic rock chart in **Figure 2.22** for each rock by following these steps:

Step 1: Determine whether the rock is foliated or nonfoliated.

Step 2: Determine the size of the mineral grains.

Step 3: If possible, determine the rock's mineral composition and note any other distinctive features.

Step 4: Use the information collected and the metamorphic rock identification key (Figure 2.21) to identify the rock.

METAMORPHIC ROCK CHART				
Sample Number	Foliated or Nonfoliated	Grain Size	Mineral Composition (if identifiable)	Rock Name

▲ **Figure 2.22** Metamorphic rock chart.

Rocks and the Rock Cycle

Name _____   Course/Section _____

Date _____   Due Date _____

1. The rock samples you encountered while completing this exercise are called *hand samples* because you can pick them up in your hand and study them. You can also easily examine hand samples with a hand lens, test them for hardness and reactivity to HCl, and feel and heft them. However, some characteristics of each rock group are so diagnostic that you can observe them at a distance and classify the rocks as igneous, sedimentary, or metamorphic. Examine the rock outcrops shown in **Figure 2.23** and determine whether each is igneous, sedimentary, or metamorphic.

   Outcrop A: _____

   Outcrop B: _____

   Outcrop C: _____

2. Match each of the metamorphic rocks listed below with one possible parent rock from the following list: bituminous coal, shale, limestone, slate, granite, sandstone.

   METAMORPHIC ROCK   NAME OF PARENT ROCK

   Marble: _____

   Slate: _____

   Phyllite: _____

   Gneiss: _____

   Quartzite: _____

   Anthracite: _____

3. Match each term or characteristic with the appropriate rock group.

   A. Foliated texture

   B. Clastic texture

   C. Gneissic texture

   D. Chemical rocks

   E. Porphyritic texture

   F. Lithification

   G. Alignment of mineral grains

   H. Silt-size particles

   I. Vesicular texture

   J. Strata or beds

   K. Felsic composition

   L. Slaty texture

   M. Glassy texture

   N. Evaporite deposits

   O. Detrital rocks

   Igneous rocks: _____

   Sedimentary rocks: _____

   Metamorphic rocks: _____

Outcrop A

Outcrop B

Outcrop C

▲ **Figure 2.23** Photos to accompany Question 1.
(Photo C courtesy of USGS)

4. Identify the foliated metamorphic rocks shown in **Figure 2.24** and list them in order from lowest to highest metamorphic grade.

A. _____
   (Name)

B. _____
   (Name)

C. _____
   (Name)

D. _____
   (Name)

Lowest _____

        _____

        _____

Highest _____

▲ **Figure 2.24** Photos to accompany Question 4.

5. Identify each of the rocks shown in **Figure 2.25** and the rock group to which each belongs.

A. _____
   (Name)

   _____
   (Rock Group)

B. _____
   (Name)

   _____
   (Rock Group)

C. _____
   (Name)

   _____
   (Rock Group)

D. _____
   (Name)

   _____
   (Rock Group)

E. _____
   (Name)

   _____
   (Rock Group)

F. _____
   (Name)

   _____
   (Rock Group)

G. _____
   (Name)

   _____
   (Rock Group)

H. _____
   (Name)

   _____
   (Rock Group)

▲ **Figure 2.25** Photos to accompany Question 5.

Plate Tectonics

LEARNING OBJECTIVES

Each statement represents an important learning objective that relates to one or more sections of this lab. After you complete this exercise you should be able to:

- **Locate the mid-ocean ridge system and major deep-ocean trenches on a world map.**
- **Describe the three types of plate boundaries and the motion associated with each.**
- **List and explain several lines of evidence that support the theory of plate tectonics.**
- **Determine the rate of seafloor spreading along a mid-ocean ridge, using paleomagnetic evidence.**
- **Determine plate velocities using the ages of a chain of volcanoes and their distances from the hot spot.**

MATERIALS

calculator
colored pencils
tracing paper

ruler
atlas, globe, or world wall map

PRE-LAB VIDEO ▶ https://goo.gl/VCjmU8

Prepare for lab! Prior to attending your laboratory session, view the pre-lab video. Each video provides valuable background that will contribute to your understanding and success in lab.

INTRODUCTION

Plate tectonics is the first theory to provide a comprehensive view of the processes that produced Earth's major surface features, including the continents and ocean basins. Within the framework of this theory, geologists have found explanations for the basic causes and distribution of earthquakes, volcanoes, and mountain belts. Further, we are now better able to explain the distribution of plants and animals in the geologic past, as well as the distribution of economically significant mineral deposits.

PART 1 Geology

3.1 From Continental Drift to Plate Tectonics

■ Locate the mid-ocean ridge system and major deep-ocean trenches on a world map.

The idea that our present-day continents were once part of a single supercontinent was proposed in 1912 by Alfred Wegener. His hypothesis, called **continental drift**, proposed that about 200 million years ago, a supercontinent called **Pangaea** began to break apart into the present continents. Over time, each landmass slowly drifted across Earth's surface to its current position.

Following World War II, advanced technologies in oceanographic research provided new evidence that supported the continental drift hypothesis. Data collected from studies of rock magnetism, the distribution of earthquakes, and mapping of the seafloor led to the development of a more comprehensive model, known as the **theory of plate tectonics**.

The theory of plate tectonics states that Earth's **lithosphere** is broken into several large, rigid slabs called **lithospheric plates**, or simply **plates** (**Figure 3.1**). Plates are continually moving, and where they move apart along the **mid-ocean ridges**, upwelling of hot mantle rock creates new seafloor. The axis (a line down the center) of some ridge segments contains deep down-faulted structures called **rift valleys**. Along other plate margins, plates slide past each other or collide to form mountains. In addition, one plate may descend beneath an overriding plate to form a **deep-ocean trench**.

The discoveries of the global mid-ocean ridge system and deep-ocean trenches were of major importance in the development of the theory of plate tectonics.

▼ **Figure 3.1** The mosaic of rigid plates that constitutes Earth's outer shell.

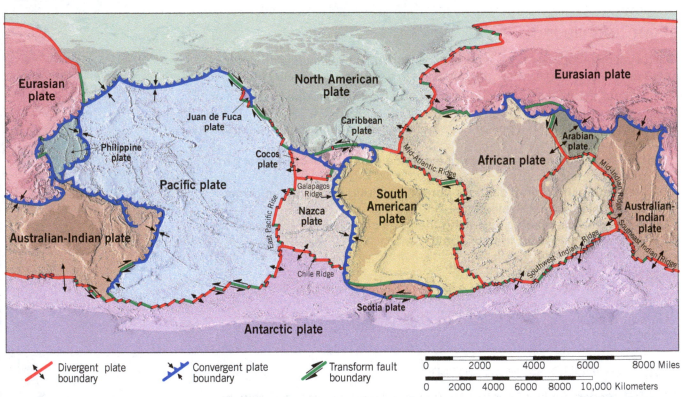

Divergent plate boundary — Convergent plate boundary — Transform fault boundary

0   2000   4000   6000   8000 Miles
0   2000 4000 6000 8000 10,000 Kilometers

ACTIVITY 3.1

Locating Ocean Ridges and Trenches

1. Using an atlas or Figure 3.1 for reference, draw the axis of the global oceanic ridge system on the map in **Figure 3.2**.

2. Use an atlas or your textbook to label the deep-ocean trenches on Figure 3.2, using the letter associated with each of the following trenches:

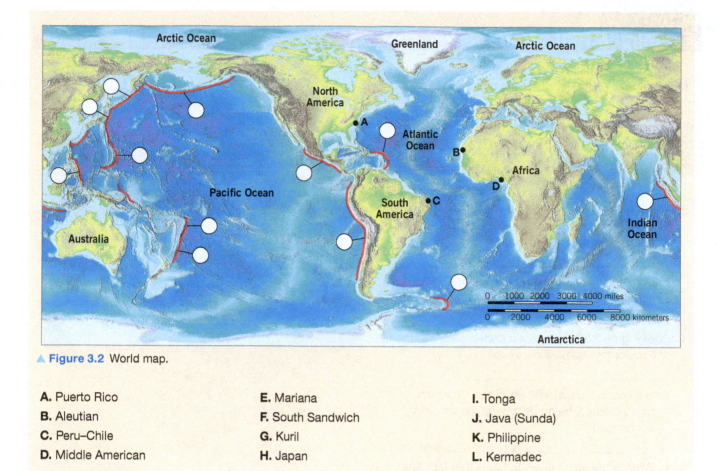

▲ **Figure 3.2** World map.

A. Puerto Rico   **E.** Mariana   **I.** Tonga
B. Aleutian   **F.** South Sandwich   **J.** Java (Sunda)
C. Peru–Chile   **G.** Kuril   **K.** Philippine
D. Middle American **H.** Japan   **L.** Kermadec

3.2 Plate Boundaries

■ **Describe the three types of plate boundaries and the motion associated with each.**

Because plates are in constant motion relative to each other, most major interactions among them (and, therefore, most of the deformation) occur along their *boundaries*. Earth scientists recognize three distinct types of plate boundaries, each distinguished by the relative movement of the plates on the opposite sides of the boundary:

- **Divergent plate boundaries** (*constructive margins*), where two plates move apart, resulting in upwelling of hot material from the mantle to create new seafloor
- **Convergent plate boundaries** (*destructive margins*), where two plates move together, resulting in oceanic lithosphere descending beneath an overriding plate, eventually to be reabsorbed into the mantle or possibly resulting in the collision of two continental blocks to create a mountain belt
- **Transform plate boundaries** (*conservative margins*), where two plates grind past each other without the production or destruction of lithosphere

ACTIVITY 3.2
Examining Plate Boundaries

Use **Figure 3.3**, which illustrates the various types of plate boundaries, to complete the following.

1. Does Figure 3.3A represent a convergent or divergent plate boundary?

   _____

   a. Are the plates along this boundary moving apart or moving together?

   _____

   b. Does this plate boundary occur at deep-ocean trenches or mid-ocean ridges?

   _____

   c. Does this plate boundary result in the construction or destruction of lithosphere?

   _____

2. Does Figure 3.3B represent a convergent, divergent, or transform plate boundary?

   _____

   a. Is lithospheric material along this boundary being created, being destroyed, or remaining unchanged?

   _____

   b. What is the name given to faults that facilitate plate movement along this type of boundary?

   _____

3. Does Figure 3.3C represent a convergent, divergent, or transform plate boundary?

   _____

   a. Are the plates along this type of boundary moving apart or moving together?

   _____

   b. Examine **Figure 3.4**, which shows the three types of convergent plate boundaries. Label each of the three diagrams with the appropriate term: *oceanic–continental convergence, oceanic–oceanic convergence,* or *continental–continental convergence.*

   d. Find one example of each type of convergent boundary on the plate map in Figure 3.1. Label each boundary with the terms from Question 3b.

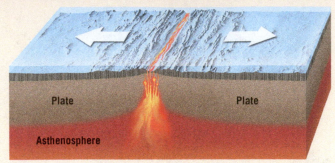

A.

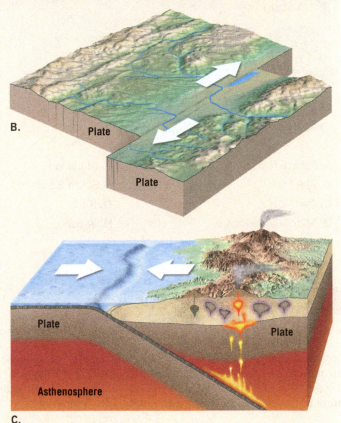

B.

C.

▲ **Figure 3.3** Illustrations of the three types of plate boundaries.

A. _____

B. _____

C. _____

▲ **Smartfigure 3.4**
Illustrations of the three types
of convergent plate boundaries.

VIDEO
https://goo.gl/eTSF6W

3.3 Evidence: The Continental Jigsaw Puzzle

■ **List and explain several lines of evidence that support the theory of plate tectonics.**

Like a few others before him, Wegener suspected that the continents might once have been joined when he noticed the remarkable similarity between the coastlines on opposite sides of the Atlantic Ocean. However, Wegener's use of present-day shorelines to fit these continents together was challenged immediately by other Earth scientists. Scientists later determined that a much better approximation of the outer boundary of a continent is the seaward edge of its continental shelf, which lies submerged a few hundred meters below sea level.

ACTIVITY 3.3A
The Continental Jigsaw Puzzle

Examine the east coast of South America and the west coast of Africa on the world map in Figure 3.2 to complete the following.

1. Does the shape of the east coast of South America conform to the shape of the east or west side of the mid-ocean ridge in the Atlantic Ocean?

   _____

2. Does the shape of the west coast of Africa conform to the shape of the east or west side of the mid-ocean ridge in the Atlantic Ocean?

   _____

3. On separate pieces of tracing paper, sketch the outlines of the continents of South America and Africa found in **Figure 3.5**. Move the tracing papers until you get the best fit of the continents. How well do they fit together?

   _____

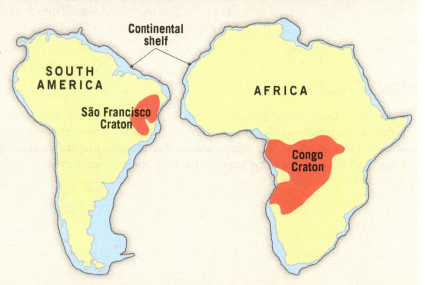

▲ **Figure 3.5** Simplified maps of South America and Africa.

Matching Rock Types

Anyone who has worked a jigsaw puzzle knows that its successful completion requires fitting the pieces together while maintaining the continuity of the picture. The "picture" that must match in the "continental drift puzzle" includes rocks of similar ages and types, as well as geologic structures, such as mountain belts. If the continents were once joined, the rocks found in a particular region on one continent should closely match, in age and type, those found in adjacent positions on the once-adjoining continent.

1. Using the same two pieces of tracing paper you used for Question 3 in Activity 3.3A, trace and shade in the areas that contain the Congo Craton and the São Francisco Craton and label each. Both of these cratons are composed of Precambrian rocks of equivalent type and age.

2. Reassemble the two continents as you did in Activity 3.3A. Do the positions of these two rock bodies support or refute the concept of continental drift?

_____

3.4   Evidence: Paleoclimates

■ **List and explain several lines of evidence that support the theory of plate tectonics.**

Because Alfred Wegener was a student of world climates, he suspected that *paleoclimate* (*ancient climate*) data might also support the idea of continental drift. His assertion was bolstered by the discovery of evidence for a glacial period dating to the late *Paleozoic era* in portions of present-day Africa, South America, Australia, and India This meant that about 300 million years ago, vast ice sheets covered extensive portions of the Southern Hemisphere as well as India

During the same span of geologic time, large tropical swamps existed in several locations in the Northern Hemisphere. The lush vegetation in those swamps was eventually buried and converted to coals that now comprise the major coal fields of the eastern United States and Northern Europe.

Based on climate studies, we know that tropical swamps form near the equator, Earth's major deserts are found centered roughly on the Tropic of Cancer and Topic of Capricorn ($23\frac{1}{2}°$ North and $23\frac{1}{2}°$ South latitude, respectively), and large continental ice sheets form in polar environments.

Figure 3.6 shows simplified paleoclimate data discovered in the rock record of the continents shown. Using tracing paper, copy these continents and their climate regions and cut them into separate pieces. Based on climate data and the shapes of the continents, restore them to their ancient location on the map in **Figure 3.7**. *Note:* Antarctica, India, and Australia have been placed in their ancient locations as a guide. (*Hint:* Begin by placing Africa adjacent to India and Antarctica.)

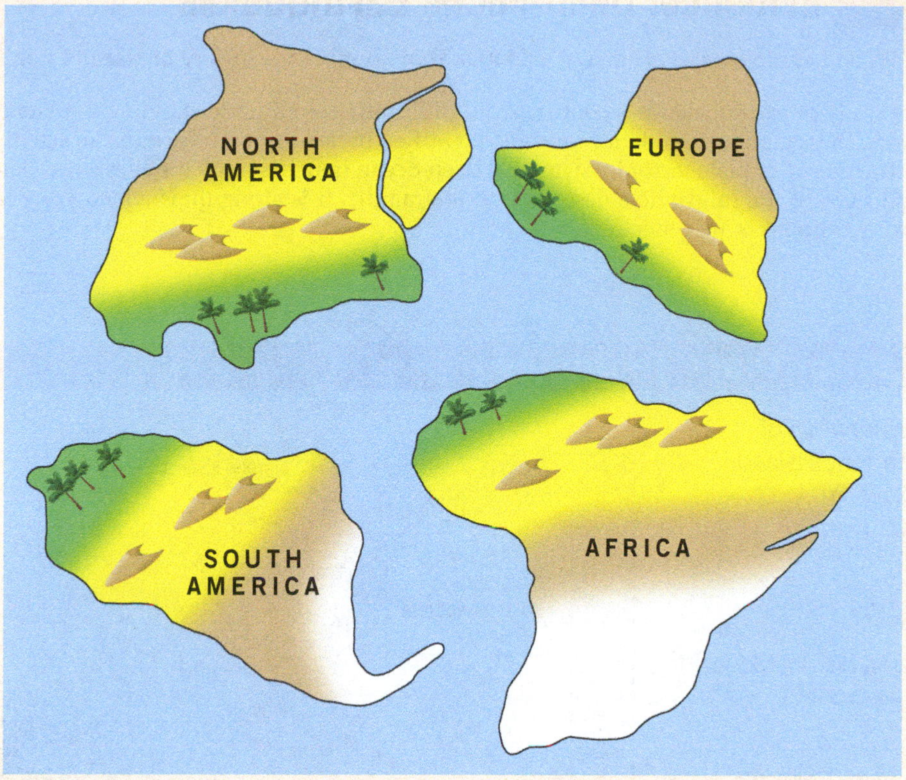
NORTH AMERICA
EUROPE
SOUTH AMERICA
AFRICA

◀ **Figure 3.6** Paleoclimate data indicate 300-million-year-old climate regions.

▼ **Figure 3.7** Map used to reconstruct continents from Figure 3.6.

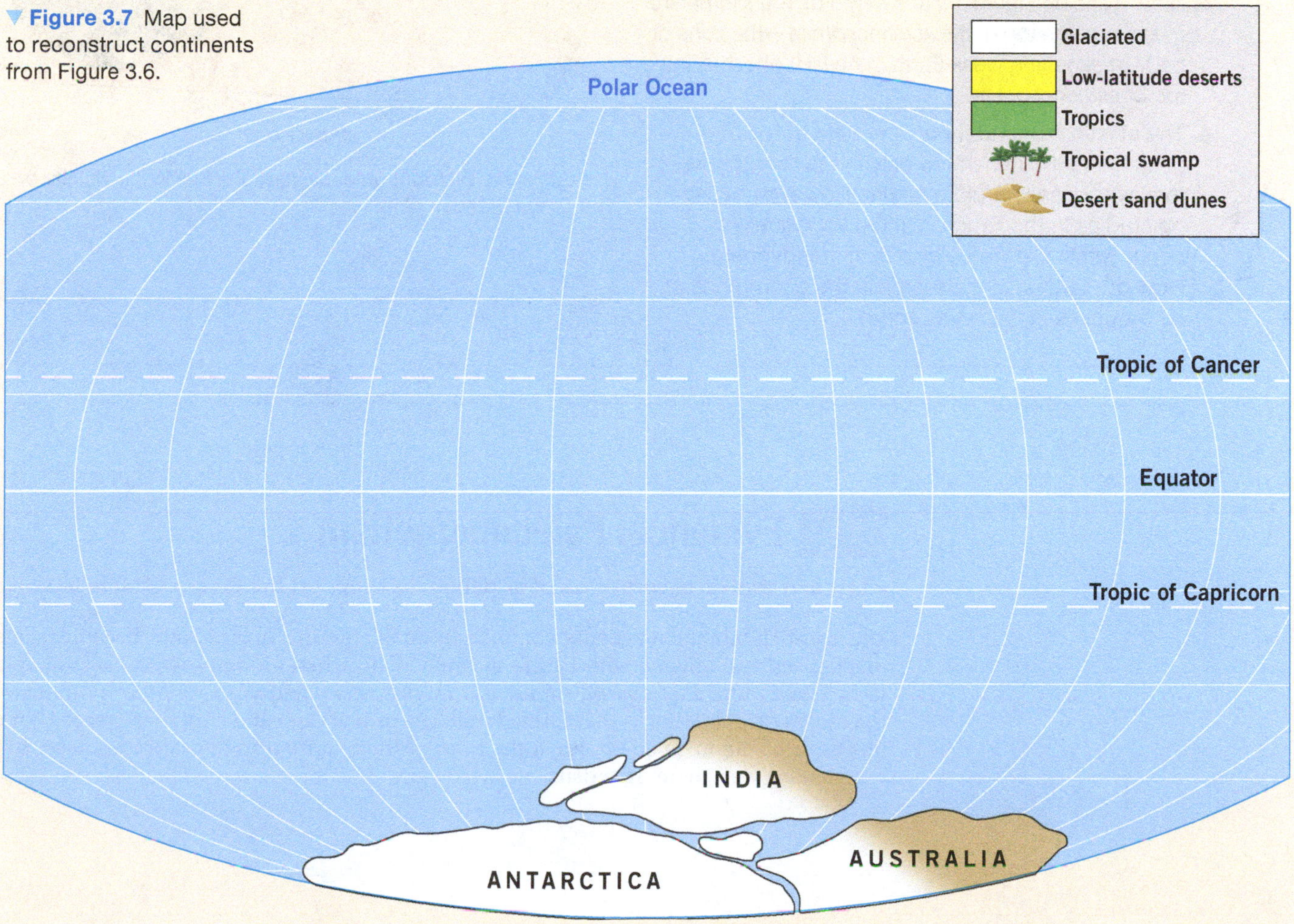
Glaciated
Low-latitude deserts
Tropics
Tropical swamp
Desert sand dunes
Polar Ocean
Tropic of Cancer
Equator
Tropic of Capricorn
INDIA
AUSTRALIA
ANTARCTICA

3.5 Evidence: Deep-Focus Earthquakes

■ List and explain several lines of evidence that support the theory of plate tectonics.

The study of earthquakes has contributed significantly to our understanding of plate tectonics. In fact, plate boundaries were first identified because of their earthquake activity.

The theory of plate tectonics states that deep-ocean trenches mark the sites where large, rigid slabs of the lithosphere descend into the mantle. It is along these zones that deep-focus earthquakes are generated.

ACTIVITY 3.5
Earthquakes: Evidence for Plate Tectonics

Figure 3.8 illustrates an idealized distribution of earthquake foci in the vicinity of the Tonga islands. Use Figure 3.8 to complete the following.

1. At approximately what depth do the deepest earthquakes occur?

   _____kilometers

2. Are the earthquake foci in the area distributed in a random manner or along a linear zone?

   _____

3. On Figure 3.8, outline the zone of earthquakes.

4. Draw a line on Figure 3.8 at a depth of 100 kilometers to indicate the top of the *asthenosphere*—the zone of partly melted and weak Earth material. Label the line *top of asthenosphere*.

5. The elastic rebound theory predicts that earthquakes can be generated only in the lithosphere—the layer of solid, rigid material. However, based on studies in the Tonga islands, earthquakes occur within the weak, ductile asthenosphere. How do Earth scientists explain the occurrence of earthquakes in this weak zone?

   _____

   _____

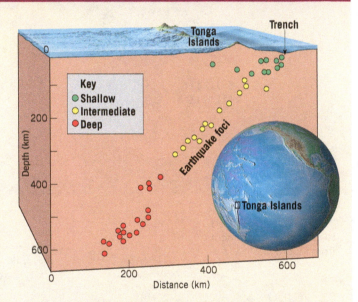

▲ **Figure 3.8** Distribution of earthquake foci in the vicinity of the Tonga islands.

3.6 Evidence: Paleomagnetism

■ List and explain several lines of evidence that support the theory of plate tectonics.

The iron-rich mineral magnetite, common in small amounts in lava flows, becomes slightly magnetized and aligned with Earth's magnetic field when molten rock cools and crystallizes. These mineral grains are similar to tiny compass needles because they point toward the magnetic poles. Rocks that formed millions of years ago and contain a "record" of the direction of the magnetic poles at the time of their formation are said to possess **fossil magnetism**, or **paleomagnetism**.

Scientists have also discovered that the polarity of Earth's magnetic field has periodically reversed, such that the magnetic north pole becomes the magnetic south pole and vice versa

As plates move apart along the mid-ocean ridge, magma from the mantle rises to the surface and creates new ocean floor. As the magma cools, the minerals align themselves with the prevailing magnetic field. When Earth's magnetic field reverses polarity, new material forming at the ridge is magnetized in an opposite (reverse) direction.

Scientists have reconstructed Earth's magnetic polarity reversals over the past several million years, as shown in **Figure 3.9**. The periods of normal polarity (when a compass needle would have had the same alignment as it does today) are shown in various colors and labeled A–F.

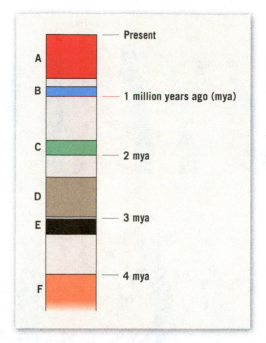

◀ **Figure 3.9** Chronology of magnetic polarity during the past 4 million years. Periods of normal polarity, when a compass needle would have been oriented as it is today, are shown in prominent colors. (Data from Allan Cox and G. Dalrymple)

ACTIVITY 3.6
Evidence: Paleomagnetism

Use Figure 3.9 to answer the following questions.

1. How many intervals—3, 5, or 7—of reverse polarity (shown in light gray color) have occurred during the past 4 million years?

   _____ intervals of reverse polarity

2. Approximately how many years ago did the current period of normal polarity (labeled A) begin?

   _____ yr ago

3. Did Earth experience normal or reverse polarity 1.5 million years ago?

   _____ polarity

4. Did the period of normal polarity, C, begin 1, 2, or 3 million years ago?

   _____ million yr ago

5. During the past 4 million years, has each interval of reverse polarity lasted more or less than 1 million years?

   _____ than 1 million yr

6. Based on the pattern of magnetic reversals shown in Figure 3.9, does it appear as though another polarity reversal will occur in the near future?

   _____

3.7   Paleomagnetism and Seafloor Spreading

■ **Determine the rate of seafloor spreading along a mid-ocean ridge, using paleomagnetic evidence.**

The pattern of magnetic polarity reversals shown in Figure 3.9 has been found in rocks of all the major ocean basins. **Figure 3.10** illustrates this pattern of reversals for sections of the mid-ocean ridges in the North Atlantic, South Atlantic, and Pacific. As new ocean crust forms along the mid-ocean ridge, it spreads out equally on both sides of the ridge. Therefore, the pattern of polarity reversals that occurs on one side of the ridge axis can be matched with the polarity of the rocks on the other side of the ridge.

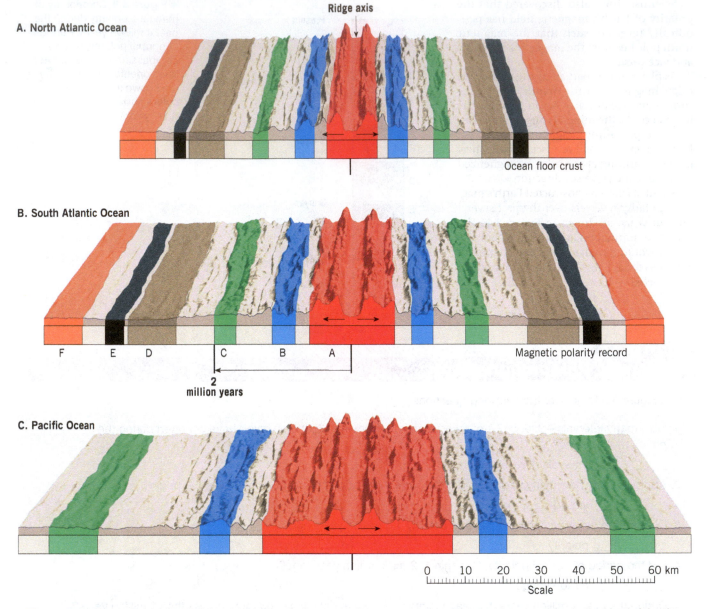

▲ **Figure 3.10** Generalized record of the magnetic polarity reversals across portions of the mid-ocean ridge in the North Atlantic, South Atlantic, and Pacific Oceans. Periods of normal polarity are shown in color and correspond to those illustrated in Figure 3.9.

ACTIVITY 3.7A
Paleomagnetism and Seafloor Spreading

Use Figures 3.9 and 3.10 to complete the following.

1. On Figure 3.10, identify and mark the periods of normal polarity with the letters A–F (as in Figure 3.9). Begin at the ridge crest and label along both sides of each ridge. (*Note:* The left side of the South Atlantic shown in Figure 3.10B has been completed as a guide.)

2. Using the South Atlantic as an example, label the *beginning* of the normal polarity period C that began *2 million years ago* on the left sides of the Pacific and North Atlantic diagrams.

3. Using the distance scale at the bottom of Figure 3.10, measure the distance from the ridge axis to the beginning of the normal polarity period C for each ridge. Place the distances on Figure 3.10. How many kilometers has the left side of each of these ocean basins spread during the past 2 million years?

South Atlantic basin: _____ km

North Atlantic basin: _____ km

Pacific basin: _____ km

4. The distances you obtained in Question 3 are for only one side of the ridge. Assuming that a ridge spreads equally on both sides, the actual distance each ocean basin has opened would be twice this amount. How many kilometers has each ocean basin opened in the past 2 million years?

South Atlantic basin: _____ km

North Atlantic basin: _____ km

Pacific basin: _____ km

ACTIVITY 3.7B
Calculating the Rate of Seafloor Spreading

When the *distance* that an ocean basin has opened and the *time* it took for it to open are known, the *rate* of **seafloor spreading** can be calculated. To determine the rate of spreading in centimeters per year for each ocean basin, convert the distance the basin has opened from kilometers to centimeters and then divide this distance by the time—2 million years in this example. Determine the rate of seafloor spreading for the Pacific and North Atlantic Ocean basins. For example, here is the calculation for the South Atlantic:

South Atlantic: distance = 72 km × 100,000 cm/km = 7,200,000 cm

$$\text{Rate of spreading} = \frac{7,200,000 \text{ cm}}{2,000,000 \text{ yr}} = 3.6 \text{ cm/yr}$$

1. North Atlantic: distance = _____ km

   ×100,000 cm/km = _____ cm

   Rate of spreading = _____ cm/yr

2. Pacific: distance = _____ km

   ×100,000 cm/km = _____ cm

   Rate of spreading = _____ cm/yr

ACTIVITY 3.7C
Determining the Ages of Ocean Basins

The following procedure will allow you to calculate the age of a portion of the North Atlantic and South Atlantic basins.

1. Using Figure 3.2, measure the distance from Point A located off the Carolina coast to Point B off the African coast. Determine the distance in kilometers and then convert that distance into centimeters.

   Distance: _____ km

   Distance: _____ cm

continued

Activity 3.7C continued

2. Divide the distance in centimeters separating the continents obtained in Question 1 by the rate of seafloor spreading for the North Atlantic basin that you calculated in Question 1 in Activity 3.7B. Your answer is the approximate age, in years, of the North Atlantic basin at that location.

Distance/Rate of seafloor spreading = _____

Age of the North Atlantic basin: _____ yr

3. Repeat the procedure above to determine the age of the South Atlantic basin from Point C along the eastern edge of Brazil to Point D off the coast of Africa.

How many years ago did South America and Africa begin to separate at that location?

Age of the South Atlantic basin: _____ yr

4. Based on your answers to Questions 2 and 3, which part of the Atlantic basin appears to have opened first?

_____

3.8 Hot Spots and Plate Velocities

■ **Determine plate velocities using the ages of a chain of volcanoes and their distances from the hot spot.**

Researchers have determined that volcanism on the Big Island of Hawaii arises from a plume of molten material moving upward from the mantle to create a **hot spot** (**Figure 3.11**). It is assumed that the position of this hot spot has remained constant during the past few million years.* In the past, as the Pacific plate moved over the hot spot, the successive

*Recent evidence indicates that mantle plumes might actually move more than previously thought, which would make these calculations somewhat inaccurate.

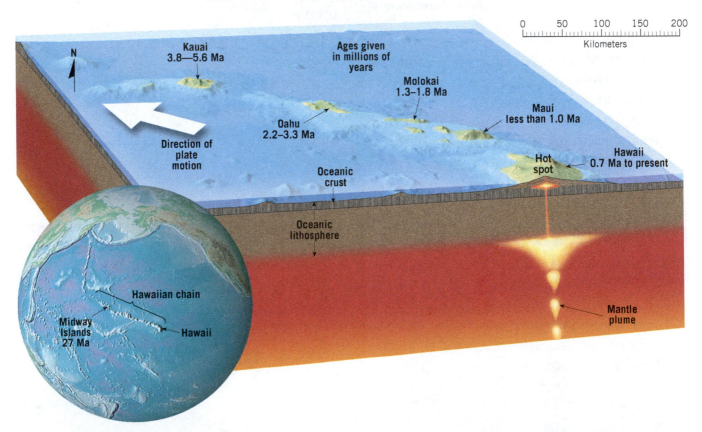

▲ **Figure 3.11** Movement of the Pacific plate over a hot spot and radiometric ages of the Hawaiian islands, in millions of years.

volcanic islands of the Hawaiian chain were built. The Big Island of Hawaii continues to grow as lava from this mantle plume flows to the sea

Radiometric dating of the volcanoes in the Hawaiian chain has revealed that the islands increase in age with increasing distance from the Big Island of Hawaii (Figure 3.11). Based on the age of an island and its distance from the hot spot, the velocity of the plate can be calculated.

ACTIVITY 3.8
Hot Spots and Plate Velocities

Use Figure 3.11 to answer the following questions.

1. What are the minimum and maximum ages of the island of Kauai?

   Minimum age: _____ million yr

   Maximum age: _____ million yr

2. What is the approximate distance (in kilometers) from the hot spot to the center of Kauai? Convert your answer to centimeters.

   _____ km

   _____ cm

3. Using the data in Questions 1 and 2, calculate the approximate maximum and minimum velocities of the Pacific plate as it moved over the Hawaiian hot spot in centimeters per year.

   Maximum velocity: _____ cm/yr

   Minimum velocity: _____ cm/yr

MasteringGeology™

Looking for additional review and lab prep materials? Go to www.masteringgeology.com for Pre-Lab Videos, Geoscience Animations, RSS Feeds, Key Term Study Tools, The Math You Need, an optional Pearson eText, and more.

Notes and calculations:

Plate Tectonics

Name _____

Date _____

Course/Section _____

Due Date _____

1. The distribution of earthquakes defines the boundaries of which major Earth feature?

   _____

2. Deep-focus earthquakes are associated with what prominent ocean-floor feature?

   _____

3. Along what prominent ocean-floor features do shallow-focus earthquakes occur?

   _____

4. Why do Earth scientists think that rigid slabs of the lithosphere are descending into the mantle near the Tonga islands?

   _____

   _____

5. Describe how paleomagnetism is used to calculate the rate of seafloor spreading.

   _____

   _____

   _____

6. Based on your calculations in Activity 3.7B, what are the rates of seafloor spreading for the following ocean basins?

   North Atlantic basin: _____ cm/yr

   Pacific basin: _____ cm/yr

7. Based on your calculations in Activity 3.7C, what are the ages for the North and South Atlantic basins at the locations examined?

   North Atlantic basin: _____ million yr old

   South Atlantic basin: _____ million yr old

8. Based on your calculations in Activity 3.8, what is the maximum velocity for the Pacific plate?

   _____ cm/yr

9. Complete the block diagrams in **Figure 3.12** to illustrate the types of plate boundaries listed below each diagram. Include arrows to indicate relative plate motion.

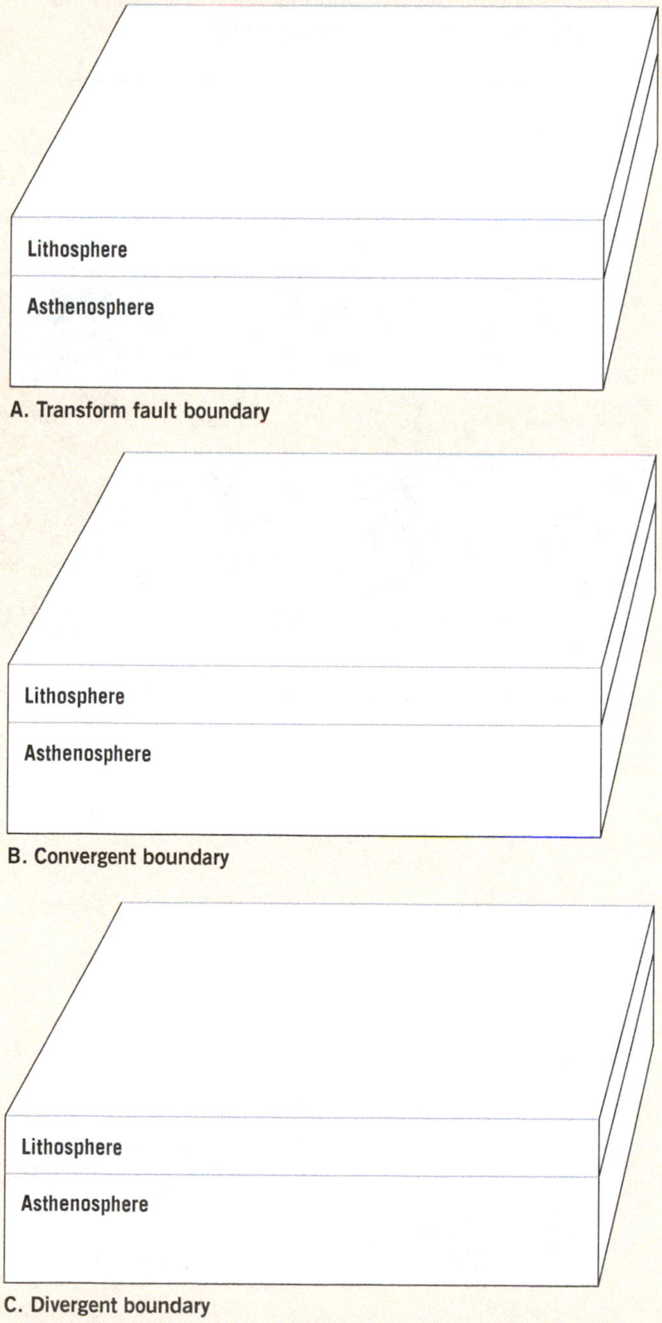

Lithosphere

Asthenosphere

A. Transform fault boundary

Lithosphere

Asthenosphere

B. Convergent boundary

Lithosphere

Asthenosphere

C. Divergent boundary

▲ **Figure 3.12** Block diagrams to accompany Question 9.

10. Use **Figure 3.13**, which illustrates a generalized cross section of the plate boundary along the western edge of South America, to answer the following questions.

   a. Label each of the following features on the block diagram.

   Asthenosphere   Continental crust

   Deep-ocean trench   Oceanic crust

   Continental lithosphere   Oceanic lithosphere

   b. What type of plate boundary is illustrated in the figure? Be as specific as possible.

   _____

11. List and explain two lines of evidence from this exercise that support the theory of plate tectonics.

   _____
   _____
   _____
   _____
   _____
   _____
   _____

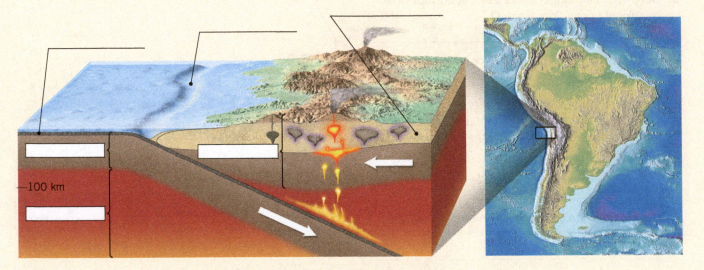

—100 km

▲ **Figure 3.13** Block diagram of the plate boundary along the western edge of South America.

Earthquakes and Earth's Interior

LEARNING OBJECTIVES

Each statement represents an important learning objective that relates to one or more sections of this lab. After you complete this exercise you should be able to:

- Examine an earthquake seismogram and identify the P waves, S waves, and surface waves.
- Use a seismogram and travel–time graph to measure the distance to the epicenter of an earthquake and determine the time it occurred.
- Locate Earth's major earthquake belts on a world map.
- List the name, thickness, and state of matter of each of Earth's layers.
- Describe the temperature gradient in Earth's interior.
- Explain why the asthenosphere likely consists of very weak material.

MATERIALS

calculator   colored pencils
drafting compass   ruler

PRE-LAB VIDEO https://goo.gl/nK9Kk6

 Prepare for lab! Prior to attending your laboratory session, view the pre-lab video. Each video provides valuable background that will contribute to your understanding and success in lab.

INTRODUCTION

Thousands of earthquakes occur around the world every day. Fortunately, most are small and cannot be detected by people. Only about 15 strong earthquakes (magnitude 7 or greater) are recorded each year, many in remote regions. Occasionally, a large earthquake occurs near a major population center. Such an event is among the most destructive natural forces on Earth. The shaking of the ground coupled with the liquefaction of soils wreaks havoc on buildings, roadways, and other structures.

Geology

4.1 Earthquakes

▼ **SmartFigure 4.1** The energy generated by earthquakes radiates outward in the form of *seismic waves* from the source of the quake, called the *hypocenter*, or *focus*. The three basic types of seismic waves generated by an earthquake are *P waves*, *S waves*, and *surface waves*. P and S waves travel through Earth's interior, and surface waves are transmitted along the surface.

VIDEO
https://goo.gl/nYjW5s

■ **Examine an earthquake seismogram and identify the P waves, S waves, and surface waves.**

An **earthquake** is ground shaking caused by the sudden and rapid movement of one block of rock slipping past another along fractures in Earth's crust called **faults**. The energy generated by an earthquake radiates outward in the form of **seismic waves** from the source of the quake, called the **hypocenter**, or focus (**Figure 4.1**).

The three basic types of seismic waves generated by an earthquake are **P waves**, **S waves**, and **surface waves**. P and S waves travel through Earth's interior, and surface waves are transmitted along the surface (see Figure 4.1). Of the three wave types, P waves have the greatest velocity and, therefore, reach the seismograph station first. Surface waves arrive at the seismograph station last.

Seismographs are instruments that detect and record seismic waves. The recordings made by seismographs are called **seismograms**. Seismograms are used to determine the time and location of earthquakes. These records also allow us to study the nature of Earth's interior.

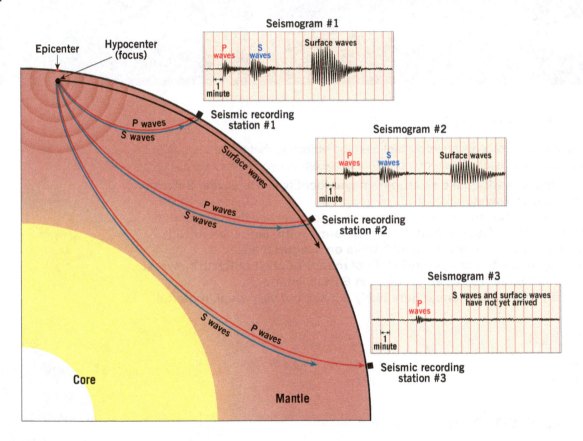

ACTIVITY 4.1
Examining a Seismogram

Use **Figure 4.2**, which shows a typical seismogram, to complete the following.

1. How many minutes elapsed between the arrival of the first P wave and the arrival of the first S wave?
   _____ min

2. How many minutes elapsed between the arrival of the first P wave and the arrival of the first surface wave?
   _____ min

3. How much time elapsed between the arrival of the first S wave and the first surface wave?

_____ min

4. Is the maximum amplitude (wave height) of the surface waves less than or greater than the maximum amplitude of the P waves?

The maximum amplitude of surface waves is _____ than that of P waves.

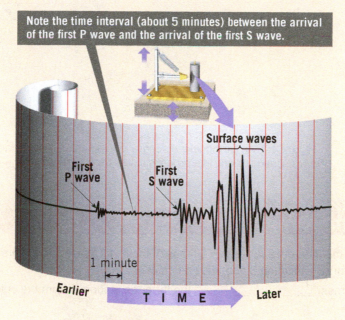

▲ **Figure 4.2** Typical earthquake seismogram.

4.2   Locating an Earthquake

■ **Use a seismogram and travel–time graph to measure the distance to the epicenter of an earthquake and determine the time it occurred.**

An earthquake's _hypocenter_, or _focus_, can occur at depths of more than 600 kilometers below Earth's surface. When locating an earthquake on a map, seismologists plot the **epicenter**, the point on Earth's surface directly above the hypocenter (see Figure 4.1).

The difference in the velocities of P and S waves provides a method for locating the epicenter of an earthquake. Both P and S waves leave the earthquake focus at the same instant. Since P waves have greater velocity, the further away the focus is from the recording instrument, the greater the difference in the arrival times of the first P wave compared to the first S wave.

To determine the distance between a recording station and an earthquake epicenter, follow these steps:

Step 1: Determine the P–S interval (the time difference in minutes between the arrival of the first P wave and the arrival of the first S wave) from the seismogram in Figure 4.2.

Step 2: Find the place on the travel–time graph (see Figure 4.4) where the vertical separation between the P and S curves is equal to the P–S interval determined in Step 1.

Step 3: From this position, draw a vertical line that extends to the bottom of the graph and read the distance to the epicenter.

To locate an earthquake epicenter, seismograms from three different stations are needed. First, determine the distance of each station from the epicenter, using the procedure just described. Then, around each station, draw a circle with a radius equal to the station's distance from the epicenter. The point where all three circles intersect is the earthquake's epicenter (**Figure 4.3**).

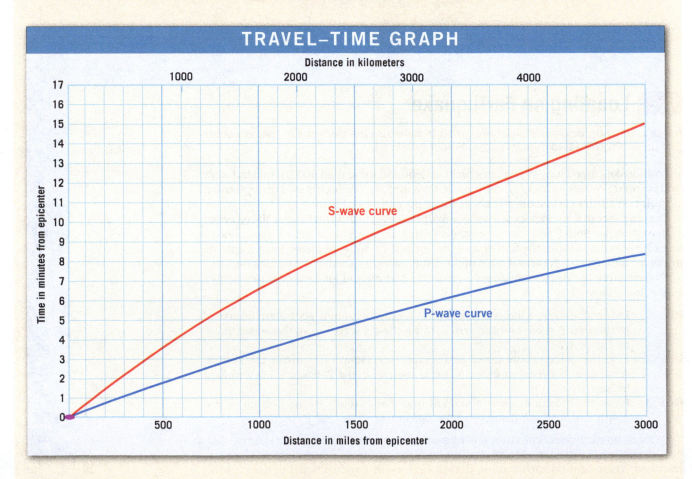

▲ **Figure 4.3** An earthquake epicenter can be located using the distances obtained from three seismic stations in a process called triangulation.

ACTIVITY 4.2A
Using a Travel–Time Graph

1. Examine **Figure 4.4**. Does the difference in the arrival times of the first P wave and the first S wave on a seismogram increase or decrease the farther a station is from the epicenter?

The difference in arrival times _____ the farther a station is from the epicenter.

TRAVEL–TIME GRAPH

Distance in kilometers

S-wave curve

P-wave curve

Time in minutes from epicenter

Distance in miles from epicenter

▲ **Figure 4.4** Travel–time graph used to determine the distance between an earthquake epicenter and a seismic station.

2. Use Figure 4.4 to determine the difference in arrival times, in minutes, between the first P wave and first S wave for stations that are the following distances from an epicenter:

1000 mi: _____ min

4000 km: _____ min

3000 mi: _____ min

On the seismogram in Figure 4.2, you determined the difference in the arrival times between the first P waves and the first S waves to be 5 minutes. Use Figure 4.2 and the travel–time graph in Figure 4.4 to complete the following.

3. What is the distance between the epicenter and the station for the earthquake recorded on the seismogram in Figure 4.2, in miles and kilometers?

_____ mi

_____ km

4. For the earthquake recorded in Figure 4.2, about how long did it take the first P wave to reach the station: 3, 6, or 14 minutes? (*Hint:* use the P-wave curve in the time-travel graph in Figure 4.4 to answer this question).

The first P wave arrived _____ minutes after the earthquake began.

5. If the first P wave was recorded at 10:39 Y.Y. local time at the station in Figure 4.2, what was the local time at the epicenter when the earthquake began. (*Hint:* The earthquake began at the epicenter.)

_____ Y.Y. local time

ACTIVITY 4.2B

Locating an Earthquake

Figure 4.5 shows seismograms for the same earthquake recorded at three locations—New York; Nome, Alaska; and Mexico City. Use this information to complete the following.

1. Use Figure 4.5 and the travel–time graph in Figure 4.4 to determine the distance between each station and the epicenter. Write your answers in **Table 4.1**.

2. Use **Figure 4.6** and a drafting compass to draw a circle around each of the three stations. Make the radius (in miles) of each circle equal to the station's distance from the epicenter as determined above. (Use the scale on the map to set the radius on the drafting compass.) The circles you draw should intersect at approximately one point. This point is the epicenter. If they do not intersect at one point, find a point that is equidistant from the edges of the three circles and use this as the epicenter. Label the epicenter on the map.

3. What are the approximate latitude and longitude of the epicenter of this earthquake?

Latitude: _____

Longitude: _____

4. The first P wave was recorded in New York at 9:01 EST. At what time (EST) did the earthquake actually occur?

_____ EST

▶ **Figure 4.5** Seismograms of the same earthquake recorded in three different cities.

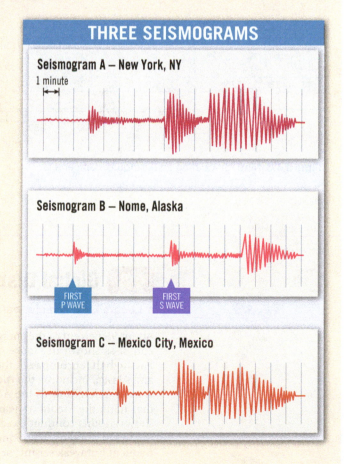

THREE SEISMOGRAMS

Seismogram A – New York, NY

1 minute

Seismogram B – Nome, Alaska

FIRST P WAVE   FIRST S WAVE

Seismogram C – Mexico City, Mexico

continued

Activity 4.2B continued

Table 4.1 Epicenter Data Table

	NEW YORK	NOME	MEXICO CITY
Elapsed time between first P waves and first S waves			
Distance from epicenter, in miles			

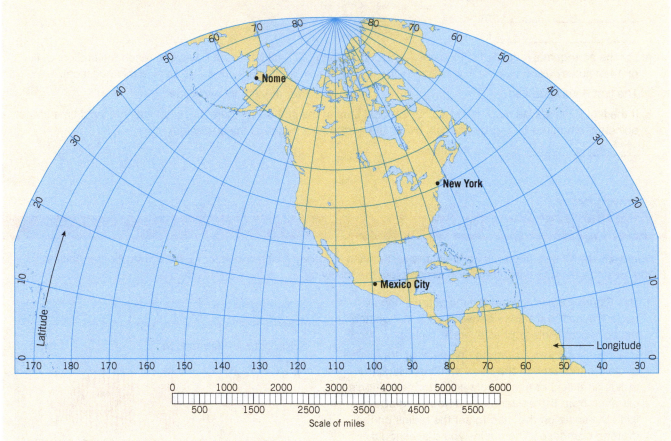

▲ **Figure 4.6** Map for locating an earthquake epicenter.

4.3 Global Distribution of Earthquakes

■ **Locate Earth's major earthquake belts on a world map.**

About 95 percent of the energy released by earthquakes originates in a few relatively narrow zones, as shown in **Figure 4.7**. The zone of greatest seismic activity, called the **circum-Pacific belt**, encompasses the coastal regions of Chile, Central America, Indonesia, Japan, and Alaska, including the Aleutian islands. Another major concentration of strong seismic activity, referred to as the *Alpine–Himalayan belt*, runs through the mountainous regions that flank the Mediterranean Sea and extends past the Himalaya Mountains. Figure 4.7 shows another continuous earthquake belt that extends for thousands of kilometers through the world's oceans. This zone coincides with the oceanic ridge system, which is an area of frequent but weak seismic activity.

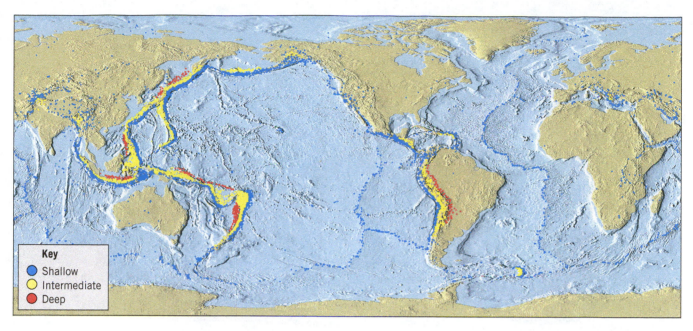

▲ **Figure 4.7** World distribution of earthquakes. (Data from NOAA)

ACTIVITY 4.3 ///
Global Distribution of Earthquakes

1. Use Figure 4.7 to answer the following questions.

   a. Do most deep-focus earthquakes occur along the margin of the Atlantic or Pacific basin?

      Most occur along the margin of the _____ basin.

   b. Do the earthquakes that occur along the western margin of South America get deeper or shallower inland from the Pacific?

      They get _____ inland from the Pacific.

   c. Are most of the earthquakes along the western margin of North America deep or shallow focus? Earthquakes along the western margin of North America have _____ focus.

2. At what depth (shallow, intermediate, or deep) do the earthquakes in the middle of the Atlantic Ocean occur?

   Earthquakes in the middle of the Atlantic Ocean occur at _____ depth.

3. With what geologic feature are the earthquakes in the mid-Atlantic associated?

   _____

4. What name is given to the zone of greatest seismic activity?

   _____

4.4 The Earth Beyond Our View

■ **List the name, thickness, and state of matter of each of Earth's layers.**

The study of earthquakes has contributed greatly to our understanding of Earth's internal structure. Variations in the travel times of P and S waves provide scientists with evidence of changes in the properties of rocks at depth.

The study of seismic waves led researchers to conclude that Earth's interior is divided into three *compositionally distinct layers*: the **crust**, **mantle**, and **core**. These layers can be further subdivided based on physical properties. Earth's major layers are illustrated in **Figure 4.8** and described below:

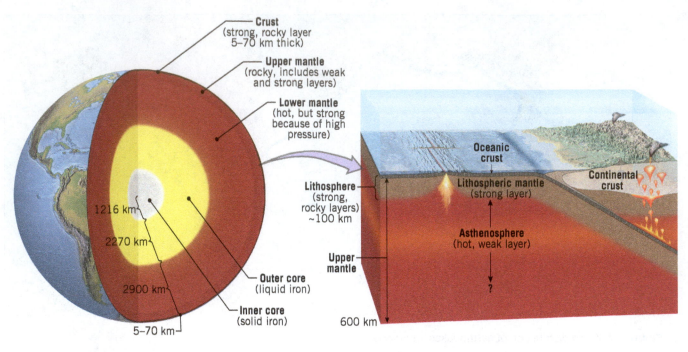

▲ **Figure 4.8** Earth's internal structure.

- **Crust** The *crust* is Earth's relatively thin, rocky outer skin. There are two types of crust: continental crust, which has an average composition of a granitic rock called granodiorite, and oceanic crust, composed of mafic (basaltic) rocks.
- **Mantle** More than 82 percent of Earth's volume is contained in the *mantle*, a rocky shell made of ultramafic rocks that extends to a depth of about 2900 kilometers (1800 miles). The mantle's physical properties vary with depth and consist of a rigid upper layer, which, combined with the continental and oceanic crust, forms the lithosphere. Beneath the lithosphere is the asthenosphere, consisting of weaker rocks. Beneath this weak layer the mantle becomes increasingly stronger (more rigid) with depth.
- **Lithosphere** The *lithosphere* consists of the entire crust and uppermost mantle and forms Earth's relatively cool, rigid outer shell that averages about 100 kilometers (62 miles) in thickness.
- **Asthenosphere** The arrow on the callout figure above implies depths closer to 500+ km. Beneath the stiff lithosphere to a depth of about 350 kilometers (200 miles) lies a soft, weak layer known as the *asthenosphere*.
- **Outer core** The *outer core* is a liquid layer 2270 kilometers (1410 miles) thick that is composed mainly of an iron–nickel alloy.
- **Inner core** The *inner core* is a solid sphere with a radius of 1216 kilometers (754 miles), and like the outer core, it is composed mainly of an iron–nickel alloy.

ACTIVITY 4.4

Discovering Earth's Interior

The study of seismic waves has improved our understanding of Earth's interior. In general, the velocities of P and S waves indicate the rigidity or stiffness of the material. Faster P and S waves indicate greater rigidity (strength). Further, S waves cannot travel through liquids because they lack rigidity. **Figure 4.9** shows the average velocities of P and S waves at various depths. Use this figure to complete the following.

1. Does the velocity of P waves and S waves increase or decrease with increased depth in the lithosphere?

   The velocity of P waves and S waves _____ with increased depth in the lithosphere.

2. What are the approximate velocities of P and S waves at the bottom of the lithosphere?

   P wave velocity: _____ km/sec

   S wave velocity: _____ km/sec

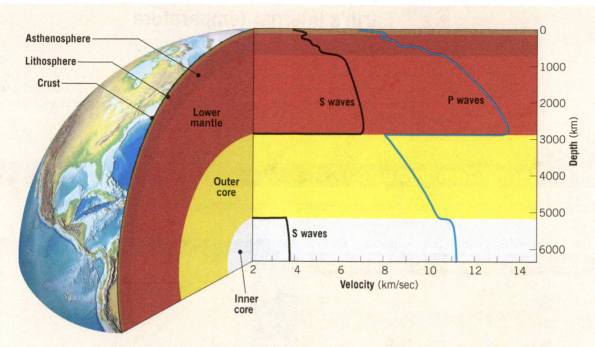

▲ **Figure 4.9** Illustration showing how P and S wave velocities vary with depth. Although S waves cannot penetrate the outer core, they can still appear in the inner core. When P waves strike the boundary between the inner core and the outer core, some of that energy is converted into S waves, which then travel though the solid inner core.

3. Does the velocity of P waves and S waves increase or decrease immediately below the lithosphere?

   The velocity of P waves and S waves _____ immediately below the lithosphere.

4. Does the change in velocity of seismic waves as they enter the asthenosphere indicate that the asthenosphere is more or less rigid than the lithosphere?

   The asthenosphere is _____ rigid than the lithosphere.

5. How does the velocity of seismic waves change with increasing depth in the lower mantle?

   _____

6. Does the change in velocity of seismic waves with increasing depth in the lower mantle indicate that the rock in the mantle becomes more or less rigid with depth?

   The rock in the mantle becomes _____ rigid with depth.

7. What happens to S waves when they reach the outer core, and what does this indicate about this layer?

   _____

   _____

8. Do P waves increase or decrease in velocity as they enter the outer core?

   The velocity of P waves _____.

9. What are the approximate velocities of P and S waves at the bottom of the mantle?

   P wave velocity: _____ km/sec

   S wave velocity: _____ km/sec

10. What are the approximate velocities of P and S waves at the bottom of the inner core?

    P wave velocity: _____ km/sec

    S wave velocity: _____ km/sec

11. Based on your answers to Questions 9 and 10, compare the rigidity of the material in Earth's inner core to that of the lowermost mantle.

    _____

4.5 **Earth's Internal Temperature**

■ **Describe the temperature gradient in Earth's interior.**

Measurements of temperatures in deep wells and mines indicate that Earth's temperatures increase with depth. The rate of temperature increase is called the **geothermal gradient**. Although the geothermal gradient varies from place to place, an average rate for a particular region can be calculated.

ACTIVITY 4.5 ///

Earth's Internal Temperature

Use Table 4.2, which shows idealized internal temperatures at various depths, to complete the following.

Table 4.2 Idealized Internal Temperatures of Earth

DEPTH (KM)	TEMPERATURE (CELSIUS)
0	20°
25	600°
50	1000°
75	1250°
100	1400°
125	1525°
150	1600°

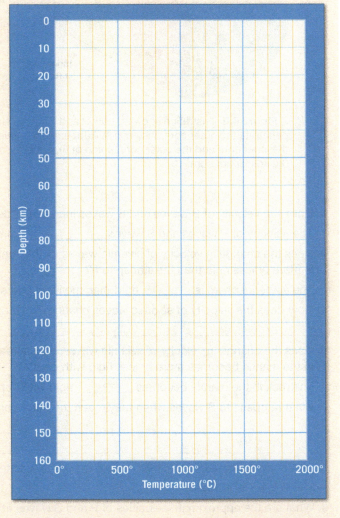

▲ **Figure 4.10** Graph for plotting temperature and melting point curves.

1. Plot the temperature values from Table 4.2 on the graph in **Figure 4.10**. Then draw a line to connect the points. Label the line *geothermal gradient*.

2. Referring to the graph, does Earth's internal temperature increase at a constant or changing rate with increasing depth?

   Earth's internal temperature increases at a _____ rate with increasing depth.

3. Is the rate of temperature increase from the surface to 100 kilometers greater or less than the rate of increase below 100 kilometers?

   The rate of temperature increase is _____.

4. Is the temperature at the base of the lithosphere, about 100 kilometers below the surface, approximately 600°C, 1400°C, or 1800°C?

   _____°C

4.6 **Melting Temperatures of Rocks**

■ **Explain why the asthenosphere likely consists of very weak material.**

The approximate melting points of the igneous rocks granite and basalt, under various pressures (depths), have been determined in the laboratory and are shown in Table 4.3. Granite that contains water and basalt were selected because they are common materials in Earth's crust.

Table 4.3 Idealized Melting Temperatures of Granite (with water) and Basalt at Various Depths Within Earth

GRANITE (WITH WATER)		BASALT (DRY)	
DEPTH (KM)	MELTING TEMP. (CELSIUS)	DEPTH (KM)	MELTING TEMP. (CELSIUS)
0	950°	0	1100°
10	700°	25	1160°
20	660°	50	1250°
30	625°	100	1400°
40	630°	150	1600°

ACTIVITY 4.6
Melting Temperatures of Rocks

1. Plot the melting temperatures for wet granite and dry basalt from Table 4.3 on the graph in Figure 4.10. Draw a line of a different color for each set of points and label them *melting curve for wet granite* and *melting curve for basalt*.

   Refer to the plots on the graph you completed in Figure 4.10 to complete the following.

2. At approximately what depth does wet granite reach its melting temperature and generate magma?

   _____ km

3. Oceanic crust and the underlying rocks to a depth of about 100 kilometers have a basaltic composition. Does the melting curve for basalt indicate that the lithosphere above approximately 100 kilometers has or has not reached the melting temperature for basalt? Therefore, at those depths, should basalt be solid or molten?

   The melting temperature _____ been reached.

   Basalt should be _____.

4. Referring to Figure 4.10, at approximately what depth does basalt reach its melting temperature?

   _____ km

5. Referring to Figure 4.9, what is the name of the layer that begins at a depth of about 100 kilometers and extends to approximately 600 kilometers?

   _____

6. Does the graph you constructed support or refute the concept of a weak asthenosphere that is capable of "flowing"?

   _____

Mastering Geology™

Looking for additional review and lab prep materials? Go to **www.masteringgeology.com** for Pre-Lab Videos, Geoscience Animations, RSS Feeds, Key Term Study Tools, The Math You Need, an optional Pearson eText, and more.

Notes and calculations:

Earthquakes and Earth's Interior

Name _____

Date _____

Course/Section _____

Due Date _____

1. Explain the difference between a seismograph and a seismogram.

   _____

   _____

2. Use **Figure 4.11** to sketch a typical seismogram in which the first P wave arrives 3 minutes ahead of the first S wave and 8 minutes ahead of the first surface wave. Label each type of wave.

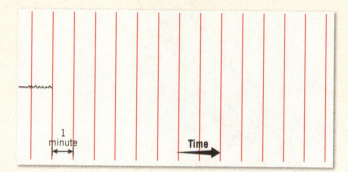

1 minute

Time

▲ **Figure 4.11** Illustration to accompany Question 2.

3. Use Figure 4.4 and Figure 4.11 to determine the distance from the earthquake epicenter to the seismic station.

   _____ mi

4. List the three zones around the globe where most earthquakes occur.

   _____

   _____

   _____

5. The velocity of S waves decreases as these waves leave the lithosphere and enter the asthenosphere. What does this suggest about the strength of the rocks in the asthenosphere compared to those in the overlying lithosphere?

   _____

   _____

6. Explain why S waves do not travel through Earth's outer core.

   _____

   _____

7. Use Figure 4.8 to complete the following about Earth's layers.

   a. The crust ranges in thickness from about _____ to _____ kilometers.

   b. The mantle is _____ kilometers thick.

   c. The outer core is _____ kilometers thick.

   d. The inner core has a radius of _____ kilometers.

8. On the graph you constructed in Figure 4.10, at what depth did you determine that wet granite would melt?

   _____ km

9. What is the location directly above an earthquake hypocenter (focus) called?

   _____

continued

10. Label Earth's major layers on **Figure 4.12**, using the following terms: core, crust, asthenosphere, mantle, outer core, lithosphere, and inner core.

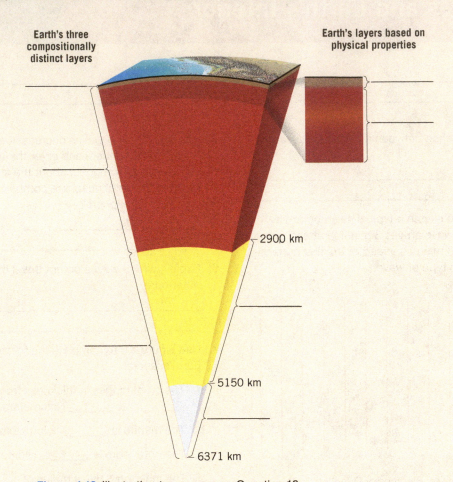

▲ **Figure 4.12** Illustration to accompany Question 10.

Volcanism and Volcanic Hazards

LEARNING OBJECTIVES

Each statement represents an important learning objective that relates to one or more sections of this lab. After you complete this exercise you should be able to:

- Explain why some volcanic eruptions are explosive and others are effusive.
- Summarize the characteristics of shield volcanoes and provide one example of this type of volcano.
- Describe the formation, size, and composition of cinder cones.
- List the characteristics of composite volcanoes and briefly describe how they form.
- Describe two major geologic hazards associated with volcanoes.

MATERIALS

string
protractor

PRE-LAB VIDEO ▶ https://goo.gl/bCcYpY

 Prepare for lab! Prior to attending your laboratory session, view the pre-lab video. Each video provides valuable background that will contribute to your understanding and success in lab.

INTRODUCTION

Volcanoes are a source of human curiosity—for scientists and nonscientists alike. The study of volcanoes provides the only means of directly observing processes that occur many kilometers below Earth's surface. Volcanic eruptions are among the most spectacular, yet destructive, natural phenomena on Earth.

PART 1 Geology

5.1 The Nature of Volcanic Eruptions

■ **Explain why some volcanic eruptions are explosive and others are effusive.**

Volcanic activity is commonly perceived as a process that produces cone-shaped structures that periodically erupt in a violent manner. However, many eruptions are not explosive. What determines the manner in which volcanoes erupt?

Recall that **magma**, molten rock that may contain some solid crystalline material and also varying amounts of dissolved gas (mainly water vapor and carbon dioxide), is the parent material of volcanic eruptions. Erupted magma is called **lava**. Most magma that erupts to form a volcano falls into one of three main compositional groups: *basaltic (mafic)*, *andesitic (intermediate)*, and *rhyolitic (felsic)*. The compositional differences among magmas also affect several other properties, as summarized in Figure 5.1.

Basaltic (mafic) magmas have the *lowest* silica (SiO_2) content, have the *least* gas content, and erupt at the *highest* temperatures. As a result, basaltic magmas have the *lowest* **viscosity** (*viscos* = sticky) and produce the most fluid lavas. Basaltic lavas can flow for several kilometers or more before congealing. Geologists refer to quiescent eruptions of fluid basaltic lava as **effusive eruptions** (*effus* = pour forth). *Shield volcanoes*, like those of the Hawaiian islands, are composed mainly of fluid basaltic lava flows.

Although basaltic magma tends to generate lava flows, sometimes gases accumulate near the top of a basaltic magma chamber. Upon escaping, these gases eject lava fragments that produce a spectacular display called a **lava fountain**. This process generates **pyroclastic material**, also called **tephra**, including pea- to walnut-sized fragments called *cinders*, and larger *volcanic bombs* that can still be partially molten when they land. When pyroclastic material accumulates around a single vent, it forms a *cinder cone*. These comparatively small volcanoes commonly form on the flanks of much larger shield volcanoes.

By contrast, rhyolitic (felsic) magmas have the *highest* silica content, have the *highest* gas content, and erupt at the *lowest* temperatures. The more silica (SiO_2) in magma, the greater its viscosity. Because felsic magmas are very viscous ("sticky" and resistant to flow), they often crystallize at depth. On those occasions when they reach Earth's surface, they are generally too stiff to flow. Instead, the magma extrudes like toothpaste coming out of a tube and piles up around the vent to form a bulbous structure called a **lava dome**. Like lava flows, lava domes typically do not have enough accumulated gas pressure to erupt explosively, although they often precede or are followed by explosive activity. Although rare, felsic magmas can also generate extraordinarily massive volcanic eruptions, as exemplified by the explosive activity that produced the landscape of Yellowstone National Park.

Properties of Magma Bodies with Differing Compositions

Composition	Silica Content (SiO_2)	Gas Content (% by weight)	Eruptive Temperature	Viscosity	Tendency to Form Pyroclastics	Volcanic Landform
Basaltic (MAFIC) High in Fe, Mg, Ca, low in K, Na	**Least** (~50%)	**Least** (0.5–2%)	**Highest** 1000–1250°C	**Least**	**Least**	Shield volcanoes, basalt plateaus, cinder cones
Andesitic (INTERMEDIATE) Varying amounts of Fe, Mg, Ca, K, Na	Intermediate (~60%)	Intermediate (3–4%)	Intermediate 800–1050°C	Intermediate	Intermediate	Composite cones
Rhyolitic (FELSIC) High in K, Na, low in Fe, Mg, Ca	**Most** (~70%)	**Most** (5–8%)	**Lowest** 650–900°C	**Greatest**	**Greatest**	Pyroclastic flow deposits, lava domes

▲ **Figure 5.1** Compositional differences of magma bodies cause their properties to vary.

Andesitic (intermediate) magmas have characteristics that range between those of basaltic (mafic) magmas and rhyolitic (felsic) magmas. These magmas tend to form *composite volcanoes*, which consist of interbedded layers of lava flows and pyroclastic material. Sometimes andesitic magmas produce effusive eruptions consisting of lava flows that tend to be thicker and much shorter than fluid basaltic lavas. They also generate explosive eruptions, like the 1980 eruption of Mount St. Helens. Because andesitic magmas (as well as the even more viscous rhyolitic magmas) are relatively resistant to flow, gas bubbles tend to remain trapped in the magma, forming a sticky froth. When these highly gaseous magmas erupt, they expel pyroclastic material at nearly supersonic speeds, producing **eruption columns** consisting mainly of volcanic ash, pumice, and gas.

ACTIVITY 5.1
The Nature of Volcanic Eruptions

1. Define *magma*.

   _____

2. List these magmas in order from the lowest to highest silica (SiO_2) content: basaltic (mafic), rhyolitic (felsic), and andesitic (intermediate).

   _____

3. Based on their silica content, which of these magmas is the most viscous? Least viscous?

   Most viscous:

   _____

   Least viscous:

   _____

4. Which of these magmas tend to produce effusive (quiescent) eruptions?

   _____

5. Which of these magmas produce the most explosive eruptions?

   _____

6. What two types of volcanoes are associated with eruptions of fluid basaltic lavas?

   _____

7. With what type of volcano are andesitic (intermediate) magmas most associated?

   _____

Use **Figure 5.2** to complete Question 8.

8. Fill in each blank with the number of the image that illustrates each of the following volcano-related phenomena.

   a. Lava fountain _____

   b. Fluid lava flow _____

   c. Volcanic dome _____

   d. Pyroclastic material (tephra) _____

   e. Eruption column _____

   f. Silica-rich lava flow _____

(1)   (2)

(3)   (4)

(5)   (6)

▲ **Figure 5.2** Images to accompany Question 8.

5.2 Shield Volcanoes

■ **Summarize the characteristics of shield volcanoes and provide one example of this type of volcano.**

Shield volcanoes are some of the largest volcanoes on Earth and exhibit the shape of a broad, slightly domed structure that resembles a warrior's shield (**Figure 5.3**). Most shield volcanoes form on the ocean floor as **seamounts** (submarine volcanoes), with some growing large enough to form volcanic islands. Examples include the Hawaiian islands in the Pacific and the Canary Islands in the Atlantic. Both volcanic island chains were generated by mantle plumes rather than forming along a plate boundary.

Although less common, some shield volcanoes form on continental crust. Included in this group are Nyamuragira, Africa's most active volcano, and Newberry Volcano, Oregon.

Horizontal

▲ **Figure 5.3** Mauna Loa, Earth's largest volcano, is one of five shield volcanoes that collectively make up the Big Island of Hawaii. Shield volcanoes are built primarily of fluid basaltic lava flows and contain only a small percentage of pyroclastic materials. (USGS)

ACTIVITY 5.2 //

Shield Volcanoes

Use Figure 5.3, which is an image of Mauna Loa, Earth's largest volcano, to complete Questions 1–3.

1. Measure the slope of Mauna Loa's flank using a protractor. (*Note:* Use the dashed horizontal line to align the bottom of your protractor.)

   _____ °

2. What factor contributes to the gentle slopes characteristic of shield volcanoes?

   _____

3. What are the small "hills" covered in vegetation in the foreground of this image? (*Hint:* See Figure 5.5.)

   _____

4. **Figure 5.4** is an image of an eruption on Kilauea, a small shield volcano located on the flanks of Mauna Loa. Would you describe the lava as having a low or high *viscosity*? Explain your reasoning.

   _____

◄ Figure 5.4 Eruption of Kilauea that includes a large lava fountain producing a fluid lava flow. (Photo by David Reggie/Getty Images)

5. Does the lava shown in Figure 5.4 most likely have a low or high *silica* content?

_____

6. Based on your answer to Question 5, what is the likely composition of the magma that formed Kilauea?

_____

5.3 Cinder Cones

■ **Describe the formation, size, and composition of cinder cones.**

Cinder cones (also called **scoria cones**) are built from ejected lava fragments that begin to harden in flight to produce chunks of pyroclastic material (**Figure 5.5**). Most of the volume of a cinder cone consists of pea- to walnut-sized fragments that are markedly *vesicular* (*vesicula* = cavities) called **scoria**. Scoria is black to reddish-brown in color and usually has a basaltic composition. Cinder cones are relatively small, usually less than 300 meters (1000 feet) high, and may form during a single, nearly continuous eruption lasting several months.

Some cinder cones produce lava flows, which generally form in the final stages of the volcano's life span, when the underlying magma body has lost most of its gas content. Because cinder cones are composed of loose fragments rather than solid rock, lava usually flows out from the unconsolidated base of the cone rather than from its crater. This is the case for the lava flow shown in Figure 5.5.

Cinder cones number in the thousands around the globe. Some occur in groups, such as the volcanic field near Flagstaff, Arizona, which consists of about 600 cones. Others are parasitic cones on the flanks of other volcanoes, usually shield volcanoes, or within large volcanic structures called *calderas*.

▶ **SmartFigure 5.5** Cinder cones, built from ejected lava fragments (mostly cinders and bombs), are relatively small— usually less than 300 meters (1000 feet) in height. (Photo by Michael Collier)

MOBILE FIELD TRIP
https://goo.gl/RddNBN

ACTIVITY 5.3
Cinder Cones

Use **Figure 5.6**, a side view of SP Crater (an average-size cinder cone), to answer Questions 1–4.

1. Using a protractor, measure the slope of SP Crater. (*Note:* Use the dashed horizontal line to align the bottom of your protractor.)

   _____ °

2. How does the slope of SP Crater compare to the slope you determined for Mauna Loa in Activity 5.2?

   _____

   _____

3. Using the scale, measure the height of SP Crater from its base to the crater rim, indicated with an arrow. Round your answer to the nearest 100 feet.

   _____ ft

4. Based on your answer to Question 3, compare the size of a cinder cone to the size of a shield volcano.

   _____

   _____

5. Based on Figure 5.5, compare the size of the *crater* of a typical cinder cone to the overall size of the volcano.

   _____

   _____

6. What is the name for the ejected lava fragments that include pea- to walnut-sized pieces?

   _____

7. Given the nature of the materials that make up a cinder cone and the steepness of its slopes, describe what it would be like to climb from the base to the rim.

   _____

   _____

8. Using **Figure 5.7** and a piece of string, measure the length of the flow from SP Crater by following the curve along the center of the flow.

   _____ km

9. Would you describe the lava that produced this flow as fluid or viscous? Explain.

   _____

   _____

▲ **Figure 5.6** Photo of SP Crater to accompany Questions 1–4. (USGS)

▲ **Figure 5.7** Satellite view of SP Crater to accompany Questions 8 and 9. (NASA)

5.4 Composite Volcanoes

■ **List the characteristics of composite volcanoes and briefly describe how they form.**

Earth's most picturesque, yet potentially most dangerous volcanoes are **composite volca-noes**—also known as **stratovolcanoes**. They consist of layers (strata) of explosively erupted cinders and ash, interbedded with hardened lava flows. Unlike shield volcanoes, composite volcanoes are characterized by their steep profile and periodic explosive eruptions that eject huge quantities of pyroclastic material.

Classic composite cones are nearly symmetrical structures. In general, composite vol-canoes are the product of viscous silica-rich magmas having an andesitic (intermediate) composition. Such magmas tend to produce relatively short lava flows and abundant pyroclastic material. They can, however, sometimes erupt highly viscous rhyolitic (felsic) lava and occasionally fluid basaltic (mafic) lava.

ACTIVITY 5.4
Composite Volcanoes

Use **Figure 5.8**, an image showing the eruption of Mount St. Helens on May 18, 1980, to complete Questions 1–3.

1. Would you describe this as an effusive or explosive eruption? Explain your reasoning.

_____

_____

▲ **Figure 5.8** Photo of Mount St. Helens on May 18, 1980, to accompany Questions 1–3. (USGS)

continued

Activity 5.4 continued

2. What name is given to the plume rising from the volcano?

_____

3. List the primary materials that compose the Mount St. Helens plume.

_____

Use **Figure 5.9**, an image of Oregon's Mount Hood as viewed from the Columbia River, to complete Questions 4–8.

4. Measure the slope (angle) near the top of Mount Hood along the red line by using a protractor. (*Note:* Use the dashed horizontal line to align the bottom of your protractor.)

_____ °

5. Measure the slope near the base of Mount Hood along the red line by using a protractor.

_____ °

6. Describe how the slope of Mount Hood changes from the base of the volcano toward its summit.

_____

_____

7. What is the primary cause of the change in slope you measured? Refer to your textbook or the Internet if needed.

_____

_____

8. Describe the differences you noticed between the shape and angle of the slope you measured for Mount Hood as compared to the slope of Mauna Loa shown in Figure 5.3.

_____

_____

▲ **Figure 5.9** Photo of Mount Hood to accompany Questions 4–8. (USGS)

5.5 Volcanic Hazards

■ **Describe two major geologic hazards associated with volcanoes.**

Nearly 300,000 people are known to have been killed by volcanic eruptions since c.e. 1600. Today, an estimated 500 million people live near active volcanoes in places such as Japan, Indonesia, Italy, and the northwestern United States. They face a number of volcanic hazards, such as pyroclastic flows, lahars, molten lava flows, and falling ash and volcanic bombs (**Figure 5.10**).

Some of the most destructive volcanic hazards are **pyroclastic flows**, which consist of hot gases mixed with glowing ash and larger lava fragments that race down steep volcanic slopes at speeds up to 700 kilometers (450 miles) per hour, laying waste to nearly everything in their path. Another major hazard is a type of fluid mudflow, known by its Indonesian name, **lahar**. Lahars occur when volcanic debris becomes saturated with water and rapidly moves down steep volcanic slopes, generally following stream valleys. Some lahars are triggered when magma approaches the surface of a glacially clad volcano, causing large volumes of ice and snow to melt. Others are generated when earthquakes trigger the movement of poorly consolidated, water-saturated volcanic deposits.

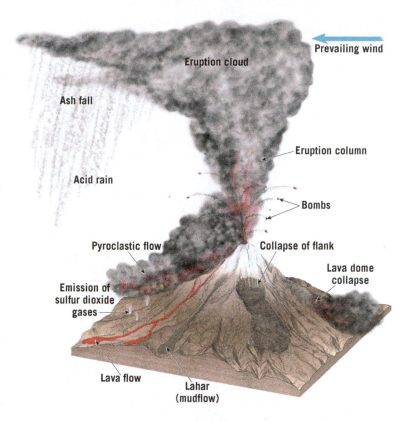

▶ **Figure 5.10** In addition to generating lahars and pyroclastic flows, volcanoes can be hazardous to human health in many other ways.

ACTIVITY 5.5A
Volcanic Hazards: Mount Rainier

Mount Rainier is an active volcano in the Cascade Range in Washington State, located about 50 kilometers (30 miles) southeast of the Seattle–Tacoma metropolitan area (**Figure 5.11**). Lahars are common at Mount Rainier because of its mantle of snow and glacial ice. A lahar can be trigged by an earthquake, which sets in motion voluminous, water-saturated rocky debris from the upper flanks of the volcano. Most lahars, however, form during an eruption when hot rock energetically mixes with snow and ice to form a sudden flood of water and rocky debris having the consistency of wet concrete. A lahar can travel down a slope at speeds exceeding 30 kilometers (20 miles) per hour.

Mount Rainier produces a major lahar every 500 to 1000 years and smaller ones more frequently. The most recent major lahar, the Electron Lahar, occurred about 600 years ago and was more than 30 meters (100 feet) thick where the community of Electron, Washington, is located today. A much more massive lahar, called the Osceola Lahar, occurred about 5600 years ago (**Figure 5.12**). The Osceola Lahar formed when the volcano's summit and northeast flank collapsed, producing a horseshoe-shaped crater. This scar was subsequently filled with lava flows.

▲ **Figure 5.11** Mount Rainier as seen from Puget Sound near Tacoma, Washington, which may be impacted by the next lahar to flow down from the volcano. (Photo by Carolyn Driedger, USGS)

continued

Activity 5.5A continued

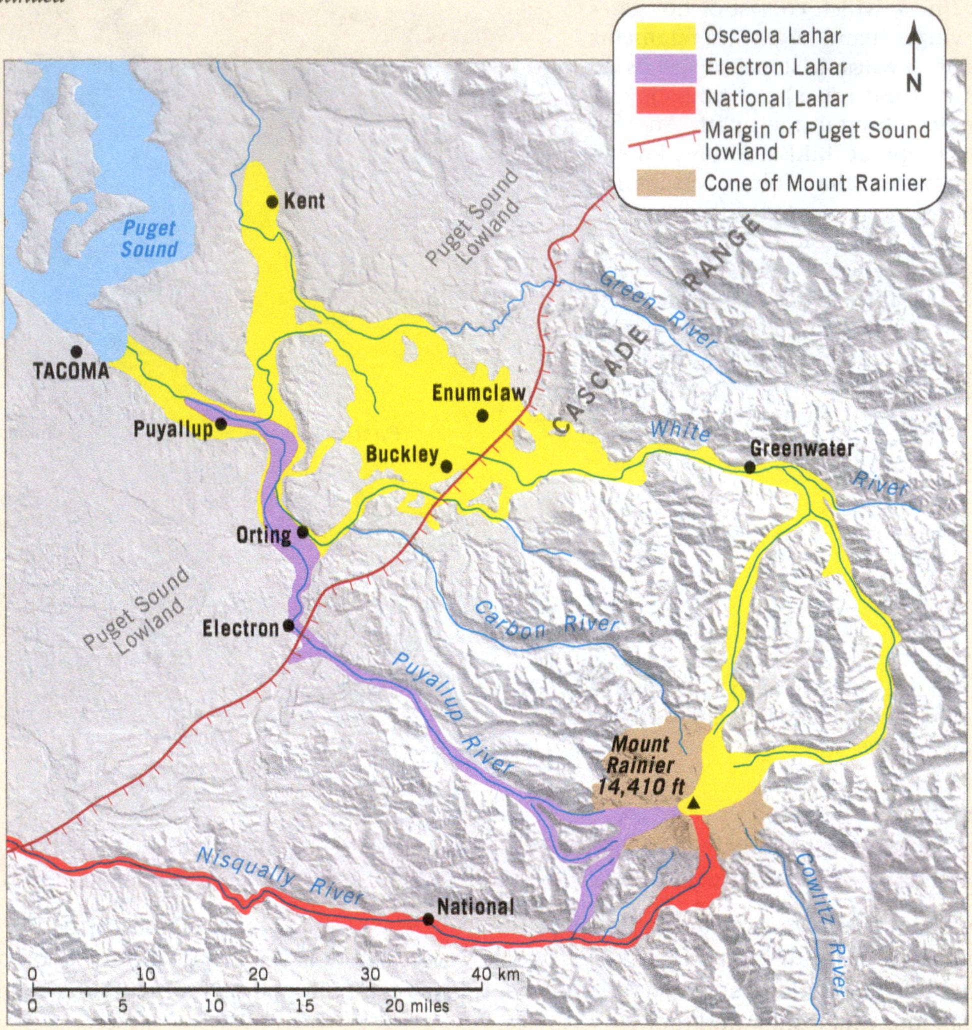

▲ **Figure 5.12** Map showing some notable lahars that flowed from Mount Rainier. The most recent major lahar, the Electron Lahar, occurred about 600 years ago and was more than 30 meters (100 feet) thick where the present community of Electron, Washington, is located. The much larger Osceola Lahar, which occurred about 5600 years ago, had a thickness of at least 90 meters (300 feet) in one location.

Osceola's deposits cover an area of about 550 km^2 (212 mi^2) in the Puget Lowland and extend at least as far as the Seattle suburb of Kent. One arm of the lahar entered Commencement Bay, which is now the site of the Port of Tacoma.

1. Based on the image in Figure 5.11, what type of volcano is Mount Rainier?

_____

2. This image also shows that the upper portion of Mount Rainier is covered with snow and glacial ice. How might this snow and ice contribute to future volcanic hazards?

_____

_____

3. What term is used for mudflows that form on volcanoes?

_____

4. Can mudflows be triggered when a volcano is not erupting? Explain.

_____

_____

5. Notice in Figure 5.12 that the Osceola Lahar was confined mainly to river valleys until it reached the Puget Lowland, where it began to spread out. What was the probable cause of this change?

_____

_____

6. Using a string, measure the distance on Figure 5.12 from the top of Mount Rainier to the city of Kent, Washington, carefully following the easternmost path of the lahar. (*Note:* See the North arrow.) What is the maximum distance traveled by the Osceola Lahar?

_____ km

7. Assuming that the Osceola Lahar traveled at an average speed of 30 kilometers (20 miles) per hour, how long would it have taken to travel from Mount Rainier to the town of Kent? Use your answer to Question 6 as the distance traveled.

_____ hr

8. If monitoring stations were established around the base of Mount Rainier to send out a warning whenever a lahar is triggered, do you think that effort would have the potential to help save lives? Explain.

_____

_____

9. What material would be incorporated into a lahar that would allow scientists to date a prehistoric mudflow using carbon-14 dating? (See the section "Dating with Carbon-14" in Chapter 11 of your textbook, if needed.)

_____

10. What value is gained from dating prehistoric occurrences of lahars?

_____

_____

ACTIVITY 5.5B

Volcanic Hazards: Mount St. Helens

On May 18, 1980, a destructive volcanic eruption occurred on Mount St. Helens in Washington. The event was triggered by an earthquake that generated a massive landslide, and the resulting loss of overlying material exposed magma in the volcano's neck to much lower pressures. This reduction in confining pressure caused the highly viscous, gas-charged molten rock to blast out the entire north flank of the volcano. The resulting lateral blast, consisting of a mixture of very hot volcanic gases, ash, and pulverized older rock, flowed northward from the volcano toward Spirit Lake at speeds that reached 1080 kilometers per hour (670 miles per hour). By the time this highly energetic pyroclastic flow reached its first human victims, it was still about 360°C (680°F). The eruption claimed 57 lives; some died from the intense heat and the suffocating cloud of ash and gases, others from the impact of the blast, and still others from being trapped by the lahars that followed (Figure 5.13).

At the same time, large quantities of hot gases and ash were propelled vertically roughly 24 kilometers (over 15 miles) into the atmosphere. During the next few days, this very fine-grained material was carried around Earth by strong upper-air winds. Crops were damaged in central Montana, and measurable deposits were reported as far away as Oklahoma and Minnesota (Figure 5.14).

▲ **Figure 5.13** The vehicle of a *National Geographic* photographer, located about 16 kilometers (10 miles) from the volcano, following the eruption Mount St. Helens on May 18, 1980. (USGS)

continued

Activity 5.5B continued

Following the initial blast, at least 17 additional pyroclastic flows occurred, all of which were products of "fresh" lava that had moved up the vent. Two weeks later, when these pyroclastic flows were safe to approach, they were still at temperatures of 300° (570°). Most contained volcanic ash, as well as boulder-size chunks of pumice. Additional pyroclastic flows, like the one shown in **Figure 5.15**, occurred sporadically during the following year.

1. What name is given to fast-moving currents of hot gases and volcanic material that move at very high speeds down the flanks of volcanoes?

   _____

2. The average chemical composition of the ash from the Mount St. Helens eruption included about 65 percent silica (SiO_2). Describe where the magma that produced this eruption would fit within the range of compositions shown on the chart in Figure 5.1.

   _____

   _____

3. Did the magma that produced the May 18, 1980, eruption of Mount St. Helens have a relatively low or high silica content? Explain.

   _____

   _____

4. Use **Figure 5.16**, which shows the area destroyed by the lateral blast, to measure the distance from the center of Mount St. Helens to the farthest point that the blast reached, indicated by the letter **A**. (*Note*: It is estimated that it took the lateral blast less than 2 minutes to reach this location.)

   _____ km, _____ m

5. Use Figure 5.14 to determine how much volcanic ash blanketed the town of Ritzville, Washington.

   _____

6. What is the furthest state from the volcano that received a measurable amount of volcanic ash during the Mount St. Helens eruption? (Use the distance scale if needed.)

   _____

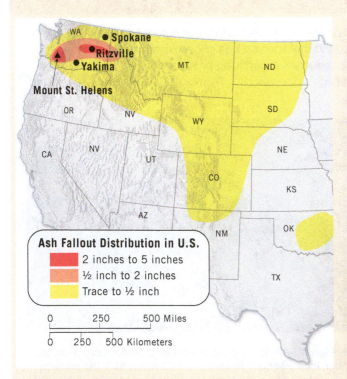

▲ **Figure 5.14** Map showing the distribution of volcanic ash from the eruption of Mount St. Helens on May 18, 1980.

▲ **Figure 5.15** Pyroclastic flow from the August 7, 1980, eruption of Mount St. Helens. (Photo by Peter Lipman/USGS)

7. In what general direction were the upper-level winds blowing during the eruption?

_____

8. Which of the three types of volcanoes discussed pose the greatest threat to property and human life?

_____

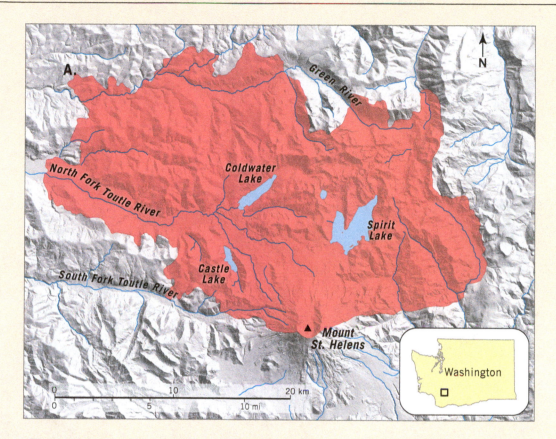

▲ **Figure 5.16** Area affected by the lateral blast of the May 18, 1980 eruption of Mount St. Helens.

MasteringGeology™   Looking for additional review and lab prep materials? Go to www.masteringgeology.com for Pre-Lab Videos, Geoscience Animations, RSS Feeds, Key Term Study Tools, The Math You Need, an optional Pearson eText, and more.

Notes and calculations:

Volcanism and Volcanic Hazards

Name _____

Date _____

Course/Section _____

Due Date _____

1. Define *magma*.

   _____

   _____

2. What type of magma has the lowest silica (SiO$_2$) content? Highest silica content?

   Lowest: _____ Highest: _____

3. What type of magma has the lowest viscosity? Highest viscosity?

   Lowest: _____ Highest: _____

4. Does the lava flow shown in **Figure 5.17** have a high or low viscosity?

   _____

5. What name is given to the nongaseous material ejected into the air by a volcanic eruption?

   _____

6. Which type of magma tends to produce effusive (quiescent) eruptions?

   _____

7. Which type of magma tends to produce lava domes?

   _____

8. What two types of volcanoes are associated with eruptions of fluid basaltic lavas?

   _____

9. Describe the basic shape of a shield volcano.

   _____

   _____

10. What type of volcano is shown in **Figure 5.18**?

    _____

11. What name is applied to the material being mined from the volcano shown in Figure 5.18?

    _____

12. Describe how the shape (slope) of a composite volcano changes from its base to its top.

    _____

    _____

    _____

▲ **Figure 5.17** Photo to accompany Question 4.

▲ **Figure 5.18** Photo to accompany Question 10 and 11.

13. What type of volcanos are Earth's largest?

_____

14. What hazard does Mount Rainier pose to the Seattle–Tacoma metropolitan area?

_____

15. Figure 5.19 shows a profile view of Newberry Volcano as seen from Pilot Butte in Bend, Oregon. Based on its profile, shown with a blue line, what type of volcano is Newberry Volcano?

_____

16. What are the small structures that are visible on the lower flanks of Newberry Volcano in Figure 5.19?

_____

▲ **Figure 5.19** Photo to accompany Question 15 and 16.

Geologic Maps, Block Diagrams, and Rock Structures

LEARNING OBJECTIVES

Each statement represents an important learning objective that relates to one or more sections of this lab. After you complete this exercise you should be able to:

- **Describe the three types of differential stress.**
- **Explain the orientation of folded rocks and faults, using the terms** *strike* **and** *dip*.
- **Draw and interpret a simple geologic block diagram and complete a geologic cross section.**
- **Recognize common structural features, including anticlines, synclines, domes, and basins.**
- **Recognize the common types of faults.**
- **Interpret a simple geologic map.**

MATERIALS

ruler

colored pencils

hand lens

protractor

PRE-LAB VIDEO ▶ https://goo.gl/nK9Kk6

 Prepare for lab! Prior to attending your laboratory session, view the pre-lab video. Each video provides valuable background that will contribute to your understanding and success in lab.

INTRODUCTION

Structural geology is the study of the architecture and evolution of Earth's crust. The geologic features that result from the interactions of Earth's plates, called *rock* or *geologic structures*, include *folds* (wave-like undulations), *faults* (fractures along which one rock body slides past another), and *joints* (fractures that display no movement). This exercise explores the forces that deform rocks and the resulting rock structures.

PART 1 Geology

6.1 Deformation

■ **Describe the three types of differential stress.**

Every body of rock, no matter how strong, has a point at which it will deform by bending or breaking. **Deformation** is a general term that refers to the changes in the shape or position of a rock body in response to differential stress. Most crustal deformation occurs along plate boundaries, where plate motions generate the tectonic forces that cause rock to deform.

Stress: The Force that Deforms Rocks

Geologists use the term **stress** to describe the forces that deform rocks. When stress is applied unequally in different directions, it is termed **differential stress**. We will consider three types of differential stress (**Figure 6.1**):

- **Compressional stress** Differential stress that squeezes a rock mass as if placed in a vise is known as *compressional stress*. Compressional stresses are most often associated with convergent plate boundaries.
- **Tensional stress** Differential stress that pulls apart or elongates rock bodies is known as *tensional stress*. Along divergent plate boundaries, where plates are moving apart, tensional stresses stretch and lengthen rock bodies.
- **Shear stress** Differential stress can also cause rock to shear, which involves the movement of one part of a rock body past another. Shearing is perhaps most easily observed along a fault where one rock body moves relative to another.

Types of Deformation

When stress is applied *gradually*, rocks initially respond by deforming elastically. Changes that result from **elastic deformation** are recoverable; that is, like a rubber band, the rock will snap back to nearly its original size and shape when the stress is removed. Once the elastic limit (strength) of a rock is surpassed, it either bends or breaks.

At low temperatures and pressures found near Earth's surface, most rocks behave like a brittle solid and fracture or break when the strength of the rock is surpassed. This type of deformation is called **brittle deformation**. If the rocks on either side of the fracture move, the geologic feature is called a **fault**. A **joint** is a fracture or break in a rock along which there has been no displacement.

▼ **Figure 6.1** Diagram showing the type of stress and resulting deformation of rock layers.

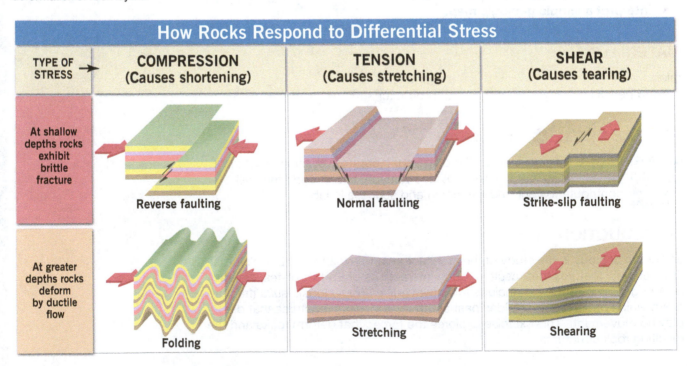

How Rocks Respond to Differential Stress			
TYPE OF STRESS →	**COMPRESSION** (Causes shortening)	**TENSION** (Causes stretching)	**SHEAR** (Causes tearing)
At shallow depths rocks exhibit brittle fracture	Reverse faulting	Normal faulting	Strike-slip faulting
At greater depths rocks deform by ductile flow	Folding	Stretching	Shearing

Ductile deformation, which occurs at high temperatures and pressures deep within Earth, is a type of solid-state flow that produces a change in the shape of an object without fracturing. The result is often wavelike undulations called **folds**.

©2019 Pearson Education, Inc.

ACTIVITY 6.1
Deformation

1. Define *rock deformation* in your own words.

_____

2. List the three types of differential stress.

_____

3. What type of plate boundary is most commonly associated with compressional stress?

_____

4. In **Figure 6.2**, describe the type of deformation you would expect if you dropped each of the items listed onto a concrete floor.

Elastic, Brittle, or Ductile Deformation	
Material	**Type of Deformation**
Wine glass	
Modeling clay	
China plate	
Tennis ball	
Pizza dough	

▲ **Figure 6.2** Common materials that exhibit either elastic, brittle, or ductile deformation when dropped on a concrete floor.

5. Are *folds* examples of brittle or ductile deformation? Are *faults* examples of brittle or ductile deformation?

Folds: _____ deformation   Faults: _____ deformation

6.2 Describing the Orientation of Geologic Structures: Strike and Dip

■ Explain the orientation of folded rocks and faults, using the terms *strike* and *dip*.

By studying the orientations of tilted sedimentary strata and faults, geologists can often reconstruct the geologic structures that lie beneath the surface. In many locations, much of the bedrock is concealed by vegetation or buried by recent sedimentation. Consequently, reconstruction must be done using data gathered from a limited number of **outcrops**, sites where bedrock is exposed at the surface.

Geologists use measurements called *strike* (trend) and *dip* (inclination) to define the orientation of a rock layer or fault (**Figure 6.3**). By knowing the strike and dip of rocks at the surface, geologists can predict the nature and structure of rock units that are hidden beneath the surface.

Strike is the compass direction of the line produced by the intersection of an inclined rock layer or fault with a horizontal plane at the surface (Figure 6.3). The strike, or compass bearing, is generally expressed as an angle relative to north. For example, "north 10° east" (N10°E) means the line of strike is 10 degrees to the east of north. The strike of the rock units illustrated in Figure 6.3 is north 42° east (N42°E).

Dip is the angle of inclination of the surface of the rock unit or fault from the horizontal plane. Dip includes both an angle of inclination and a direction toward which the rock

▶ **Figure 6.3** Illustration of the strike and dip of a rock layer.

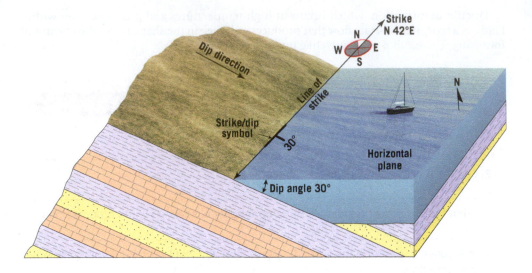

is inclined. In Figure 6.3 the dip angle of the rock layer is 30°. The direction of dip will always be at a 90° angle to the strike. (To illustrate, hold a closed book at an angle to the tabletop. The upper edge of your book represents the strike. Regardless of the way you point the book, the direction of dip of the book is always at 90°, or at a right angle, to the strike.)

Typically, the strike and dip of rock units and faults are shown on geologic maps with standard symbols that provide descriptive information about them—for example, $\overline{_{20°}}$. The long line shows the strike direction, the short line indicates the direction of the dip, and the dip angle is noted. In Figure 6.3, the strike-dip symbol indicates that the rocks are dipping toward the southeast at a 30° angle from the horizontal plane (30°SE).

ACTIVITY 6.2

Strike and Dip

Figure 6.4 illustrates geologic map views (views from directly overhead) of two hypothetical areas, showing a strike-dip symbol for a rock layer in each area.

1. Complete the information requested below each map.
2. Draw a single large arrow on each map to indicate the direction of dip of the rock layer.

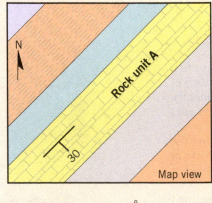

Strike: North _____° _____

Direction of dip: _____

Angle of dip: _____°

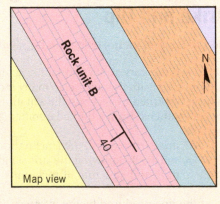

Strike: North _____° _____

Direction of dip: _____

Angle of dip: _____°

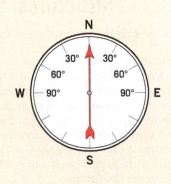

▲ **Figure 6.4** Geologic map views of two hypothetical areas.

6.3 Geologic Maps and Block Diagrams

■ Draw and interpret a simple geologic block diagram and complete a geologic cross section.

To interpret a region's geologic history, geologists often construct geologic maps and block diagrams of the area (**Figure 6.5**). A **geologic map** is a representation of Earth's surface, as viewed from above, that shows the locations and orientations of the rock units that outcrop at the surface (Figure 6.5B). Geologists use block diagrams to illustrate both the **map view** and usually two **cross-sectional views** of an area Hence, a **block diagram** is a small three-dimensional view of a portion of Earth's crust that allows you to visualize rock layers at the surface as well as underground (Figure 6.5C).

To construct these models, geologists first go into the field and measure the strike (trend) and dip (inclination) of sedimentary strata at as many outcrops as is practical. These data, and a description of each rock unit, are then plotted on a map or an aerial photograph, using T-shaped strike and dip symbols. From the orientation of the strata at the surface, geologists try to determine the orientation and shape of the buried structures (see Figure 6.5C).

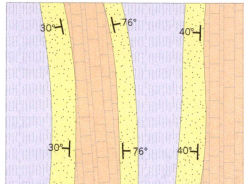

A. Study area

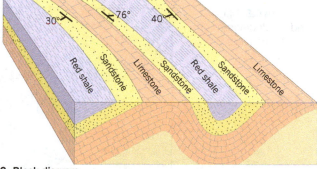

B. Map view

C. Block diagram

▶ **Figure 6.5** Diagram illustrating **A.** a surface view of the study area; **B.** a map view of the area as it would appear on a geologic map; and **C.** a block diagram. By establishing the strike and dip of outcropping sedimentary beds on a map, geologists infer the orientation of the structure below ground.

ACTIVITY 6.3

Using Block Diagrams

Use the block diagram in Figure 6.5C as a guide to complete Questions 1–3.

1. Complete the cross sections on the front and side of the block diagram in **Figure 6.6**.

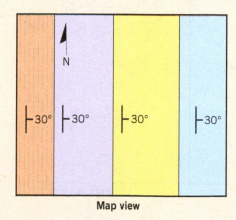

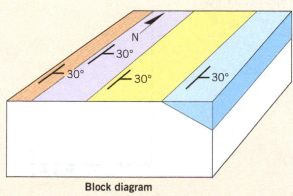

▶ **Figure 6.6** Map view and block diagram of a hypothetical area.

Map view   Block diagram

continued

Activity 6.3 continued

2. Complete the cross section of the side of the block diagram in **Figure 6.7**.

▶ **Figure 6.7** Map view and block diagram of a hypothetical area.

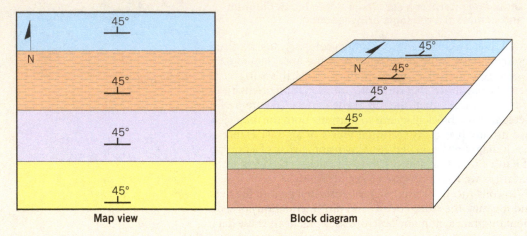

Map view   Block diagram

3. Complete the front and side of the block diagram in **Figure 6.8**.

▶ **Figure 6.8** Map view and block diagram of a hypothetical area.

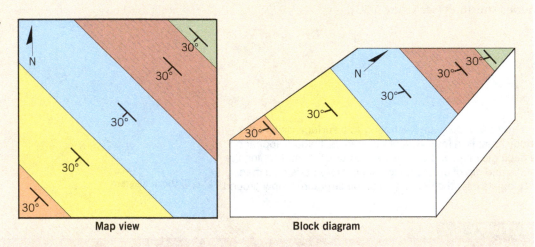

Map view   Block diagram

4. Use the guidelines on the top of page 99 to sketch both a map view and a block diagram on **Figure 6.9**.

▶ **Figure 6.9** Map view and block diagram of a hypothetical area.

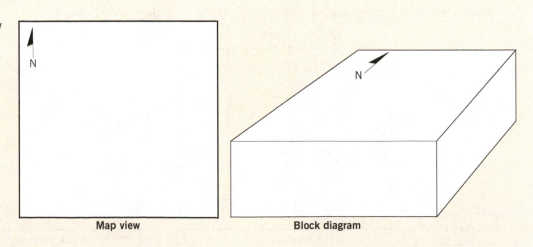

Map view   Block diagram

a. Draw four sedimentary layers of equal thickness that have a strike of N90°E. (*Hint:* Sketch the map view first.)

b. Each layer is dipping to the south at an angle of 60°.

c. On the block diagram complete the map view and both cross-sectional views.

d. Add strike-dip symbols for each rock layer on both parts of Figure 6.9.

6.4 Folds: Rock Structures Formed by Ductile Deformation

■ **Recognize common structural features including anticlines, synclines, domes, and basins.**

During mountain building, formerly flat-lying layers of sedimentary or volcanic rocks are often deformed into a series of waves called *folds*. Anticlines and synclines are the two most common types of folds (**Figure 6.10**). Rock layers that fold upward, forming an arch, are called **anticlines**. Often associated with anticlines are downfolds, or troughs, called **synclines**.*

To understand folds and folding, it is important to become familiar with certain terminology. During deformation, each layer can be described as being bent around an imaginary axis called a **hinge line** (**Figure 6.11A**). Folds are also described by their **axial plane**, which is an imaginary plane that divides the fold as equally as possible into two halves, called *limbs*. In a *symmetrical fold*, the limbs are mirror images of each other and diverge at the same angle (see Figure 6.10). In an *asymmetrical fold*, the limbs each have different angles of dip. A fold in which one limb is tilted beyond the vertical is referred to as an *overturned fold* (see Figure 6.10).

*By strict definition, an *anticline* is a structure in which the oldest strata are found in the center. This most typically occurs when strata are upfolded. Furthermore, a *syncline* is strictly defined as a structure in which the youngest strata are found in the center. This occurs most commonly when strata are downfolded.

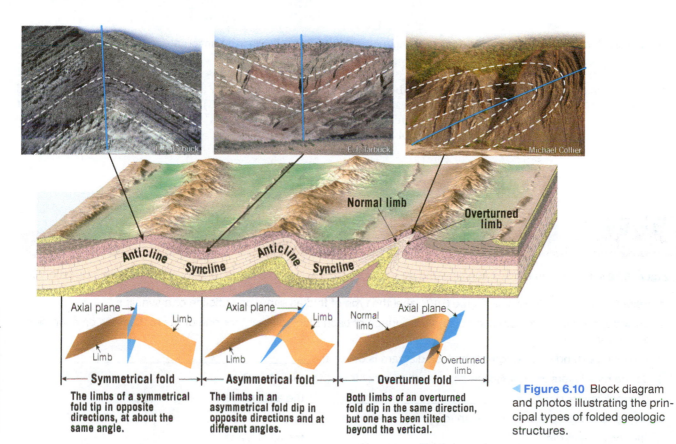

◄ **Figure 6.10** Block diagram and photos illustrating the principal types of folded geologic structures.

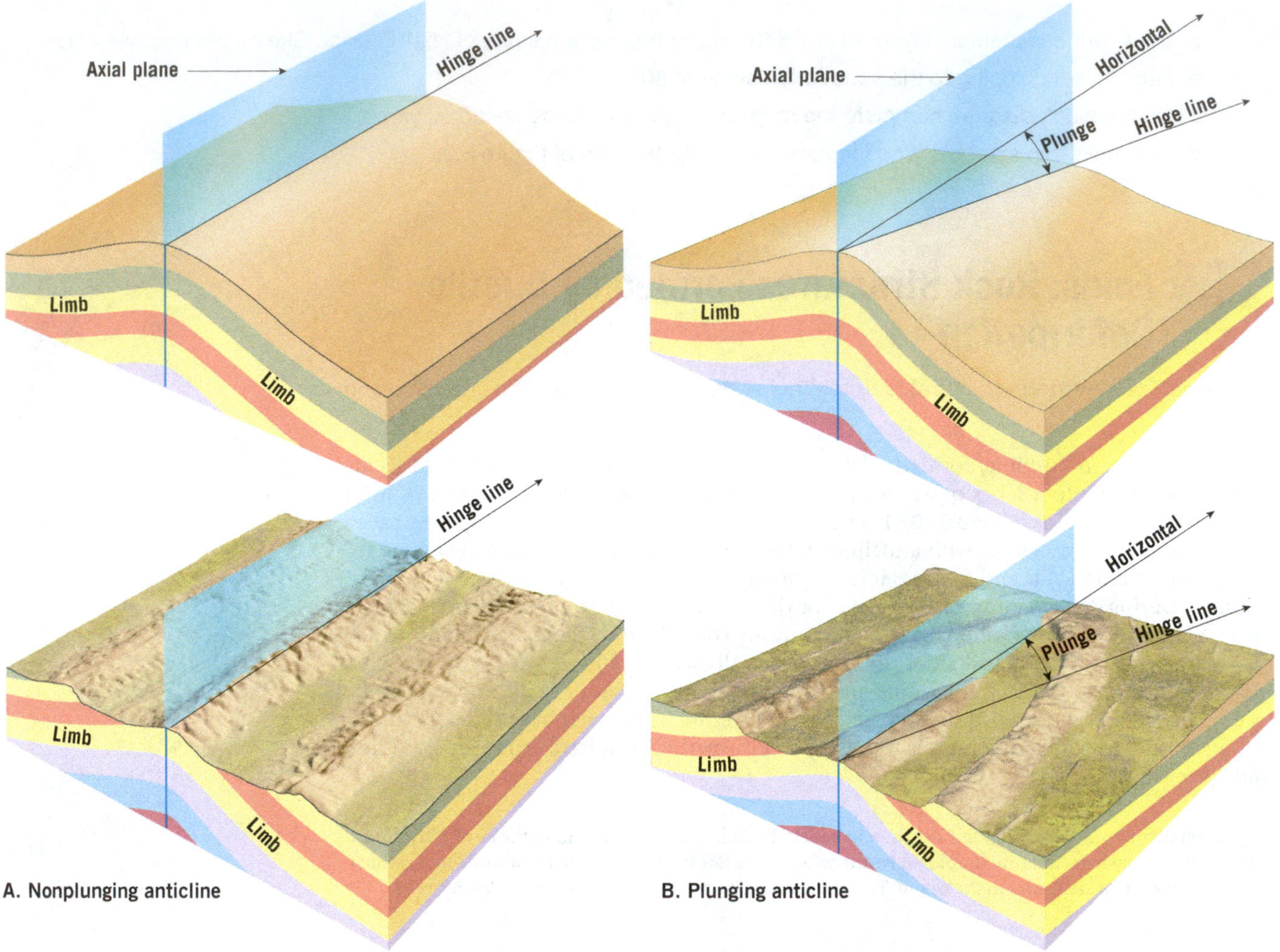

▲ **Figure 6.11** Features of simple folds. **A.** Nonplunging (horizontal) anticline. **B.** Plunging anticline.

Folds can also be tilted by tectonic forces that cause their hinge lines to slope downward. Folds of this type are said to be *plunging* (**Figure 6.11B**). As shown in Figure 6.11B, the outcrop pattern of an eroded plunging anticline "points" in the direction it is plunging. The opposite is true of a plunging syncline.

ACTIVITY 6.4A

Examining Folds

Figure 6.12 illustrates an eroded anticline and an eroded syncline. Use this figure to answer the following questions.

1. Below each block diagram in Figure 6.12, label the type of fold as either an anticline or a syncline.
2. Draw appropriate strike-dip symbols on the surface of each block diagram for one of the rock layers on each side of the axial plane.
3. The plane extending through each block diagram is called the_____.
4. Do the rock layers in an anticline dip toward or away from the axial plane?

   The rock layers dip _____ the axial plane.
5. Do the rock layers in a syncline dip toward or away from the axial plane?

   The rock layers dip _____ the axial plane.

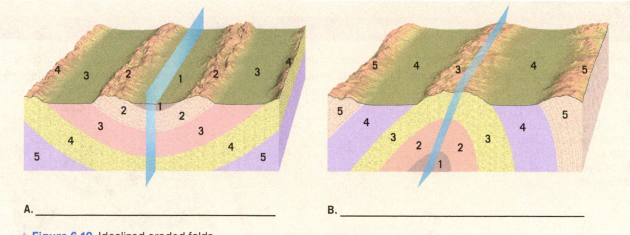

A. _____

B. _____

▲ **Figure 6.12** Idealized eroded folds.

6. Is the syncline shown in Figure 6.12 symmetrical or asymmetrical? Is the anticline symmetrical or asymmetrical?

The syncline is _____.

The anticline is _____.

7. Are these folds plunging or nonplunging folds?

The folds are _____.

8. The *principle of superposition* (see Exercise 10) states that, in strata that have not been overturned, the oldest rocks are at the bottom of the stack. With this in mind, list the rock layers in each block diagram, by number, from oldest to youngest.

Syncline:

Oldest _____ Youngest

Anticline:

Oldest _____ Youngest

9. Label the oldest exposed layers on the surface of both block diagrams, on both sides of the axial plane, with X.

10. Complete the following statements that describe what happens to the ages of the surface rocks as you move further away from the axial plane of each of the following structures.

   a. On an *eroded syncline*, do the rocks exposed at the surface get older or younger further from the axial plane?

   _____

   b. On an *eroded anticline*, do the rocks exposed at the surface get older or younger further from the axial plane?

   _____

11. On **Figure 6.13**, complete the block diagram, using the information provided on the map view. Rocks illustrated with the same pattern/colors are part of the same rock layer.

12. Write the names of the two types of geologic structures illustrated at the appropriate place on the block diagram in Figure 6.13.

▼ **Figure 6.13** Map view and block diagram of a hypothetical area.

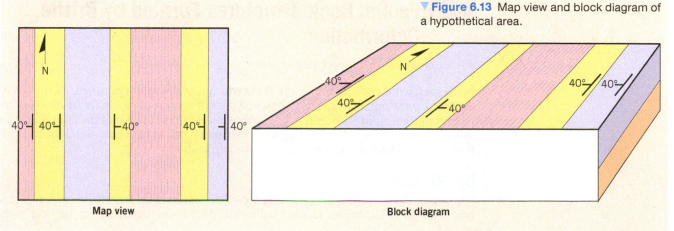

Map view

Block diagram

ACTIVITY 6.4B
Examining Domes and Basins

Anticlines and synclines are somewhat linear features caused by compressional forces. Two other types of folds, **domes** and **basins**, are roughly circular features that result from vertical displacement. Upwarping of sedimentary rocks produces a dome, whereas a basin is a downwarped structure (**Figure 6.14**).

▶ **Figure 6.14** A typical eroded dome has an outcrop pattern that is roughly circular, with the rock layers dipping away from its center.

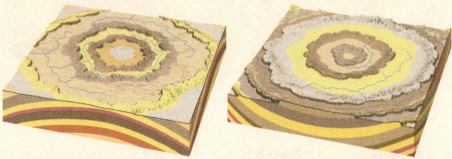

A. Upwarping produces a *dome*.   B. Downwarping produces a *basin*.

1. Draw several strike-dip symbols on Figure 6.14 that would be appropriate for a dome and a basin, respectively.

2. In a dome, are the oldest surface rocks found near the center or at the flanks?

_____

3. In a basin, are the oldest surface rocks found near the center or at the flanks?

_____

4. **Figure 6.15** is a simplified geologic map of the Black Hills in western South Dakota. Based on the ages of the rocks in the area, is this area a dome or basin?

_____

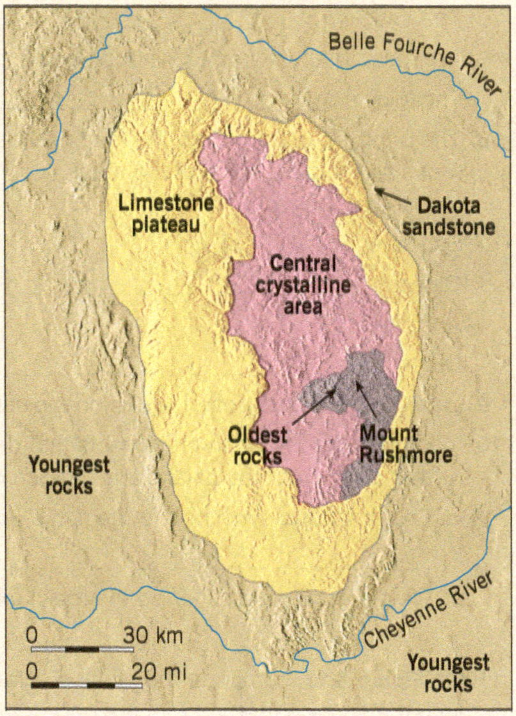

▶ **Figure 6.15** Simplified geologic map of the Black Hills, South Dakota.

6.5 Faults: Rock Structures Formed by Brittle Deformation

■ **Recognize the common types of faults.**

Faults are fractures or breaks in rocks along which movement has occurred. As with folds, the terms *strike* and *dip* are used to describe the orientation of faults. Faults in which the relative movement of rock units is primarily vertical (along the dip of the fault plane) are called **dip-slip faults**. When the dominant motion of rock units is horizontal (in the direction of the strike), the faults are called **strike-slip faults**.

Dip-Slip Faults
Dip-slip faults are described by noting the relative motion of the blocks on top of and below the fault. Geologists identify the rock surface immediately above the fault as the

hanging wall block and the rock surface below as the **footwall block** (**Figure 6.16**). These names were first used by miners who excavated metallic ore deposits, such as gold that had deposited along inactive fault zones. The miners would walk on the rocks below the mineralized fault zone (*the footwall block*) and hang their lanterns on the rocks above (*the hanging wall block*).

If the hanging wall block moves down relative to the footwall block, the fault is classified as a **normal fault** (Figure 6.16A). Normal faults result from tensional forces. Compressional forces may produce a **reverse fault**, where the hanging wall moves up relative to the footwall (Figure 6.16B).

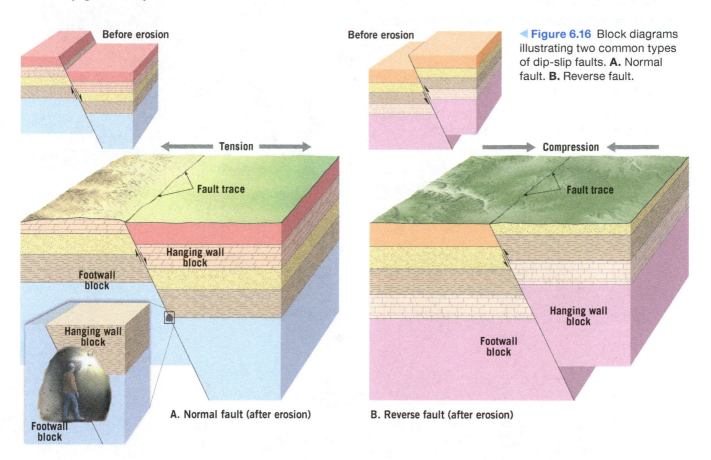

◄ **Figure 6.16** Block diagrams illustrating two common types of dip-slip faults. **A.** Normal fault. **B.** Reverse fault.

Strike-Slip Faults

Strike-slip faults result from horizontal displacement of rock units along the strike or trend of a fault. Some strike-slip faults, such as the famous San Andreas Fault system in California, are nearly vertical faults that exhibit tens or hundreds of kilometers of displacement.

A strike-slip fault is described as a **right-lateral fault** or a **left-lateral fault**, depending on the relative motion of the blocks. When facing the fault, if the crustal block on the opposite side is displaced to the *right*, it is a *right-lateral fault*. Likewise, if the opposite side is displaced to the *left*, it is a *left-lateral fault*.

<hr>

ACTIVITY 6.5

Examining Faults

1. On each of the block diagrams in Figure 6.16, determine the relative ages of the surface rocks on both sides of the fault trace (after erosion) and write the word *older* or *younger* on the appropriate side of each fault trace.

2. Do dip-slip faults exhibit mainly vertical or horizontal displacement?

   Dip-slip faults exhibit mainly _____ displacement.

continued

Activity 6.5 continued

3. Is the fault shown in **Figure 6.17** a dip-slip or strike-slip fault?

   It is a _____ fault.

4. Complete **Figure 6.18** by illustrating an eroded reverse fault on Part B. Place arrows on both sides of the fault to illustrate relative motion.

5. Is the fault illustrated in **Figure 6.19** a right-lateral or left-lateral fault?

   This is a _____ fault.

6. Complete the block diagram in **Figure 6.20** by drawing a left-lateral fault on Part B.

7. Draw arrows on both sides of the fault in Figure 6.20B to illustrate the relative motion of each block.

▲ **Figure 6.17** This fault is located along a road cut in **Nevada.** (Photo by E. J. Tarbuck)

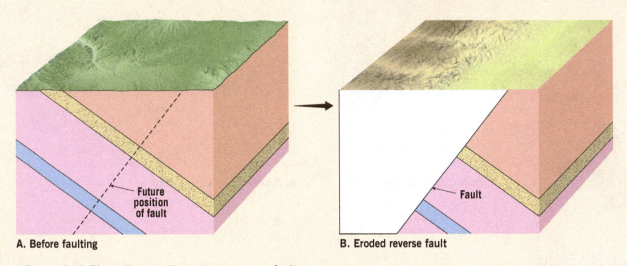

A. Before faulting

Future position of fault

B. Eroded reverse fault

Fault

▲ **Figure 6.18** Block diagram illustrating a reverse fault.

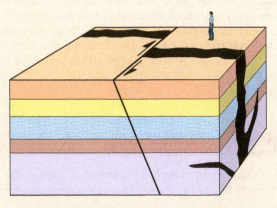

▲ **Figure 6.19** Block diagram illustrating a strike-slip fault.

©2019 Pearson Education, Inc.

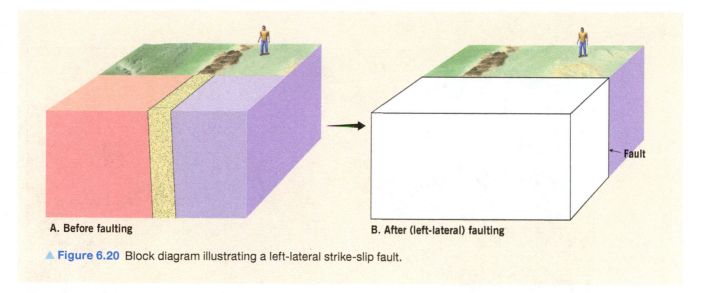

A. Before faulting   B. After (left-lateral) faulting

▲ **Figure 6.20** Block diagram illustrating a left-lateral strike-slip fault.

6.6 Examining Geologic Maps

■ **Interpret a simple geologic map.**

Geologic maps show the ages, distribution, and orientation of the basic rock structures. These maps allow geologists to interpret the nature of the rocks below the surface and determine the forces that caused their deformation.

ACTIVITY 6.6
Interpreting a Geologic Map

Examine Figure 6.21, which is a simplified geologic map of Devils Fence, Montana, and Figure 6.22, a key describing the rock layers on the Devils Fence map. Then answer the following questions. (*Note:* You may find the geologic time scale in Figure 10.14, page 175, a helpful reference.)

1. What are the names and approximate ages, in years, of the youngest and oldest *sedimentary* rock units shown in the map key in Figure 6.22 (page 107)?

   Youngest sedimentary rock unit: _____

   _____ million yr

   Oldest sedimentary rock unit: _____

   _____ million yr

2. On the map in Figure 6.21, write the word *oldest* where the oldest sedimentary rock unit is exposed at the surface and the word *youngest* where the youngest sedimentary rocks occur.

3. Are the intrusive igneous rocks near the center of the structure younger or older than the adjacent sedimentary rocks? (*Hint:* See Figure 6.22.)

   The igneous rocks are _____.

4. Examine the strike and dip of the rock units on the Devils Fence geologic map in Figure 6.21. Draw multiple large arrows on the map, pointing in the direction of dip on several rock units.

   a. Do the rock layers located near the center of the map in Section 14 dip toward the northwest or southeast? (Note: A section is a plot of land and the middle of each section is numbered in red.)

   _____

   b. The same rocks in Section 14 are also found in Section 18. Do the rocks in Section 18 dip toward the east or west?

   _____

continued

Activity 6.6 continued

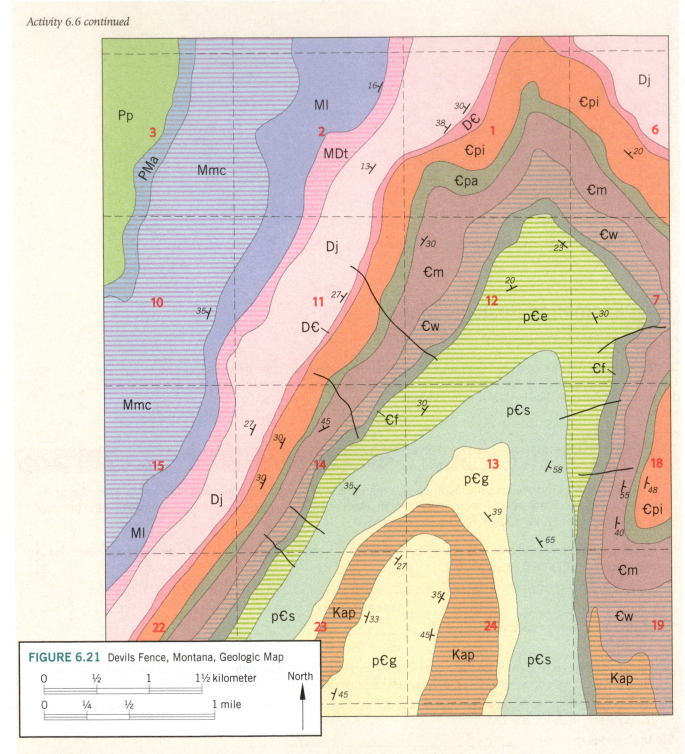

FIGURE 6.21 Devils Fence, Montana, Geologic Map

0 ½ 1 1½ kilometer

0 ¼ ½ 1 mile

North

▲ Simplified geologic map of Devils Fence, Montana.

5. What is the approximate angle of dip of the units in Section 18?

Angle: _____ °

6. Draw a dashed line representing the hinge line of the large geologic structure that occupies most of the map. Label the *hinge line*.

7. Are the rocks getting older or younger farther from the hinge line?

_____

8. Is this geologic structure a plunging anticline or syncline?

   This is a plunging _____.

9. Provide two lines of evidence that support your answer to Question 8.

   _____

   _____

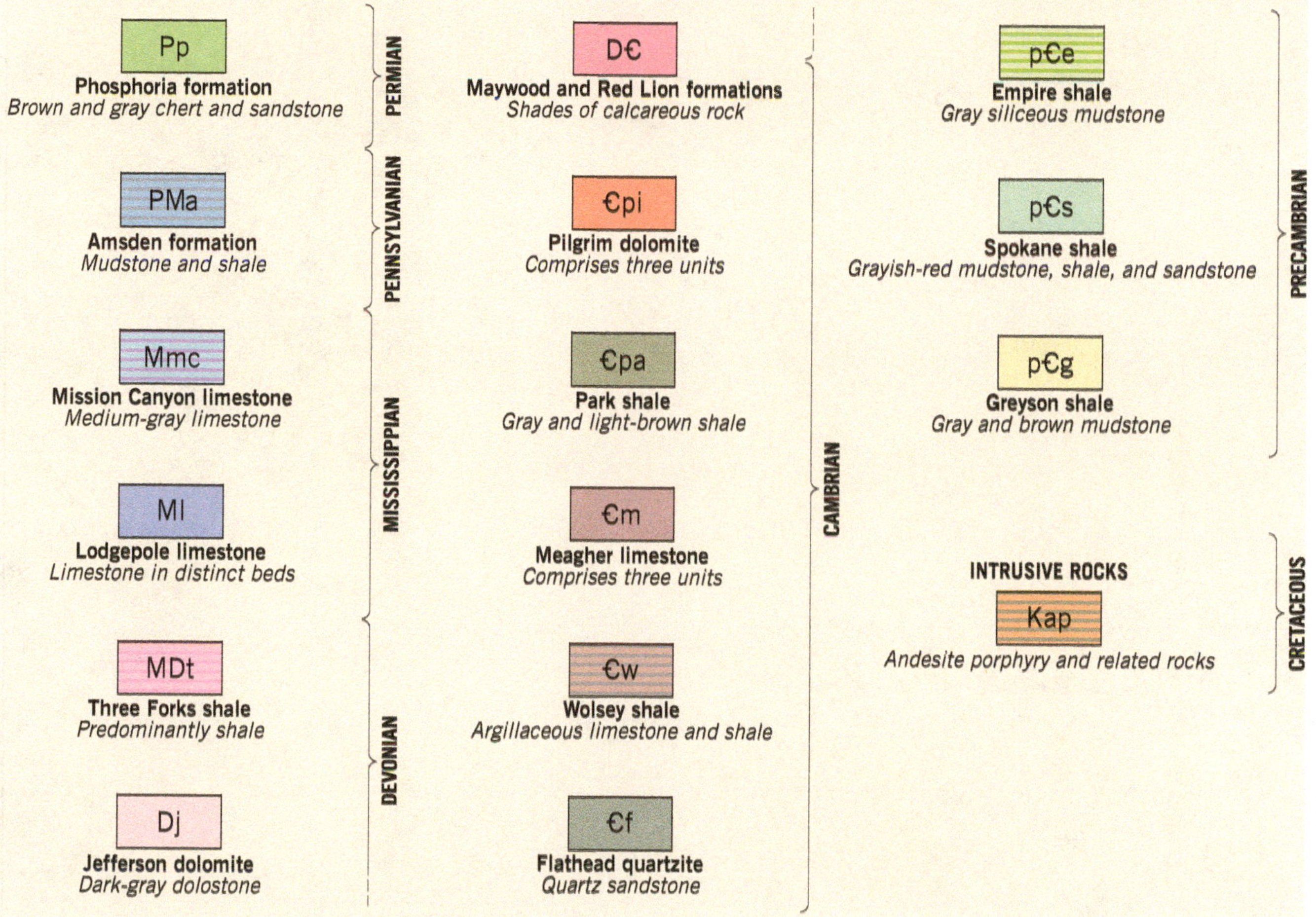

PERMIAN		
Pp	**DꞒ**	**pꞒe**
Phosphoria formation	**Maywood and Red Lion formations**	**Empire shale**
Brown and gray chert and sandstone	*Shades of calcareous rock*	*Gray siliceous mudstone*

PERMIAN

Pp
Phosphoria formation
Brown and gray chert and sandstone

PENNSYLVANIAN

PMa
Amsden formation
Mudstone and shale

MISSISSIPPIAN

Mmc
Mission Canyon limestone
Medium-gray limestone

Ml
Lodgepole limestone
Limestone in distinct beds

MDt
Three Forks shale
Predominantly shale

DEVONIAN

Dj
Jefferson dolomite
Dark-gray dolostone

DꞒ
Maywood and Red Lion formations
Shades of calcareous rock

Ꞓpi
Pilgrim dolomite
Comprises three units

Ꞓpa
Park shale
Gray and light-brown shale

Ꞓm
Meagher limestone
Comprises three units

Ꞓw
Wolsey shale
Argillaceous limestone and shale

Ꞓf
Flathead quartzite
Quartz sandstone

CAMBRIAN

pꞒe
Empire shale
Gray siliceous mudstone

pꞒs
Spokane shale
Grayish-red mudstone, shale, and sandstone

pꞒg
Greyson shale
Gray and brown mudstone

PRECAMBRIAN

INTRUSIVE ROCKS

Kap
Andesite porphyry and related rocks

CRETACEOUS

▲ **Figure 6.22** Key for the geologic map in Figure 6.21.

Mastering Geology™

Looking for additional review and lab prep materials? Go to www.masteringgeology.com for Pre-Lab Videos, Geoscience Animations, RSS Feeds, Key Term Study Tools, The Math You Need, an optional Pearson eText, and more.

Notes and calculations:

Name _____

Date _____

Course/Section _____

Due Date _____

1. Refer to **Figure 6.23** to describe the strike and dip of the limestone layer.

   Strike: North _____ ° _____

   Direction of dip: _____

   Angle of dip: _____ °

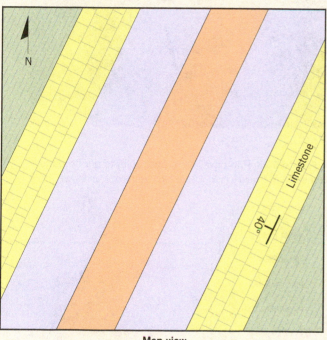

Map view

▲ **Figure 6.23** Map view to accompany Questions 1 and 2.

2. On Figure 6.23 add dip and strike symbols on all unlabeled beds to produce a map view of an *asymmetrical syncline*.

3. Where are the youngest rocks located in an eroded syncline?

   _____

4. Name the geologic structure shown in **Figure 6.24**. What was the dominant type of stress that acted on this rock body?

   Structure: _____

   Stress: _____

▲ **Figure 6.24** Photograph to accompany Question 4.
(Photo by E.J. Tarbuck)

5. Was compressional stress or tensional stress responsible for producing the geologic structure shown on the Devils Fence geologic map (Figure 6.21)?

   _____

6. Complete the block diagram in **Figure 6.25**. Show at least four sedimentary rock layers and add arrows to show the relative movement on both sides of the normal fault.

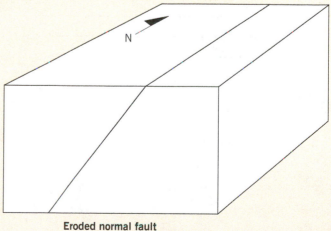

Eroded normal fault

▶ **Figure 6.25** Block diagram of an eroded normal fault.

7. Identify which type of strike-slip fault (right-lateral or left-lateral) is shown in **Figure 6.26**.

This is a _____ strike-slip fault.

▲ **Figure 6.26** Photo to accompany Question 7. (Photo courtesy of USGS)

8. Label the hanging wall block and the footwall block on each of the faults illustrated in **Figure 6.27**. On each photo, draw arrows showing the relative movement on each side of the fault. Identify the type of fault illustrated in each photo and describe the type of stress that produced it.

A. Fault type: _____

    Stress type: _____

B. Fault type: _____

    Stress type: _____

9. **Figure 6.28** is a simplified geologic map showing a large geologic structure that underlies Michigan and parts of the surrounding states. What name is given to this type of geologic structure? How did you figure this out?

_____

_____

_____

A.

B.

▲ **Figure 6.27** Photographs of two faults to accompany Question 8. (Photo A by E. J. Tarbuck; Photo B by Marli Miller)

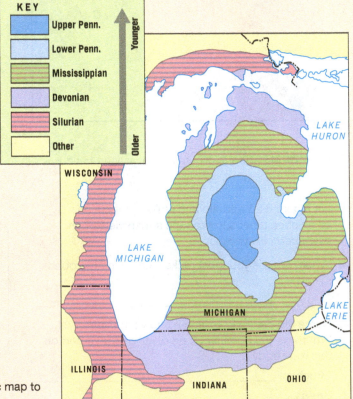

KEY

- Upper Penn.
- Lower Penn.
- Mississippian
- Devonian
- Silurian
- Other

Younger → Older

WISCONSIN

LAKE HURON

LAKE MICHIGAN

MICHIGAN

LAKE ERIE

ILLINOIS

INDIANA   OHIO

▲ **Figure 6.28** Simplified geologic map to accompany Question 9.

Aerial Photographs, Satellite Images, and Topographic Maps

LEARNING OBJECTIVES

Each statement represents an important learning objective that relates to one or more sections of this lab. After you complete this exercise you should be able to:

- **Investigate Earth's surface features using aerial photos and satellite images.**
- **Explain what a topographic map is.**
- **Distinguish between fractional scale and graphic scale. Measure distances using a graphic map scale.**
- **Describe the location of a parcel of land using the Public Land Survey system.**
- **Interpret contour lines to determine the elevation, relief, and shapes of landforms on a topographic map.**
- **Construct a simple topographic map by drawing contour lines.**
- **Prepare a topographic profile.**
- **Analyze a USGS topographic quadrangle.**

MATERIALS

ruler	stereoscope
hand lens	topographic maps

PRE-LAB VIDEO https://goo.gl/zXE74F

Prepare for lab! Prior to attending your laboratory session, view the pre-lab video. Each video provides valuable background that will contribute to your understanding and success in lab.

INTRODUCTION

Geologists use many tools to study Earth's surface, including aerial photographs, satellite images, and maps. The maps that geologists work with most often, called *topographic maps*, use contour lines to show landforms. This exercise will show you how to read a topographic map, a skill that is important when studying Earth's varied landforms in the exercises that follow.

PART 1 Geology

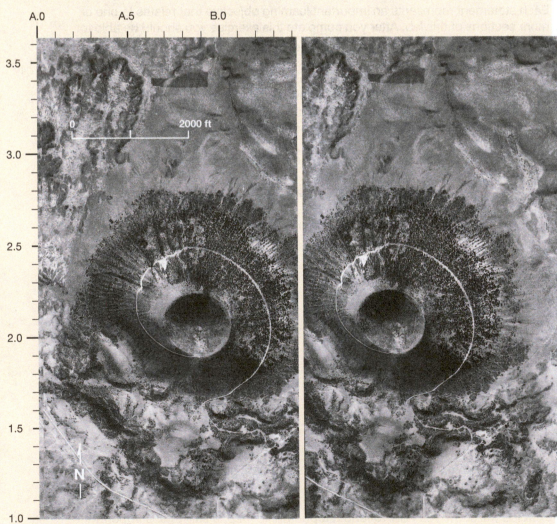

Distance between eyes of user

Stereoscope

Left line of sight   Center line   Right line of sight

Stereogram

7.1 Aerial Photographs

■ **Investigate Earth's surface features using aerial photos and satellite images.**

Aerial photographs are important tools for studying Earth's surface. When overlapping aerial photographs are combined, they produce **stereograms** that allow you to see Earth's surface in three dimensions. To view a stereogram, a **stereoscope** is placed directly over the line that separates the two images (**Figure 7.1**). As you look through a stereoscope, you may have to adjust it slightly until the image appears in three dimensions. The topography you see through the stereoscope will be vertically exaggerated—making slopes and other features appear steeper than they actually are.

◄ **Figure 7.1** How to align a stereoscope to view a stereogram.

Aerial Photographs

Obtain a stereoscope from your instructor, unfold it, and center it over the line that separates the two aerial photographs of the volcanic cone in **Figure 7.2**. Use this stereogram to complete the following.

▶ **Figure 7.2**
Stereogram of Mt. Capulin, a volcanic cinder cone located in northeastern New Mexico. The stereogram consists of two overlapping aerial photographs taken from an altitude of approximately 4800 meters (16,000 feet). To view the three-dimensional image of the volcano, center a stereoscope over the line that separates the two photographs. Then, while looking through the stereoscope, adjust the stereoscope until the image appears in three dimensions. The inset photo shows a traditional ground-level view of Mt. Capulin.
(Jeffrey M. Frank/ Shutterstock)

2000 ft

N

1. Identify and label the crater at the summit of the volcano.

2. Outline and label the lava flow, located directly north of the volcano.

3. What is the white, curved feature that extends from the base of the cone to its summit?

_____

4. Mark the highest point on the volcano with an X.

7.2 Satellite Images

■ **Investigate Earth's surface features using aerial photos and satellite images.**

Since the early 1970s, hundreds of satellites have been launched to systematically collect images of Earth's surface, using a variety of remote-sensing techniques. Computer enhancement of these images provides Earth scientists with data that allow them to study phenomena such as volcanic eruptions, hurricanes, and oil spills in real time (**Figure 7.3A,B**). These data have been invaluable in estimating the potential hazards associated with events such as these.

Satellite images may also reveal features that were previously unknown or unidentified. For example, heavily eroded impact craters that struck Earth millions of years ago have recently been identified using satellite images.

▼ **Figure 7.3** Satellite images are invaluable to our study of Earth. **A.** Satellite image of an ash plume produced by Iceland's Eyjafjallajökull volcano, April 2010. **B.** Image of oil slick produced by a leak on the Deepwater Horizon platform located south of the Mississippi Delta. (Photos courtesy of NASA)

A. Ash plume, Iceland

B. Oil slick in the Gulf of Mexico

ACTIVITY 7.2
Satellite Images

Use the satellite images in **Figure 7.4A,B** to complete the following.

A. Gosses Bluff Crater

B. Goat Paddock Crater

▲ **Figure 7.4** Satellite images to go with Activity 7.2. **A.** Gosses Bluff crater is an eroded remnant of an impact crater located near the center of Australia, about 170 kilometers (100 miles) west of Alice Springs. **B.** Goat Paddock crater is an impact crater, located in northwestern Australia, which is largely covered by sediment deposited during the Pleistocene epoch. (Photos courtesy of NASA)

continued

Activity 7.2 continued

1. Draw a line that shows the outermost extent of the impact crater in each of these images.

2. Which of these satellite images also captured a wildfire?

_____

7.3 Topographic Maps

■ **Explain what a topographic map is.**

Topography refers to the general configuration of Earth's surface—that is, the "shape of the land." **Topographic maps**, also called **quadrangles**, are two-dimensional scale models of Earth's three-dimensional surface viewed from directly above. The third dimension, elevation (height) of the landscape, is shown with contour lines that connect points of equal elevation relative to sea level. You will learn more about how to interpret and draw contour lines later in this exercise. In addition to showing elevations, topographic maps also show water features, geographic place names, and a variety of cultural features.

The U.S. Geological Survey (USGS) is the principal government agency that provides topographic maps of the United States. The USGS has been producing topographic maps since the late 1800s, following a standard format. In addition to using standard colors and symbols, each map contains information about the location of the mapped area, the date the map was drawn and/or revised, a map scale, a north arrow, and the names of adjoining quadrangles.

Topographic maps are bounded by parallels of latitude on the north and south and meridians of longitude on the east and west. The lines of latitude and longitude that identify the boundaries of a quadrangle are labeled at the four corners of the map in degrees (°), minutes ('), and seconds (") and are indicated at intervals along the margins. Maps that cover 15 minutes of latitude and 15 minutes of longitude are called *15-minute series topographic maps*. A $7\frac{1}{2}$-minute series topographic map covers $7\frac{1}{2}$ minutes of latitude and $7\frac{1}{2}$ minutes of longitude. (*Note:* There are 60 minutes of arc in 1 degree and 60 seconds of arc in 1 minute of arc. Therefore, $\frac{1}{2}$ minute is the same as 30 seconds.) A more complete examination of latitude and longitude can be found in Exercise 23, "Location and Distance on Earth."

ACTIVITY 7.3
Examining a Topographic Map

Obtain a copy of a topographic map from your instructor and examine it. Use the map to answer the questions that follow.

PLEASE DO NOT WRITE OR MARK ON THE MAP.

1. A topographic map is assigned a name that is found in the upper-right corner of the map. What is the name of your map?

_____

2. Notice the small reference map and compass arrow in the lower margin of the map. In what part of the state (northern, southwestern, etc.) is the area located?

_____

3. What is the name of the map that adjoins the western edge of your map?

_____

4. When was the area surveyed? When was the map published?

   Surveyed: _____   Published: _____

5. Refer to the inside cover of this manual to access the standard topographic map symbols used on USGS quadrangles. Using the standard map symbols as a guide, locate examples of various types of roads, buildings, woodland areas, and streams on the topographic map supplied by your instructor.

6. In general, what colors are used for the following types of features?

Highways and roads: _____

Buildings: _____

Urban areas: _____

Wooded areas: _____

Water features: _____

7. How much area is covered, in latitude and longitude, on the topographic map you are using? (*Hint:* Look for the map series, listed in minutes.)

_____

7.4 Map Scales

■ **Distinguish between fractional scale and graphic scale. Measure distances using a graphic map scale.**

You are probably familiar with scale model airplanes or cars that are miniature representations of the actual objects. Maps are similar in that they are "scale models" of Earth's surface. Each map has a **map scale** that expresses the relationship between distance on the map and actual distance on Earth's surface. Different map scales depict an area on Earth with varying degrees of detail. On a topographic map, scale is usually indicated in the lower margin and is expressed in two ways: fractional scales and graphic scales.

Fractional scales indicate how much the portion of Earth's surface represented on the map has been reduced from its actual size. For example, a map with a scale of 1/24,000 (or 1:24,000) tells the viewer that a distance of 1 unit on the map represents a distance of 24,000 of the *same* units on Earth (e.g., 1 inch equals 24,000 inches). The U.S. Geological Survey publishes maps of various scales with expansive as well as highly detailed coverage.

A **graphic scale**, or **bar scale**, is a bar that is divided into segments that show the relationship between distance on the map and actual distance on Earth (**Figure 7.5**). Bar scales showing miles, feet, and kilometers are generally provided on most topographic maps. The left side of the bar is often divided into fractions to allow for more accurate measurements. You can use a graphic scale to make your own "map ruler" by transferring the bar scale to a piece of paper.

▼ **Figure 7.5** Typical graphic or bar scale.

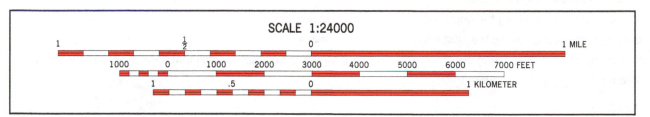

Map Scales

1. Depending on the map scale, 1 inch on a topographic map represents various distances on Earth. Convert the following scales and round to the nearest mile. (*Hint:* 5280 feet = 1 mile.)

MAP SCALE	FRACTIONAL SCALE		
1:62,500	1 inch on the map represents _____ inches, or _____ miles, on Earth.		
1:250,000	1 inch on the map represents _____ inches, or _____ miles, on Earth.		

continued

Activity 7.4 continued

2. Examine **Figure 7.6** and complete the map description by choosing the appropriate terms: (smaller *or* larger) (more *or* less). Maps with small fractional scales (e.g., 1:250,000) cover a _____ area and provide _____ detail. Maps with large fractional scales (e.g., 1:24,000) cover a _____ area and provide _____ detail.

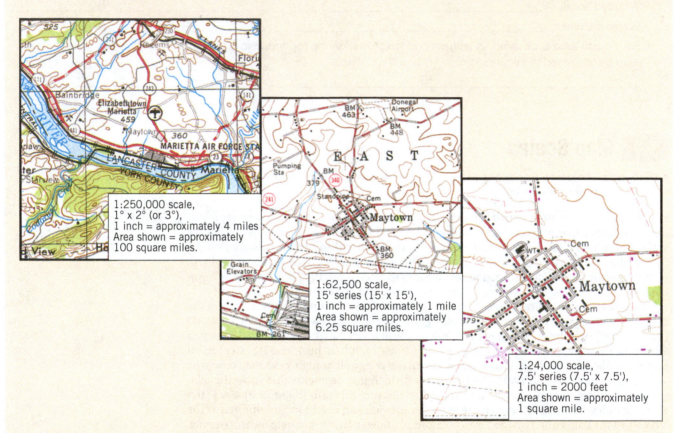

1:250,000 scale,
1° x 2° (or 3°),
1 inch = approximately 4 miles
Area shown = approximately
100 square miles.

1:62,500 scale,
15' series (15' x 15'),
1 inch = approximately 1 mile
Area shown = approximately
6.25 square miles.

1:24,000 scale,
7.5' series (7.5' x 7.5'),
1 inch = 2000 feet
Area shown = approximately
1 square mile.

▲ **Figure 7.6** Portions of three topographic maps of the same area, showing the effect that different map scales have on the detail illustrated.

3. Use the graphic scale on the topographic map provided by your instructor to construct a "map ruler" in miles and measure the following distances.

   a. Width of the map along the south edge = _____ miles

   b. Length of the map along the east edge = _____ miles

4. How many square miles (area) are represented on your topographic map? (Recall that area = width × length.)

   Map area = _____ square miles

7.5 **Public Land Survey System**

■ **Describe the location of a parcel of land using the Public Land Survey system.**

The Public Land Survey (PLS) began shortly after the Revolutionary War in 1785, when the federal government became responsible for large land areas beyond the 13 original colonies (**Figure 7.7A**). Its purpose was to provide accurate maps of most of what would become the United States of America This arduous task involved teams of surveyors—working mostly on foot—dividing almost 1.5 billion acres into parcels of land called *congressional townships* and *sections*. This survey system provided a means for the government to sell land in order to raise public funds. As in the past, the Public Land Survey

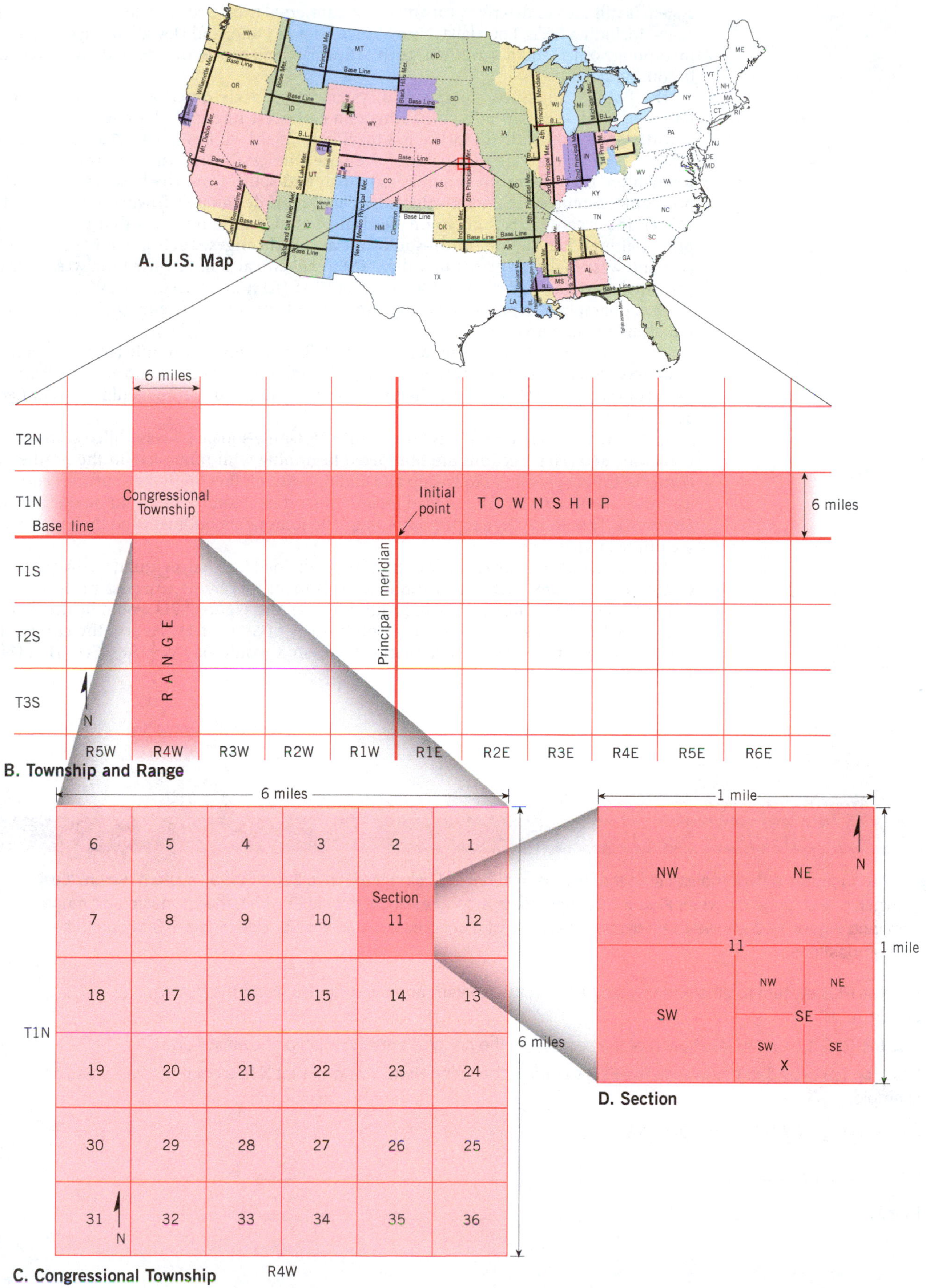

A. U.S. Map

6 miles

T2N

T1N

Congressional
Township

Initial
point

T O W N S H I P

6 miles

Base line

T1S

T2S

R A N G E

Principal   meridian

T3S

N

R5W   R4W   R3W   R2W   R1W   R1E   R2E   R3E   R4E   R5E   R6E

B. Township and Range

6	5	4	3	2	1
7	8	9	10	Section 11	12
18	17	16	15	14	13
19	20	21	22	23	24
30	29	28	27	26	25
31	32	33	34	35	36

6 miles

T1N

6 miles

N

C. Congressional Township   R4W

1 mile

NW   NE   N

11   1 mile

SW   SE

NW   NE

SW   SE

X

D. Section

▲ **Figure 7.7** The Public Land Survey (PLS) system.

system is still used to describe property being transferred from one party to another. Several states, including Texas, Louisiana, New Mexico, Maine, Ohio, and Hawaii, use systems that incorporate other land descriptions, such as Spanish land grants, that existed prior to their becoming states.

The Public Land Survey uses a grid system to precisely describe the location of a parcel of land. It begins at a point where an east–west line, called a **base line**, meets a north–south line, called a **principal meridian**, as shown in **Figure 7.7B**. Horizontal lines at 6-mile intervals that parallel the base line establish east–west tracts, called **townships**. Each township is numbered north and south from the base line. The first horizontal 6-mile-wide tract north of the base line is designated Township One North (T1N), the second T2N, and so on. Vertical lines at 6-mile intervals that parallel the principal meridian define north–south tracts, called **ranges**. Each range is numbered east and west of the principal meridian. The first vertical 6-mile-wide tract west of the principal meridian is designated Range One West (R1W), the second R2W, and so on. On a topographic map, the townships and ranges covered by the map are printed in red along the margins.

The intersection of a township and a range defines a 6-mile-by-6-mile rectangle, called a **congressional township**, which is identified by referring to its township and range numbers. For example, in **Figure 7.7B**, the shaded congressional township would be identified as T1N, R4W.

Each congressional township is divided into 36 1-mile-square parcels, called **sections**—each with 640 acres. Sections are numbered beginning with number 1 in the northeast corner of the congressional township and ending with number 36 in the southeast corner, as shown in **Figure 7.7C**. The shaded section of land in Figure 7.7C would be designated Section 11, T1N, R4W. On a topographic map, the sections are outlined, and their numbers are printed in red.

To describe smaller areas, sections may be subdivided into halves, quarters, or quarters of a quarter (**Figure 7.7D**). Each subdivision is identified by its compass position. For example, the 40-acre area designated with the letter X in Figure 7.7D would be described as the SW $\frac{1}{4}$ (southwest $\frac{1}{4}$) of the SE $\frac{1}{4}$ (southeast $\frac{1}{4}$) of Section 11. Hence, the complete locational description of the area marked with an X would be SW $\frac{1}{4}$, SE $\frac{1}{4}$ Sec. 11, T1N, R4W.

ACTIVITY 7.5A
Public Land Survey System

Figure 7.8 illustrates a hypothetical area that has been surveyed using the Public Land Survey system. **Figure 7.8A** is a township and range diagram, **Figure 7.8B** represents a congressional township within the township and range system, and **Figure 7.8C** shows three different sections of a congressional township. Use Figure 7.8 to complete the following questions.

1. Use the PLS system to label the townships along the western edge and ranges along the bottom of Part A.

2. Use the PLS system to label each of the sections in the congressional township shown in Part B.

3. In the space provided below, use the PLS system to describe Plots Y and Z. Plot X has been completed as an example.

Plot X: <u>NW</u> $\frac{1}{4}$, <u>SW</u> $\frac{1}{4}$, Sec. <u>8</u>, T<u>3N</u>, R<u>4W</u>

Plot Y: _____ $\frac{1}{4}$, _____ $\frac{1}{4}$, Sec. _____, T _____, R _____

Plot Z: _____ $\frac{1}{4}$, _____ $\frac{1}{4}$, Sec. _____, T _____, R _____

▶ **Figure 7.8** Hypothetical Public Land Survey system map, showing the locations of various parcels of land.

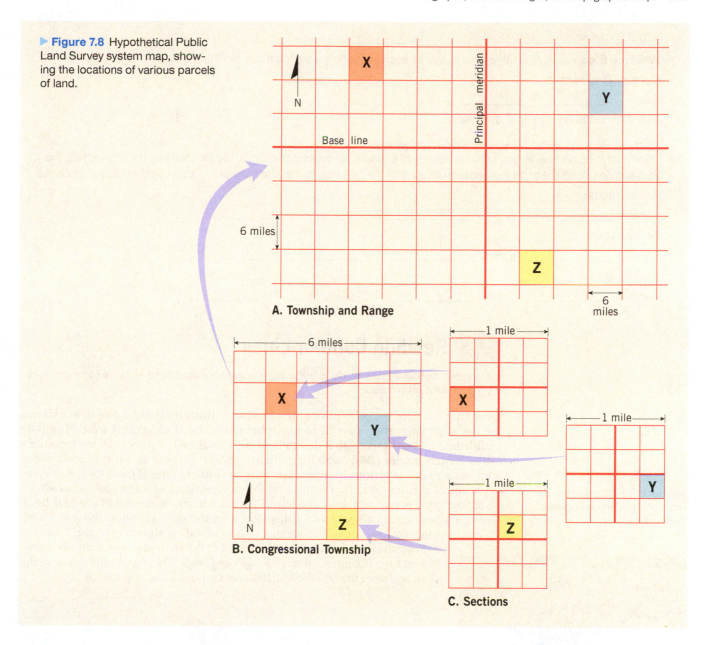

Public Land Survey System

Use the topographic map supplied by your instructor to complete the following.

PLEASE DO NOT WRITE OR MARK ON THE MAP.

1. List the townships and ranges shown on the map.

Townships: _____

Ranges: _____

2. Complete the following statement by choosing the appropriate term: (eastward *or* westward) (northward *or* southward).

To arrive at the principal meridian, people living in this area should travel _____, and to arrive at the base line, they should travel _____.

continued

Activity 7.5B continued

3. What are the section, township, and range at each of the following locations on the map?

Center of the map:

Sec. _____, T _____, R _____

Extreme northeast corner of the map:

Sec. _____, T _____, R _____

4. Your instructor will supply you with the names of three features (school, church, etc.) located on the map. Using the PLS system, write the complete location of each of the features, to the nearest $\frac{1}{4}$ of a section, in the following spaces.

Feature name: _____

Location: _____ $\frac{1}{4}$, Sec. _____, T _____, R _____

Feature name: _____

Location: _____ $\frac{1}{4}$, Sec. _____, T _____, R _____

Feature name: _____

Location: _____ $\frac{1}{4}$, Sec. _____, T _____, R _____

7.6 Reading Contour Lines

■ **Interpret contour lines to determine the elevation, relief, and shapes of landforms on a topographic map.**

Depicting the height or elevation of the land and illustrating the shape of landforms are the most important uses of topographic maps. Land elevations were originally established by land surveys that determined the elevations of some selected locations, called **bench marks (BM)**, which are highly accurate. These elevations are marked with brass plates at various locations around the United States (**Figure 7.9**). Surveyors work outward from these locations to establish elevations of numerous other sites.

Topographic maps, miniature models of Earth's surface, make use of contour lines. **Contour lines** connect all points on a map that have the same elevation above sea level. To visualize this concept, imagine a small volcanic island, as shown in **Figure 7.10**. The point at which sea level intersects the land at the shoreline represents the 0-foot contour line. The intersection of imaginary planes 50 feet and 100 feet above the ocean represent the 50-foot and 100-foot contour lines, respectively.

▼ **Figure 7.9** Photo of a bench mark (BM) in Bryce National Park, Utah.

(Photo by E. J. Tarbuck)

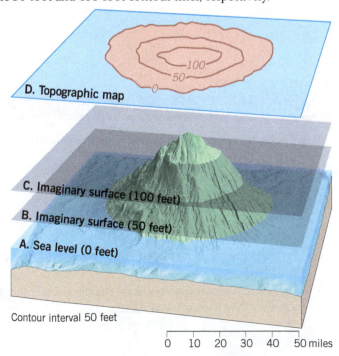

D. Topographic map

C. Imaginary surface (100 feet)

B. Imaginary surface (50 feet)

A. Sea level (0 feet)

Contour interval 50 feet

0 10 20 30 40 50 miles

▶ **Figure 7.10** How topographic maps are constructed. A contour line is drawn where an imaginary horizontal plane intersects the land surface. Where the ocean surface (plane A) intersects the land, it forms the 0-foot contour line. Plane B is 50 feet above sea level, and it intersects the land to form the 50-foot contour line. Plane C is 100 feet above the ocean, and it intersects the land to form the 100-foot contour line. The topographic map that results when the contour lines are drawn on a map is shown in D.

ACTIVITY 7.6
Reading Contour Lines

To effectively use topographic maps, familiarity with the rules for reading contour lines is necessary. Use Figure 7.11, a simplified contour map of a small volcanic cone and nearby landforms, to complete the following. (General rules for reading contour lines are provided in Figure 7.11.)

1. The difference in elevation between adjacent contour lines is called the **contour interval**. Look on the bottom of this map to identify the contour interval.

   Contour interval: _____

2. What is the difference in elevation between Points A and B?

   Difference in elevation: _____ feet

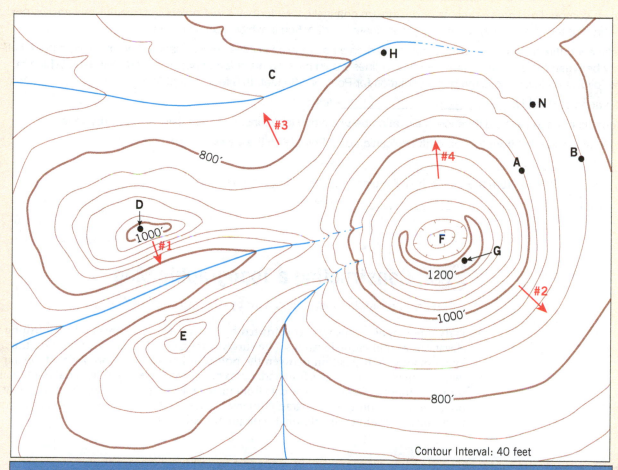

Contour Interval: 40 feet

GENERAL RULES FOR READING CONTOUR LINES

1. A contour line connects points of equal elevation.

2. A contour line never branches or splits.

3. Steep slopes are shown by closely spaced contours.

4. Hills are represented by closed contour lines.

5. Closed contours with hachure marks on the downhill side represent a closed depression.

6. When contour lines cross streams or dry stream channels, they form a "V" that points upstream.

7. Contour lines that occur on opposite sides of a valley always occur in pairs.

▲ **Figure 7.11** Topographic map to be used to answer questions in Activity 7.6.

continued

Activity 7.6 continued

3. Notice that every fifth contour line, called an **index contour**, is printed as a bold brown line, and the elevation of that line is provided (in feet). List the elevations for each index contour shown on this map.

_____ feet, _____ feet, _____ feet

4. Closely spaced contours indicate steep slopes. Which of the four slopes shown with red arrows labeled 1–4 is the steepest? Which is the least steep?

Steepest: _____   Least steep: _____

5. One or more roughly circular closed contours indicates a hill. Which of the landforms labeled B–E are hills?

Hill landforms: _____

6. Closed contours with hachures (short lines) that point downslope indicate depressions (basins without outlets). Which of the landforms labeled D–G are depressions?

Depressions: _____

7. When contour lines cross streams or dry stream channels, they form a V *that points upstream*. Draw arrows next to three of the streams (shown in blue) to indicate the direction in which each is flowing.

8. Estimating the elevations of places not located on a contour line involves extrapolation. For example, a point halfway between the 500- and 600-foot contour lines would have an elevation of approximately 550 feet. Which of the following elevations is the best estimation for Point N: 830 feet, 870 feet, or 890 feet?

Elevation of Point N: _____ feet

9. **Relief** is defined as the difference in elevation between two locations, such as a hill and a nearby valley.

a. What is the relief from the top of the volcanic cone (Point G) to the valley (Point H) below?

Relief: _____ feet

b. What is the relief from the top of the hill (Point D) to the valley (Point H) below?

Relief: _____ feet

7.7 Constructing a Topographic Map

■ **Construct a simple topographic map by drawing contour lines.**

Early topographic maps were constructed by first surveying an area and establishing the elevations at numerous locations. The surveyor then sketched contour lines on the map by estimating their position between the points of known elevation. Today, topographic maps are made by computers from radar data or a series of stereoscopic aerial photographs similar to those in Figure 7.2. The data from these images is processed by computers that are programmed to construct topographic maps. However, considerable field work is necessary to establish elevations and eliminate computational errors.

ACTIVITY 7.7
Constructing a Topographic Map

The process of constructing a simple topographic map will prove useful in future labs, when you will be asked to interpret landforms on topographic maps.

1. Use the elevations in **Figure 7.12** as a guide for drawing contour lines. The 100-foot contour line is provided for reference. Using a 20-foot contour interval, use a pencil to draw a contour line for each 20-foot change in elevation below and above 100 feet (e.g., 60 feet, 80 feet, 120 feet). You will have to estimate the elevations between the points. Label each contour line with its elevation.

2. Does the land shown on the topographic map you constructed generally slope downward toward the north or south?

Land slopes downward toward the _____.

3. Show the direction each stream is flowing by drawing arrows on the map.

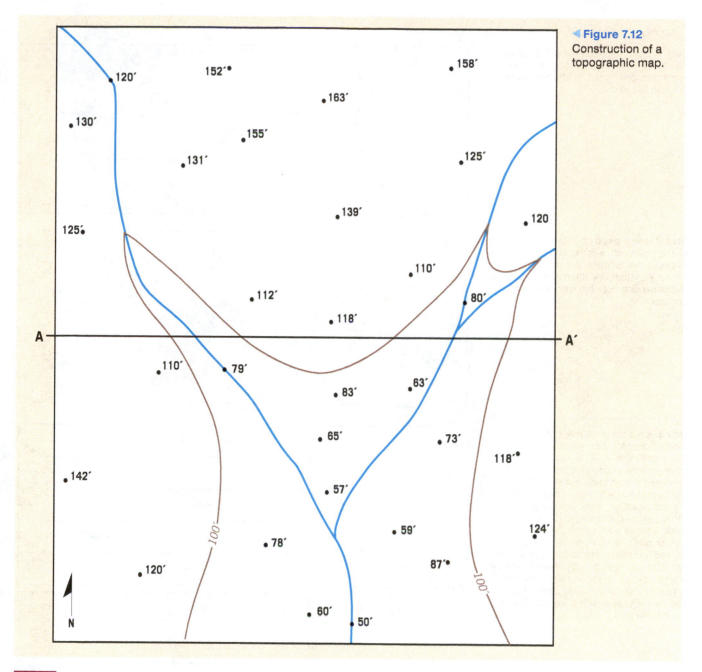

◀ **Figure 7.12**
Construction of a topographic map.

7.8 Drawing a Topographic Profile

■ **Prepare a topographic profile.**

A topographic map depicts Earth's surface as it would appear when viewed from an airplane directly above the area. In some circumstances, a topographic profile (side view) provides a more valuable representation of an area. **Figure 7.13** shows the steps to follow to draw a topographic profile.

ACTIVITY 7.8
Drawing a Topographic Profile

Follow the steps illustrated in Figure 7.13 to draw a topographic profile.

1. Use the *profile graph* in **Figure 7.14** to construct a west–east profile along the line A–A' on the contour map you completed in Figure 7.12.

continued

Activity 7.8 continued

Step 1. Our sample profile will be along line A–A'. When constructing profiles on maps provided by your instructor, do not draw on the map.

Step 2. Lay a piece of paper along the line of the profile you want to construct. In this example this is line A–A'. Mark each place where a contour line intersects the edge of the paper and note the elevation of the contour line.

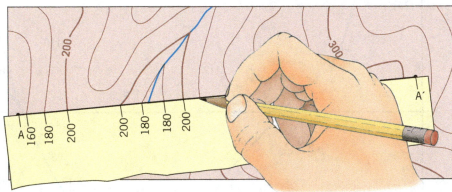

Step 3. On a separate piece of paper, draw a horizontal line slightly longer than your profile line, A–A'. Select a vertical scale for your profile that begins slightly below the lowest elevation along the profile and extends slightly beyond the highest elevation. Mark this scale on either side of the horizontal line. Lay the marked paper edge (from Step 2) along the horizontal line. Wherever you have marked a contour line on the edge of the paper, place a dot directly above the mark at an elevation on the vertical scale equal to that of the contour line. Connect the dots on the profile with a smooth line to see the finished product.

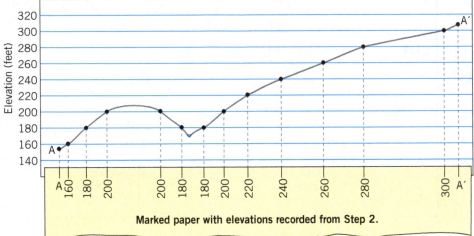

Marked paper with elevations recorded from Step 2.

▲ **Figure 7.13** How to construct a topographic profile.

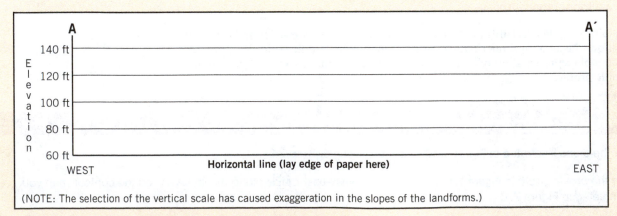

(NOTE: The selection of the vertical scale has caused exaggeration in the slopes of the landforms.)

▲ **Figure 7.14** Graph to be used to construct a topographic profile.

7.9 Analysis of a Topographic Map

■ **Analyze a USGS topographic quadrangle.**

Figure 7.15 is a portion of a quadrangle that shows an area near Leadville, Colorado, which is located about 75 miles southwest of Denver in the Rocky Mountains.

ACTIVITY 7.9

Analysis of a Topographic Map

Use the portion of the Leadville North quadrangle provided in Figure 7.15 (page 126) to answer the following questions.

1. What is the contour interval of this map?

   Contour interval: _____ feet

2. What is the difference in elevation from one *index contour* to the next?

   Difference in elevation: _____ feet

3. On most quadrangles, each section is numbered in red and outlined in red or, occasionally, dashed black lines. Find Section 9, located near the center of the map, and measure its width and length in miles, using the bar scale provided.

   Sections are _____ mile(s) wide and _____ miles(s) long.

4. Locate the small intermittent stream (blue dashed line) just below the red number 9 that denotes Section 9. Toward what general direction does the stream flow? Explain how you arrived at your answer.

   Direction of stream flow: _____

   Explain: _____

5. What is the approximate elevation of the point marked with an X in Section 8?

   Elevation of X: _____ feet

6. What is the approximate relief between point X and the surface of Turquoise Lake?

   Relief: _____ feet

7. Which of the following phrases best describes the topography of Tennessee Park: steep, rugged terrain; hilly terrain; or gradually sloping to flat terrain?

   _____

8. The area on both sides of Tennessee Creek in Section 9 is marked with blue symbols that do not represent streams. What do these symbols indicate? (See the inside front cover of the lab book, if necessary.)

   _____

9. Describe the change in elevation you would encounter as you walked from the boat ramp (Point A) on the northeast shore of Turquoise Lake, directly northward along the dashed line to Point B. (Use terms such as *steep*, *gradual*, *moderate*, *uphill*, *downhill*, and *relatively flat*.)

   _____

   _____

10. What is the elevation of the town of Leadville, Colorado, to the nearest 1000 feet?

    Elevation of Leadville: _____ feet

11. Examine the numbers of each section shown in red on this map. Portions of how many congressional townships are represented?

    Portions of _____ congressional townships

MasteringGeology™

Looking for additional review and lab prep materials? Go to www.masteringgeology.com for Pre-Lab Videos, Geoscience Animations, RSS Feeds, Key Term Study Tools, The Math You Need, an optional Pearson eText, and more.

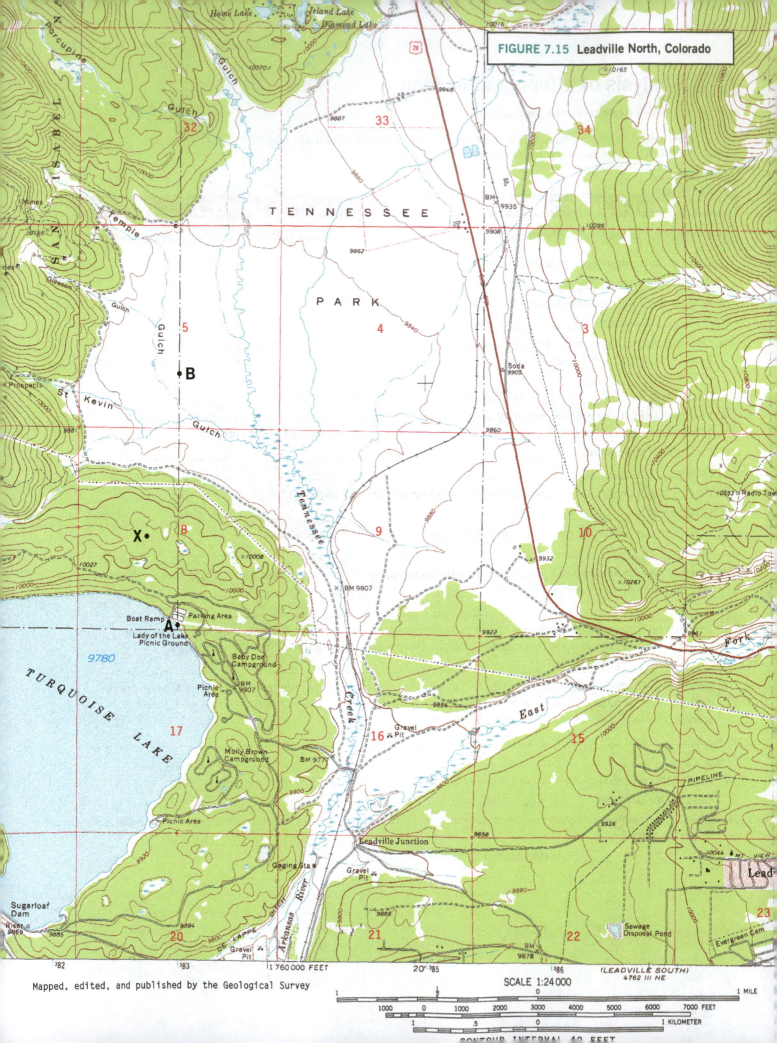

FIGURE 7.15 Leadville North, Colorado

Aerial Photographs, Satellite Images, and Topographic Maps

Name _____

Date _____

Course/Section _____

Due Date _____

Use **Figure 7.16**, which shows both a perspective view and a topographic map of a hypothetical area, to answer Questions 1–10.

1. Determine the contour interval used on this map.

   Contour interval: _____ feet

2. List the elevation of each *index contour* on the map.

   _____

3. What are the elevations of the locations designated A, B, and C?

   A: _____ feet

   B: _____ feet

   C: _____ feet

A. Perspective aerial view

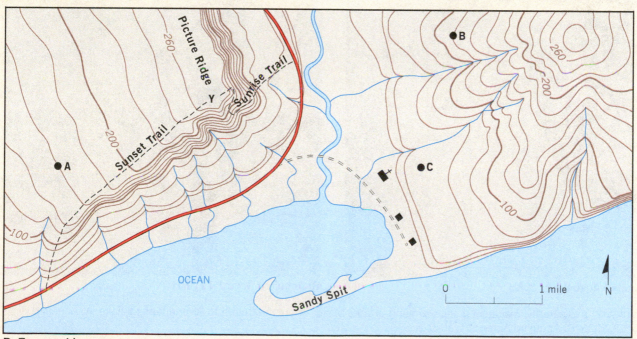

B. Topographic map

▲ **Figure 7.16** Two views of a hypothetical coastal area. (After U.S. Geological Survey)

4. What is the elevation of the church?

Elevation of the church: _____ feet

5. Indicate two areas on this map that have steep slopes by writing "steep" on the map. What characteristic of contour lines indicates steep terrain?

_____

6. Describe the relief of Sandy Spit with one of the following terms: low, medium, or high.

The relief of Sandy Spit is _____

7. Mark the top of the hill in the northeast corner of the map with a small x. What is the *maximum* elevation of this hill, to the nearest foot?

Maximum elevation: _____ feet

8. Is the top of Picture Ridge, designated Y, at a higher or lower elevation than the top of the hill that you labeled in Question 7?

Point Y is _____ than Point X.

9. What is the maximum relief shown on this contour map?

Maximum relief: _____ feet

10. Imagine that you are taking a group of friends hiking to Picture Ridge. Which trail (Sunrise Trail *or* Sunset Trail) would you select for those who want to take the easiest (more gradual) route?

Easiest route to the top of Picture Ridge: _____

11. Use the satellite images in **Figure 7.17**, which show a small portion of Lake Mead, to complete the following.

   a. Describe the width of the Colorado River in Image A as compared to its appearance in Image B.

   The Colorado River is _____ in Image A than in Image B.

   b. Describe how this portion of Lake Mead has changed between 1985, when Image A was acquired, and 2010, when Image B was acquired.

   _____
   _____

   c. Suggest a likely cause of the change(s).

   _____
   _____

A. Acquired August 22, 1985

B. Acquired August 11, 2010

▲ **Figure 7.17** Two satellite images of the eastern end of Lake Mead, where the Colorado River enters the lake. **A.** Acquired August 22, 1985. **B.** Acquired August 11, 2010. (Images courtesy of NOAA)

Shaping Earth's Surface: Running Water and Groundwater

LEARNING OBJECTIVES

Each statement represents an important learning objective that relates to one or more sections of this lab. After you complete this exercise you should be able to:

- **Sketch, label, and discuss the hydrologic cycle.**
- **Explain the relationship between infiltration and runoff.**
- **Discuss the effect of urbanization on runoff and infiltration.**
- **Identify river and valley features on a topographic map, including rapids, meanders, cut banks, point bars, floodplains, oxbow lakes, and backswamps.**
- **Describe the occurrence and uses of groundwater.**
- **Construct a contour map of the water table to determine the direction and rate of groundwater movement.**
- **Explain the relationship between groundwater withdrawal and land subsidence.**
- **Identify on a topographic map karst landscape features, including sinkholes, disappearing streams, and solution valleys.**

MATERIALS

calculator	graduated cylinder	colored pencils
ruler	coarse sand, fine sand, soil	string
stereoscope	ring stand	stop watch
beaker	small funnel	
hand lens	cotton ball	

PRE-LAB VIDEO https://goo.gl/12gFgh

Prepare for lab! Prior to attending your laboratory session, view the pre-lab video. Each video provides valuable background that will contribute to your understanding and success in lab.

INTRODUCTION

Water is continually on the move, from the ocean to land and back again in an endless cycle. This exercise deals with the part of the hydrologic cycle in which water returns

PART 1 Geology

to the sea. Some water travels quickly via rushing streams, and some moves much more slowly, below the surface. In this exercise, we examine the factors that influence the distribution and movement of water, as well as how water sculpts landscapes.

▼ SmartFigure 8.1 Earth's water balance, a quantitative view of the hydrologic cycle.

VIDEO
https://goo.gl/8foF4E

8.1 Examining the Hydrologic Cycle

■ **Sketch, label, and discuss the hydrologic cycle.**

Earth's water is constantly moving between Earth's surface and atmosphere. The **hydrologic cycle** describes the continuous movement of water from the oceans to the atmosphere, from the atmosphere to the land, and from the land back to the sea. Over most of Earth, the quantity of precipitation that falls on the land must eventually be accounted for by the sum total of **evaporation**, **transpiration** (the release of water vapor by vegetation), **runoff**, and **infiltration**.

A portion of the precipitation that falls on land will soak into the ground through a process called *infiltration*. If the rate of rainfall exceeds the ability of the surface to absorb it, the additional water flows over the surface and becomes *runoff*. Runoff initially flows in broad sheets that form tiny channels called *rills*. The rills merge to form gullies, which eventually join to create streams. **Erosion** by both groundwater and running water wears down the land and shapes Earth's surface.

Figure 8.1 illustrates Earth's *water balance*, a quantitative view of the hydrologic cycle. The figure implies a globally uniform exchange of water between Earth's atmosphere and surface, but factors such as climate, steepness of slope, surface materials, vegetation, and degree of urbanization produce local variations.

Hydrologic Cycle

Precipitation 284,000 km³
Precipitation 96,000 km³
Evaporation/Transpiration 60,000 km³
Evaporation 320,000 km³
Runoff 36,000 km³
Infiltration
Oceans

ACTIVITY 8.1
Examining the Hydrologic Cycle

Use Figure 8.1 as a reference to complete the following:

1. Globally, from which source does more water evaporate into the atmosphere: oceans or land?
_____

2. Approximately what percentage of the total water evaporated into the atmosphere comes from the oceans?

$$\text{Percentage from oceans} = \frac{\text{Ocean evaporation}}{\text{Total evaporation}} \times 100\% = \underline{\hspace{1cm}} \%$$

3. Notice in Figure 8.1 that more water evaporates from the oceans than is returned directly by precipitation. If sea level is not dropping, identify a source of water for the oceans in addition to precipitation.
_____

4. Worldwide, about how much of the precipitation that falls on the land becomes runoff: 35, 55, or 75 percent?
About _____ % becomes runoff.

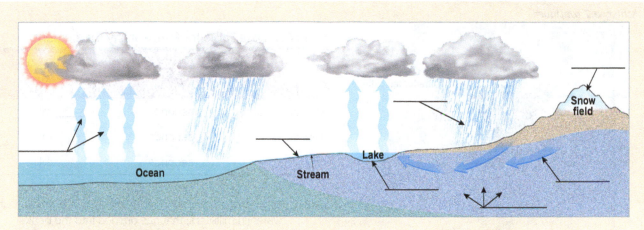

▲ **Figure 8.2** Illustration (cross section) of the hydrologic cycle.

5. Much of the water that falls on land does not immediately return to the ocean via runoff. Instead, it is temporarily stored in reservoirs such as lakes. In some mountainous and polar regions, what features serve as reservoirs to temporarily store water?

_____

6. Label the drawing in **Figure 8.2** with the letters that correspond to the following terms:

   A. Runoff   D. Evaporation

   B. Infiltration   E. Precipitation

   C. Groundwater   F. Reservoir

8.2   Infiltration and Runoff

■ **Explain the relationship between infiltration and runoff.**

When it rains, most of the water that reaches the land surface will infiltrate or run off. The balance between infiltration and runoff is influenced by factors such as the _permeability of the surface material, slope of the land, intensity of the rainfall,_ and _type and amount of vegetation._ When the ground becomes saturated—that is, when it contains all the water it can hold—runoff occurs on the surface.

ACTIVITY 8.2
Permeability Experiment

This experiment is designed to help you understand how the **permeability** (i.e., the ability of a material to transmit fluids) of various Earth materials affects how rapidly rainwater can be absorbed into the ground. Examine the equipment setup in **Figure 8.3** and complete the experiment.

Step 1:   Obtain the following equipment and materials from your instructor:

graduated cylinder	cotton ball	soil
beaker	coarse sand	ring stand
small funnel	fine sand	stop watch

Step 2:   Place a small wad of cotton in the neck of the funnel.

Step 3:   Fill the funnel above the cotton about two-thirds full with coarse sand.

Step 4:   With the bottom of the funnel placed in the beaker, pour in 50 milliliters of water and measure the time it takes until the water stops draining though the funnel. Record the time in Table 8.1.

continued

Activity 8.2 continued

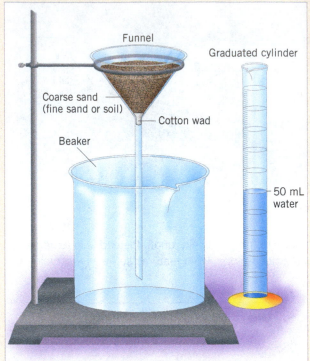

▲ **Figure 8.3** Equipment setup for permeability experiment.

Table 8.1 Data Table for Permeability Experiment

	TIME IT TAKES FOR WATER TO STOP DRAINING	MILLILITERS OF WATER DRAINED INTO BEAKER
Coarse sand	_____ seconds	_____ mL
Fine sand	_____ seconds	_____ mL
Soil	_____ seconds	_____ mL

Step 5: Using the graduated cylinder, measure the amount of water (in milliliters) that has drained into the beaker and record the measurement in Table 8.1.

Step 6: Empty and clean the graduated cylinder, funnel, and beaker.

Step 7: Repeat the experiment two additional times, using fine sand and then soil. (*Note:* In each case, fill the funnel with the material to the *same level* that was used for the coarse sand and use the *same size* piece of cotton.) Record the results of each experiment in the appropriate place in Table 8.1.

Step 8: Clean the glassware and return it to your instructor, along with any unused sand and soil.

The following questions refer to the permeability experiment above.

1. Of the three materials you tested (coarse sand, fine sand, and soil), which has the greatest permeability?
_____

2. Suggest a reason different amounts of water were recovered in the beaker for each material that was tested.
_____

3. Write a brief statement summarizing the results of your permeability experiment.
_____
_____
_____

4. Describe how each of the following conditions influences infiltration and runoff.
Highly permeable material: _____
_____

Steep slope: _____
_____

Gentle rainfall: _____
_____

Dense vegetation: _____
_____

8.3   Infiltration and Runoff in Urban Areas

■ **Discuss the effect of urbanization on runoff and infiltration.**

In urban areas, much of the surface is covered with buildings, streets, and parking lots—all of which alter the amount of water that runs off compared to the amount that soaks in.

Figure 8.4 shows two hypothetical **hydrographs** (plots of streamflow, or runoff, over time) for an area before and after urbanization. Runoff is measured by stream **discharge**, which is the volume of water flowing past a given point per unit of time, usually measured in cubic feet per second.

▼ **Figure 8.4** The effect of urbanization on streamflow before urbanization (top) and after urbanization (bottom).

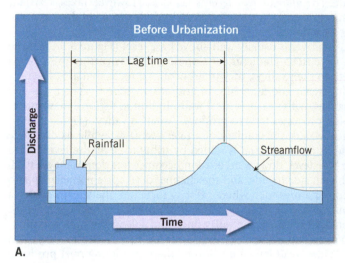

A.

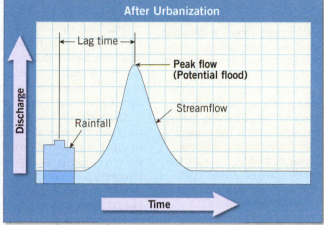

B.

ACTIVITY 8.3

Infiltration and Runoff in Urban Areas

Use Figure 8.4 to complete the following.

1. Does urbanization increase or decrease the peak streamflow?

   _____

2. What is the effect of urbanization on *lag time* (the span between when rainfall occurred and when peak stream discharge occurred)?

   _____

3. Does total runoff occur over a longer or shorter time span after an area has been urbanized?

   Runoff occurs over a _____ time after urbanization.

4. Based on what you have learned from the hydrographs in Figure 8.4, explain why urban areas often experience more flash-flooding than do rural areas during intense rainfalls.

   _____

   _____

8.4 # Running Water

■ Identify river and valley features on a topographic map, including rapids, meanders, cut banks, point bars, floodplains, oxbow lakes, and backswamps.

Of all the agents that shape Earth's surface, running water is the most important. Rivers and streams are responsible for producing a vast array of erosional and depositional landforms in both humid and arid regions. As illustrated in **Figure 8.5**, many of these features are associated with the *headwaters* of a river, while others are found near the *mouth*.

An important factor that governs the flow of a river is its **base level**—the lowest point to which a stream can erode. The *ultimate base level* is *sea level*. However, lakes, resistant rocks, and main rivers often act as *temporary*, or *local*, *base levels* that control the erosional and depositional activities of a stream for a period of time.

The *head*, or source area, of a river is usually well above base level (Figure 8.5A). At the headwaters, rivers typically have steep gradients, or slopes, and downcutting prevails. As these headwater streams deepen their valleys, they may encounter rocks that are especially resistant to erosion—forming *rapids* and *waterfalls*. In arid areas where rivers are eroding bedrock, valleys are often narrow, with nearly vertical walls. In humid regions, mass movements and other slope erosion processes tend to produce V-shaped valleys (Figure 8.5A).

Downstream from the headwaters, the gradient of a river decreases, while its discharge increases because of the additional water that is added by tributaries. As the level of the channel begins to approach base level, downward erosion becomes less dominant. Instead, the river's energy is directed from side to side, and the channel develops a *meandering* path (Figure 8.5B). On the outside of a meander, where velocity increases, the stream erodes a *cutbank*. On the inside of a meander, velocity slows and deposition of a *point bar* often occurs. Lateral erosion by a meandering river widens the valley floor, and a *floodplain* begins to form.

Near the mouth of a river where the channel is very near base level, meandering often becomes very pronounced. Widespread lateral erosion by the meandering river produces a wide floodplain. Features such as *oxbow lakes*, *natural levees*, *backswamps* or *marshes*, and *yazoo tributaries* commonly develop on broad floodplains (Figure 8.5C).

▼ **SmartFigure 8.5** Common features of stream valleys. A. Near its headwaters, a stream is well above base level, its gradient is steep, and downcutting is the dominant activity. B. In the middle portion of a stream, the channel often exhibits a meandering pattern, directing energy from side to side. The Condor Video illustrates this process. C. Near its mouth, a stream frequently flows on a wide floodplain and erodes its valley walls in only a few places.

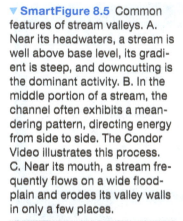

CONDOR VIDEO
https://goo.gl/uWWyPC

A. Headwaters of a river

B. Downstream from the headwaters

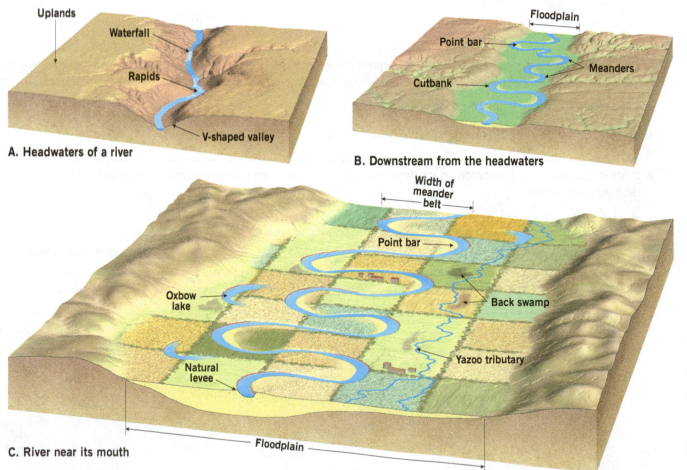

C. River near its mouth

ACTIVITY 8.4A
Examining the Portage, Montana, Topographic Map

Use the Portage, Montana, topographic map in **Figure 8.6** (on page 136) and the stereogram in **Figure 8.7** (on page 137) to complete the following.

1. Compare the stereogram in Figure 8.7 to the map in Figure 8.6. Then, on the topographic map, outline the area shown in the photo.

2. Is the terrain in Section 14, located on the west side of this map, relatively flat or hilly? Explain.

   _____

3. Label the areas that topographically resemble Section 14 on the topographic map as "upland."

4. Describe the topography in the lower half of Section 17, located three sections east of Section 14.

   _____

5. Section 17 contains a portion of the valley occupied by the Missouri River. Approximately what percentage of the area shown on the map is stream valley (similar to the lower half of Section 17) and what percentage is upland?

   Stream valley: _____ %

   Upland: _____ %

6. Which of the following best describes the shape of the Missouri River Valley along the line labeled D–D′: wide valley with a floodplain or steep-sided V-shaped valley with no floodplain?

   The valley is a _____.

7. Calculate the gradient, or slope, of this portion of the Missouri River by following the steps below:

   Step 1: Locate a contour line that crosses the river. *Note:* An *index contour* crosses the river in Section 26 (lower left portion of the map). Follow the index contour to the right to establish its elevation.

   Elevation: _____ feet

   Step 2: Locate a second index contour that crosses the river. *Note:* Look in Section 11, between the letters *V* and *E* in the word *RIVER*. What is its elevation?

   Elevation: _____ feet

   Step 3: Use a string and the bar scale to measure the approximate distance (to the nearest mile) between these two index contours.

   Distance: _____ miles

   Step 4: Calculate the gradient of the river by using the following formula:

   $$\text{Gradient} = \frac{\text{Difference in elevation (feet)}}{\text{Distance (miles)}}$$

   Gradient: _____ feet per mile

8. Approximately how many feet is the Missouri River above sea level?

   Height above sea level: _____ feet

9. What are the features in the river labeled with the letter A? (*Hint:* See the inside front cover.)

   _____

10. Based on your answer to Question 9, are the Missouri River and its tributaries actively eroding or depositing in this region?

    The Missouri River and its tributaries are _____.

11. Over time, as tributaries erode and lengthen their courses in their headwaters, what will happen to the upland areas?

    _____

    _____

continued

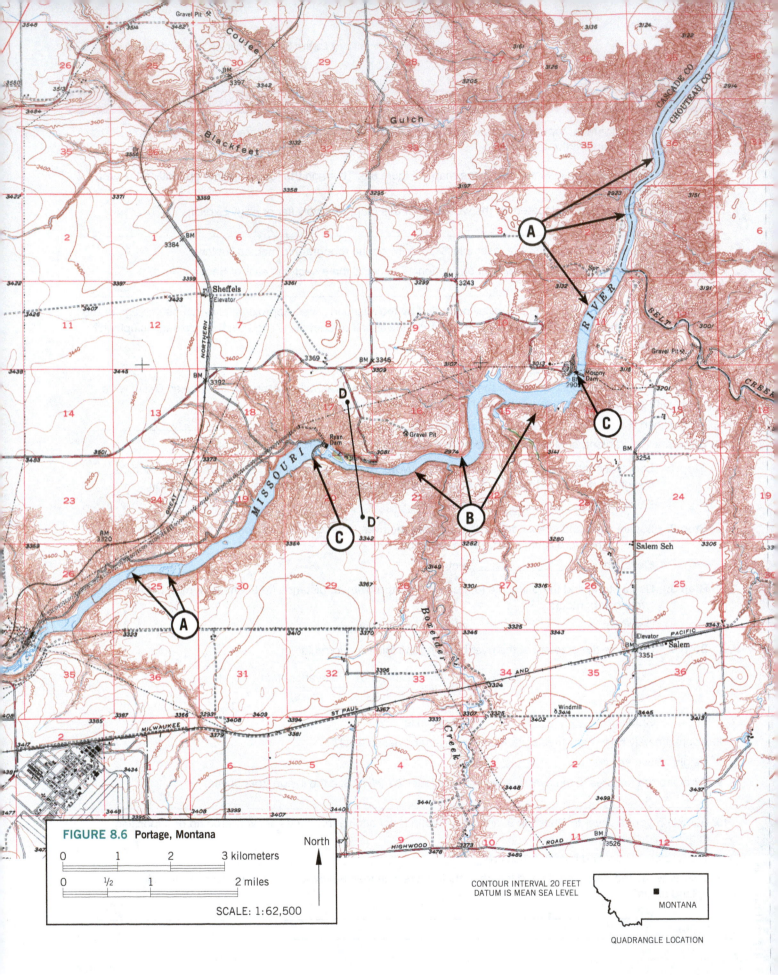

FIGURE 8.6 Portage, Montana

North

0	1	2	3 kilometers

0	½	1	2 miles

SCALE: 1:62,500

CONTOUR INTERVAL 20 FEET
DATUM IS MEAN SEA LEVEL

MONTANA

QUADRANGLE LOCATION

(Courtesy of U.S. Geological Survey)

Activity 8.4A continued

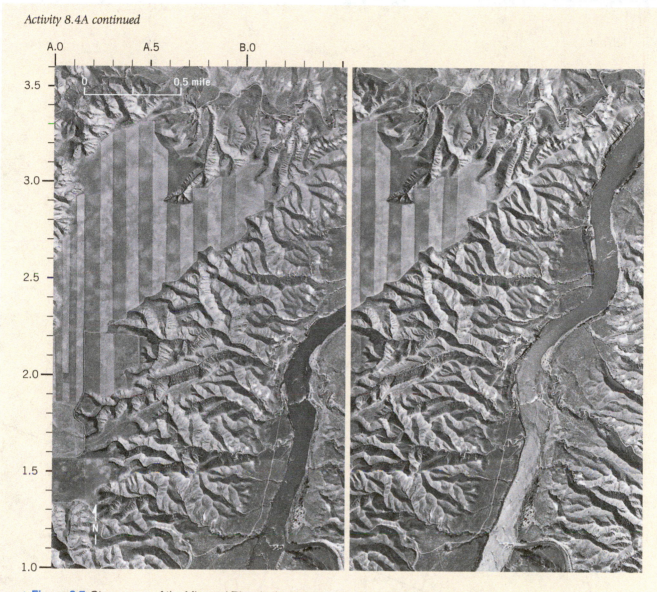

▲ **Figure 8.7** Stereogram of the Missouri River in the vicinity of Portage, Montana. (Courtesy of U.S. Geological Survey)

ACTIVITY 8.4B
Examining the Angelica, New York, Topographic Map

Refer to the Angelica, New York, topographic map in **Figure 8.8** (on page 138) to complete the following.

1. Draw an arrow on the map to indicate the direction that the main river, the Genesee, is flowing. (*Hint:* Use the elevations of the two bench marks [BM] next to the river near the top and bottom of the map to determine your answer.)

2. Use the BM elevations from Question 1 to calculate the approximate gradient of the Genesee River. (*Hint:* See Question 7 in Activity 8.4A.)

   Gradient: _____ feet per mile

3. Use **Figure 8.9** to draw a profile along line A–A′. Use only index contour lines. (*Note:* Figure 7.13 on page 124 contains a detailed explanation of how to construct topographic profiles.)

4. Approximately how many feet is the Genesee River above sea level?

   Height above sea level: _____ feet

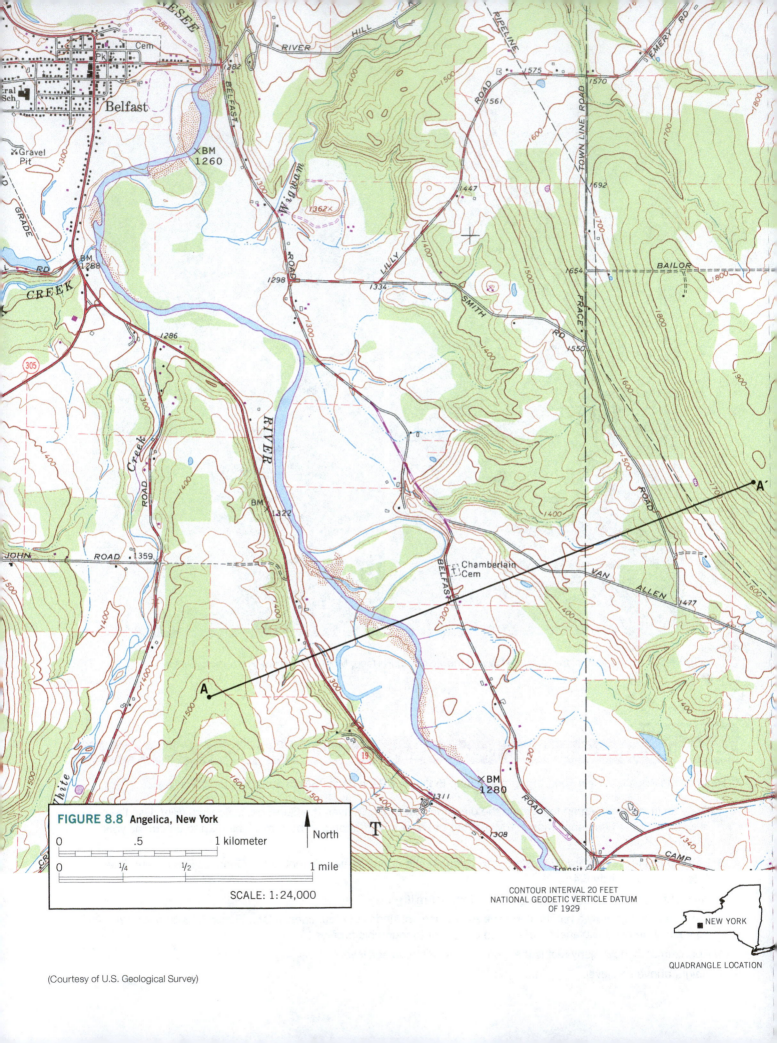

FIGURE 8.8 Angelica, New York

0   .5   1 kilometer

North

0   1/4   1/2   1 mile

SCALE: 1:24,000

CONTOUR INTERVAL 20 FEET
NATIONAL GEODETIC VERTICLE DATUM
OF 1929

NEW YORK

QUADRANGLE LOCATION

(Courtesy of U.S. Geological Survey)

Activity 8.4B continued

▲ **Figure 8.9** Profile of the Genesee River Valley.

5. The path of the Genesee River can best be described as which of the following: a straight course or a meandering course?

The Genesee River follows a _____ course.

6. Which phrase most accurately describes most of the areas beyond the Genesee River Valley: very broad and flat or relatively hilly and dissected?

Most of this area is _____.

7. Use the fractional scales on the Angelica topographic map and the Portage, Montana, map (page 136) to answer the following.

   a. What is the fractional scale for the Angelica map?

   _____

   b. How does this compare to the fractional scale for the Portage, Montana, map?

   _____

   c. What distance (in miles or fractions of a mile) does 1 inch represent on each map?

   Angelica, New York, map: _____ miles

   Portage, Montana, map: _____ miles

   d. Does the Angelica map cover *more* or *less* area than the Portage map?

   _____

   e. Explain why the Angelica map lacks numbered sections that we associate with the Public Land Survey system.

   _____

   _____

ACTIVITY 8.4C
Examining the Campti, Louisiana, Topographic Map

Refer to the Campti, Louisiana, topographic map in **Figure 8.10** (on page 140) and the stereogram of the same area in **Figure 8.11** (on page 141) to complete the following. On the map, A indicates the width of the Red River floodplain, and the dashed lines, B, mark the two sides of the meander belt of the river.

1. Approximately what percentage of the map area is flat and part of the Red River floodplain?

   Floodplain = _____ % of the map area

2. Find the index contour adjacent to the Red River. (Look above the word *RIVER* in the upper portion of the map.) The value of this contour line will indicate that the Red River is less than how many feet above sea level?

   Height above sea level: less than _____ feet

FIGURE 8.10 Campti, Louisiana

0   1   2   3 kilometers

0   ½   1   2 miles

North

SCALE: 1:62,500

CONTOUR INTERVAL 20 FEET
DATUM IS MEAN SEA LEVEL

LOUISIANA

QUADRANGLE LOCATION

Activity 8.4C continued

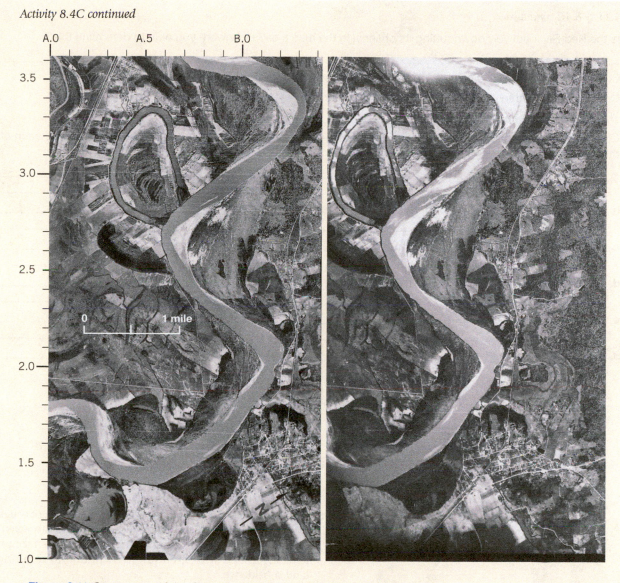

▲ **Figure 8.11** Stereogram of the Campti, Louisiana, area. (Courtesy of U.S. Geological Survey)

3. Using Figure 8.5C as a reference, identify the type of feature found at each of the following letters on the map in Figure 8.10:

Letter C (*Old River*): _____

Letter D: _____

Letter E: _____

Letter F: _____

4. If no contour lines cross the Red River and the contour interval is 20 feet, what is the maximum gradient of this section of the Red River? (*Note:* Although the red lines that cross the river resemble contour lines, they are land boundaries established by the French, using the long-lot survey system.)

Maximum gradient: _____ feet per mile

5. Identify and label examples of a point bar, a cutbank, and an oxbow lake on the stereogram in Figure 8.11.

6. Write a statement that compares the width of the meander belt of the Red River to the width of its floodplain.

_____

_____

continued

Activity 8.4C continued

7. Is the Red River actively downcutting its channel in the map area? How were you able to determine this?

_____

Complete the following by comparing the Portage, Angelica, and Campti topographic maps.

8. On which of the three maps is the gradient of the main river steepest?

The gradient is steepest on the _____ map.

9. Choosing from Figure 8.6, Figure 8.8, or Figure 8.10, write the name of the map that is best described by each of the following statements.

  a. Primarily floodplain: _____
  _____

  b. River valleys separated by broad, relatively flat upland areas: _____
  _____

  c. Greatest number of streams and tributaries: _____
  _____

  d. Poorly drained lowland area with marshes and swamps: _____
  _____

  e. Downcutting is the dominant process: _____
  _____

  f. Nearest to base level: _____
  _____

8.5 Groundwater

■ **Describe the occurrence and uses of groundwater.**

Groundwater is one of Earth's most valuable resources. It is an important source of water for human use, irrigation, and industry. In many areas, overuse and contamination threaten the groundwater supply. Furthermore, many regions face problems associated with groundwater withdrawal resulting in land subsidence.

Groundwater is water that has soaked into Earth's surface and completely fills the pore spaces in the soil and bedrock in the layer called the **zone of saturation** (**Figure 8.12**). The upper surface of this saturated zone is called the **water table**. Above the water table is the **unsaturated zone**, where the pore spaces are mainly filled with air.

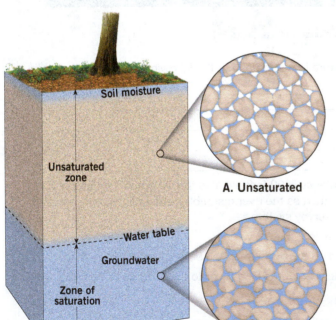

A. Unsaturated

B. Saturated

◀ **SmartFigure 8.12**
Idealized distribution of groundwater.

VIDEO
https://goo.gl/dgYbAw

ACTIVITY 8.5
Groundwater

Use **Figure 8.13**, which illustrates the subsurface of a hypothetical area, to complete the following questions.

1. Label the zone of saturation, the unsaturated zone, and the water table.

2. Describe the shape of the water table in relationship to the shape of the land surface.

_____

3. Whenever a substantial amount of water is withdrawn from a well, the water table forms a **cone of depression**. Label the cone of depression on Figure 8.13. What factors might cause a cone of depression to become larger or smaller?

_____

4. Use a pencil to shade the area between the dashed lines labeled A. This zone represents an impermeable lens of clay. Describe what will happen to water that infiltrates to the depth of the clay lens at point A.

_____

_____

5. How does the drop in the water table during a drought affect the operation of the well? The blue dashed line in Figure 8.13 represents the level of the water table during a drought.

_____

_____

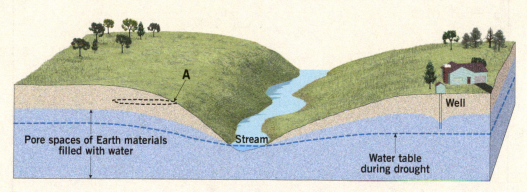

▲ **Figure 8.13** Subsurface of an area, showing saturated and unsaturated materials.

8.6   Groundwater Movement

■ **Construct a contour map of the water table to determine the direction and rate of groundwater movement.**

The movement of water on the land surface is relatively easy to visualize, and the movement of groundwater is less obvious. A common misconception is that groundwater occurs in underground rivers that resemble surface streams. Although subsurface streams exist, they are rare. Rather, as you learned in the preceding sections, groundwater exists in the pore spaces and fractures in rock and sediment. Thus, contrary to any impressions of rapid flow, the movement of most groundwater is exceedingly slow, from pore to pore.

One method of measuring the rate of groundwater movement involves introducing dye into a well. Time is measured until the coloring agent appears in another well at a known distance from the first.

ACTIVITY 8.6
Groundwater Movement

Use **Figure 8.14**, a hypothetical topographic map showing the location of several water wells, to complete the following. The numbers in parentheses indicate the depth of the water table below the surface in each well.

continued

Activity 8.6 continued

1. Calculate the elevation of the water table at each well location and write the approximate elevation on the line next to each well. Next, use a colored pencil to draw smooth 10-foot contours that show the shape of the water table. (Start with the 1160-foot contour.) Use a pencil of a different color to draw arrows on the map to indicate the direction of the slope of the water table.

   a. Toward which direction (downward) does the water table slope?

   _____

   b. Referring to the site of the proposed water well, at approximately what depth below the surface should the proposed well intersect the water table?

   _____

2. Assume that a dye was put into well A on May 10, 2017, and detected in well B on May 25, 2018. What was the rate of groundwater movement between the two wells, in centimeters per day? (*Hint:* Convert feet to centimeters.)

   Velocity: _____ centimeters per day

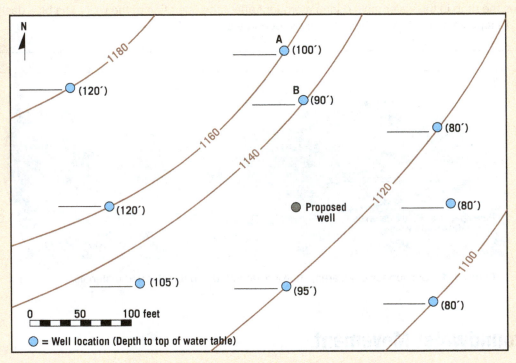

▲ **Figure 8.14** Hypothetical topographic map showing the locations of several water wells. The numbers in parentheses indicate the depth of the water table below the surface in each well.

8.7 Ground Subsidence

■ **Explain the relationship between groundwater withdrawal and land subsidence.**

As demand for freshwater increases, surface subsidence caused by the withdrawal of groundwater from **aquifers** presents a serious problem in many areas. Several major urban areas, such as Las Vegas, Houston, Mexico City, and the Central Valley of California are experiencing subsidence due to overpumping of wells (**Figure 8.15**). Parts of Mexico City have experienced as much as 7 meters (23 feet) of subsidence due to compaction of the subsurface material resulting from the reduction of fluid pressure as the water table is lowered. Once compaction occurs, porosity and permeability are permanently reduced. A subsequent rise in the water table will not restore porosity and permeability to pre-subsidence values.

Another classic example of land subsidence due to groundwater withdrawal occurred in California's Santa Clara Valley, which borders the southern part of San Francisco Bay. We will explore this episode of subsidence in the next activity.

Level of land before heavy groundwater pumping began

9 meters (30 feet)

Level of land after 52 years of heavy pumping

◀ **Figure 8.15** The marks on this utility pole indicate the level of the surrounding land in preceding years. Between 1925 and 1977, this part of the San Joaquin Valley, California, subsided almost 9 meters (30 feet) because of the withdrawal of groundwater and the resulting compaction of sediments. (Photo courtesy of U.S. Geological Survey)

ACTIVITY 8.7
Ground Subsidence

Figure 8.16 is a graph that shows the relationship between ground subsidence in Santa Clara Valley and the water level of a well in the same area. Use this figure to answer the following questions.

1. What is the general relationship between ground subsidence and the level of water in the well?

_____

2. What was the total ground subsidence, and what was the total drop in the level of water in the well during the period shown on the graph?

   Total ground subsidence: _____ feet

   Total drop in well level: _____ feet

3. During the period shown on the graph, on average, about how much land subsidence occurred with each 20-foot decrease in the water level in the well: 1 foot, 5 feet, or 10 feet?

   Subsidence: about _____ foot/feet

4. Was the ground subsidence that occurred between 1930 and 1950 less or more than the subsidence that occurred between 1950 and 1970?

   Ground subsidence from 1930 to 1950 was _____ than that from 1950 to 1970.

▲ **Figure 8.16** Ground subsidence and changing water levels in a well in the Santa Clara Valley, California. (Courtesy of U.S. Geological Survey)

continued

Activity 8.7 continued

5. Notice that minimal subsidence occurred from 1935 to 1950. Refer to the well water level during the same period of time and suggest a possible reason for the reduced rate of subsidence.

_____

_____

8.8 Karst Topography

■ **Identify on a topographic map karst landscape features, including sinkholes, disappearing streams, and solution valleys.**

Landscapes dominated by features that result from groundwater dissolving and removing the underlying soluble rock, usually limestone, are said to exhibit **karst topography**, named for the *Krs* region in the border area between Slovenia and Italy where such topography is strikingly developed. On the surface, karst topography can be identified by *sinking streams*, *solution valleys*, and depressions called *sinkholes* (**Figure 8.17**). Beneath the surface, removal of soluble rock may result in *caves* and *caverns*. A well-known karst region in the United States is in the vicinity of Kentucky's Mammoth Cave National Park. The bedrock in this area is mainly composed of soluble limestone.

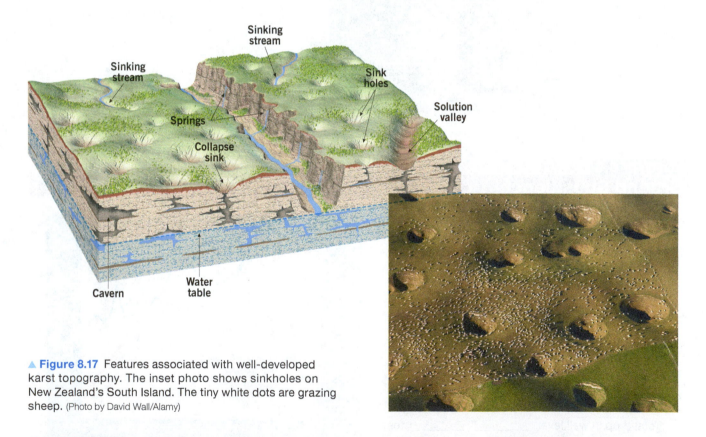

▲ **Figure 8.17** Features associated with well-developed karst topography. The inset photo shows sinkholes on New Zealand's South Island. The tiny white dots are grazing sheep. (Photo by David Wall/Alamy)

ACTIVITY 8.8
Examining a Karst Landscape

Use the Park City, Kentucky, topographic map in **Figure 8.18** (on page 147) and the stereogram of the same area in **Figure 8.19** (on page 148) to answer the following.

1. Locate three sinkholes (depressions) on the map and mark each with an X. (*Hint:* Look for closed contour lines with hachures.)

©2019 Pearson Education, Inc.

continued

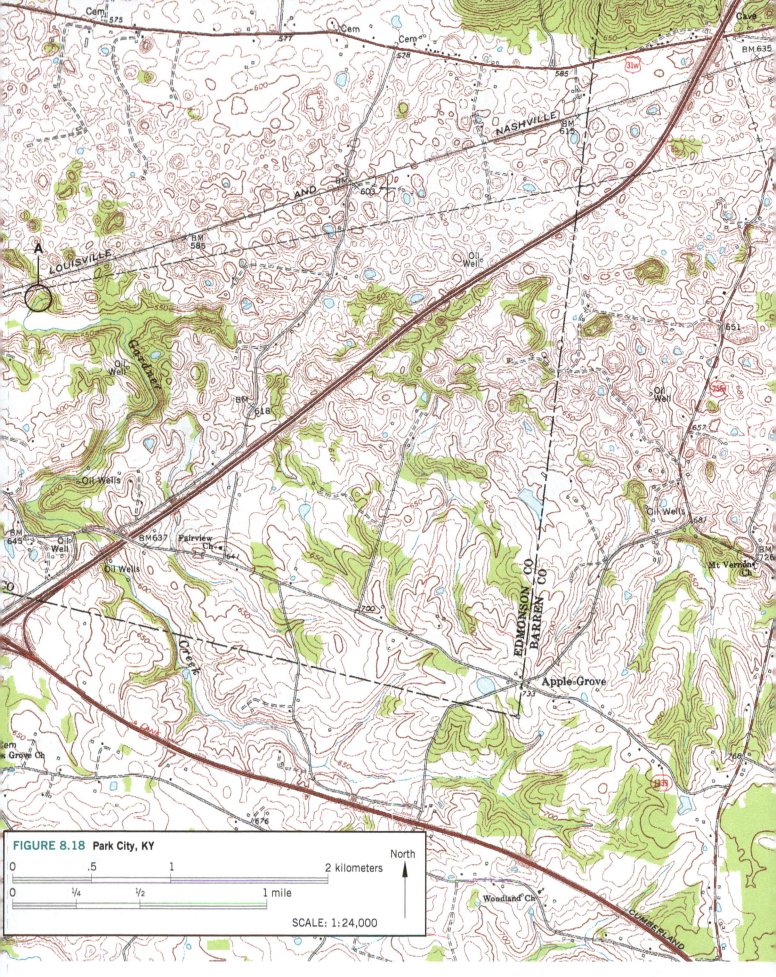

FIGURE 8.18 Park City, KY

North

| 0 | .5 | 1 | 2 kilometers |

| 0 | ¼ | ½ | 1 mile |

SCALE: 1:24,000

(Courtesy of U.S. Geological Survey)

▲ **Figure 8.19** Stereogram of the Park City, Kentucky, area. (Courtesy of U.S. Geological Survey)

2. Notice that several sinkholes have water in them. What does this indicate about the depth of the water table in this area?

_____

3. Describe what is happening to Gardner Creek in the area indicated with the letter A on the map.

_____

MasteringGeology™

Looking for additional review and lab prep materials? Go to **www.masteringgeology.com** for Pre-Lab Videos, Geoscience Animations, RSS Feeds, Key Term Study Tools, The Math You Need, an optional Pearson eText, and more.

Shaping Earth's Surface: Running Water and Groundwater

Name _____   Course/Section _____

Date _____   Due Date _____

1. Write a statement or two describing the movement of water through the hydrologic cycle, including several of the processes that are involved.

   _____

   _____

   _____

2. Assume that you need to determine the rate at which infiltration occurs in an area. What are the variables you must consider in order to formulate an answer?

   _____

   _____

3. Write a brief paragraph summarizing the results of your permeability experiment.

   _____

   _____

   _____

4. The diagrams in **Figure 8.20** show lag time between rainfall and peak flow (flooding) for an urban area and a rural area. Which diagram—(A or B)—represents a rural area? Explain your choice.

   Diagram _____ represents a rural area.

   _____

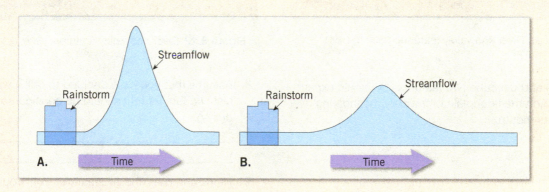

▲ **Figure 8.20** Diagrams to be used with Question 4.

continued

5. On **Figure 8.21**, identify and label as many features of the river and valley as possible.

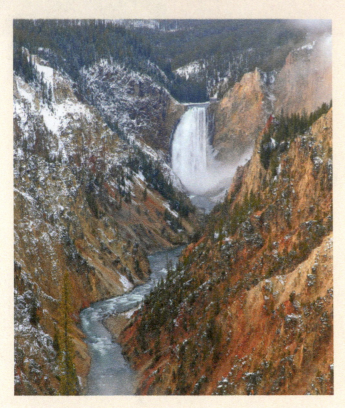

▲ **Figure 8.21** River and valley features. (Photo by Lane V. Erickson/Shutterstock)

6. Name and describe two features you would expect to find on the floodplain of a widely meandering river near its mouth.

FEATURE	DESCRIPTION
_____	_____
	_____
_____	_____
	_____

7. On **Figure 8.22**, identify and label as many features of the river and valley as possible. Is this stream more likely well above base level or near base level? Explain how you figured this out.

_____

_____

_____

▲ **Figure 8.22** River and valley features. (Photo by Michael Collier)

8. Assume that you have decided to drill a water well. List two factors that should be considered prior to drilling.

_____

_____

9. Name and describe two features you would expect to find in a region with karst topography.

FEATURE	DESCRIPTION
_____	_____
	_____
_____	_____
	_____

Shaping Earth's Surface: Arid and Glacial Landscapes

LEARNING OBJECTIVES

Each statement represents an important learning objective that relates to one or more sections of this lab. After you complete this exercise you should be able to:

- **Locate the desert and steppe regions of North America.**
- **Describe the evolution of mountainous desert landforms found in the Basin and Range region of the western United States.**
- **Contrast alpine (valley) glaciers and ice sheets.**
- **Describe the types of glacial deposits.**
- **Identify and explain the formation of depositional features commonly found in areas once covered by ice sheets.**
- **Describe the evolution of glaciated mountainous areas and the erosional features created by alpine glaciation.**

MATERIALS

calculator   ruler
hand lens   stereoscope
string

PRE-LAB VIDEO ▶ https://goo.gl/jRDTmf

Prepare for lab! Prior to attending your laboratory session, view the pre-lab video. Each video provides valuable background that will contribute to your understanding and success in lab.

INTRODUCTION

Climate has an undeniable influence on the nature and intensity of Earth's external processes. The existence and extent of deserts are largely controlled by Earth's changing climate. Another example of a prominent link between climate and geology is evident when we examine glacial landscapes. Glaciers and wind, like the running water and groundwater studied in Exercise 8, are significant erosional processes. They are responsible for creating many different landforms and are an integral link in the rock cycle.

PART 1 Geology

9.1 Deserts

■ **Locate the desert and steppe regions of North America.**

Arid (**desert**) and semiarid (**steppe**) climates cover about 30 percent of Earth's land area (Figure 9.1). Despite the arid conditions, running water is the dominant agent of erosion in deserts. Wind erosion, although more significant in dry areas than elsewhere, is only of secondary importance.

Precipitation in dry climates is not only meager but sporadic. When rain occurs, it is often in the form of short, heavy downpours. Consequently, **flash floods** are relatively common. Even when desert streams are full, they die out quickly because the flowing water soaks into the ground.

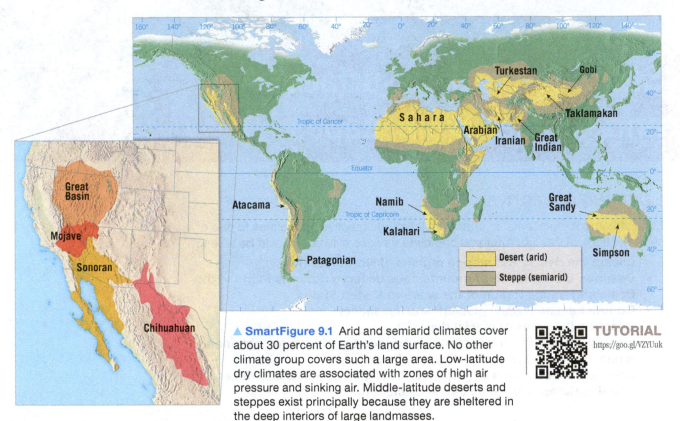

▲ **SmartFigure 9.1** Arid and semiarid climates cover about 30 percent of Earth's land surface. No other climate group covers such a large area. Low-latitude dry climates are associated with zones of high air pressure and sinking air. Middle-latitude deserts and steppes exist principally because they are sheltered in the deep interiors of large landmasses.

TUTORIAL
https://goo.gl/VZYUuk

ACTIVITY 9.1

Deserts

1. Indicate whether you think the following statements about the world's dry lands are *true* (T) or *false* (F).

   a. _____ Deserts are always hot.

   b. _____ Deserts are typically covered with sand dunes.

   c. _____ Deserts are rare, encompassing only a small percentage of Earth's land surface.

   d. _____ Deserts are practically lifeless.

   e. _____ Deserts are located away from the ocean.

Many people are surprised to learn that these statements are all common misconceptions. Midlatitude deserts experience low winter temperatures; sand dunes are widespread in only a few locations; plants and animals thrive where water is available; and the driest desert on Earth is located in coastal Chile.

2. Name the four deserts found in the United States.

   _____

   _____

9.2 Evolution of a Mountainous Desert Landscape

■ Describe the evolution of mountainous desert landforms found in the Basin and Range region of the western United States.

A classic region for studying the effects of running water in dry areas is the **Basin and Range** region, which includes portions of California, Nevada, Utah, Oregon, Arizona, and New Mexico (**Figure 9.2**). In this region, the erosion of mountain ranges and subsequent deposition of sediment in adjoining basins have produced a landscape characterized by several unique landforms.

The Basin and Range is characterized by **fault-block mountains** that formed when tensional forces elongated and fractured the crust into numerous blocks (**Figure 9.2A**). The region is characterized by interior drainage in which streams carry sediment from the mountains into adjacent basins. **Alluvial fans** and **bajadas** (features that result from the coalescence of alluvial fans) often form as streams deposit sediment on the less steep slopes at the base of the mountains (**Figure 9.2B**). Occasionally, a shallow **playa lake** may develop near the center of a basin. Playa lakes are temporary features that usually last a few weeks before evaporation and infiltration remove the water. The dry, flat lake bed that remains is called a **playa**.

As the front of a mountain is worn down by erosion and the basin fills with sediment, relief diminishes. Continuing erosion in the mountains and deposition in the basins may eventually fill the basin, leaving a nearly flat surface dotted with isolated peaks called **inselbergs** (**Figure 9.2C**).

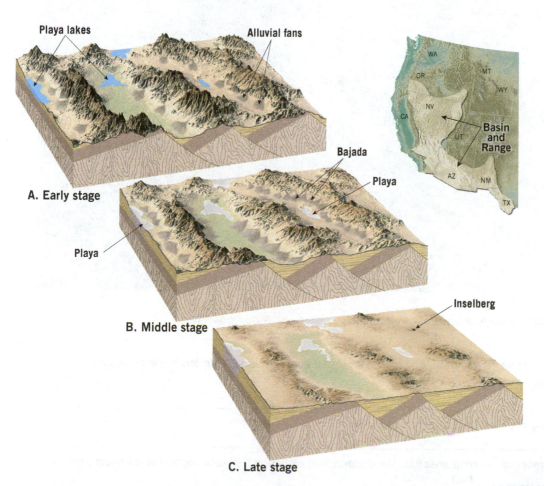

SmartFigure 9.2 Stages of landscape evolution in a block-faulted, mountainous desert such as the Basin and Range region of the West. Notice that as time advances, relief diminishes. **A.** Early stage; **B.** middle stage; **C.** late stage.

CONDOR VIDEO
https://goo.gl/6oy97G

©2019 Pearson Education, Inc.

ACTIVITY 9.2

Evolution of a Mountainous Desert Landscape

Figure 9.3 is a stereogram showing a portion of the area covered by the Antelope Peak, Arizona, topographic map (**Figure 9.4**). Use the stereogram and the map to complete the following. You may also find the diagrams in Figure 9.2 helpful.

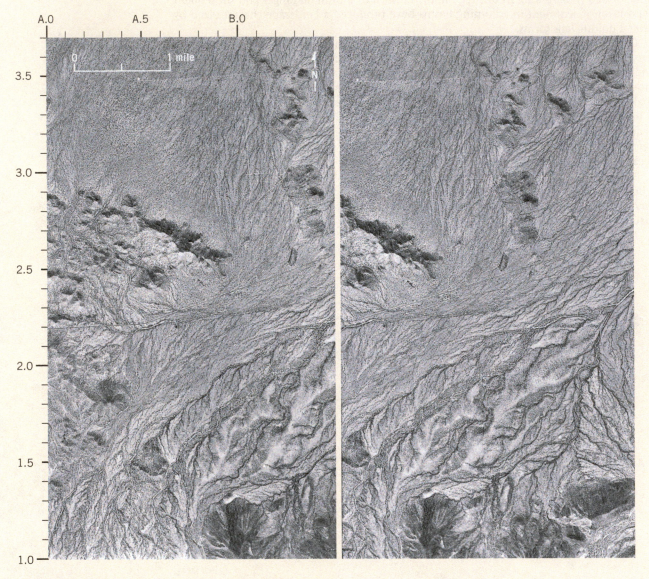

▲ **Figure 9.3** Stereogram of the Antelope Peak, Arizona, area. (Courtesy of U.S. Geological Survey)

1. On the Antelope Peak topographic map, draw a line around the area that is illustrated in the stereogram.

2. Is the vegetation in the area dense or sparse?

_____

3. Are there few or many dry stream channels? How did you figure this out?

_____

4. Determine the total relief of the map area (i.e., the difference in elevation between highest and lowest points).
   Total relief: _____ feet

continued

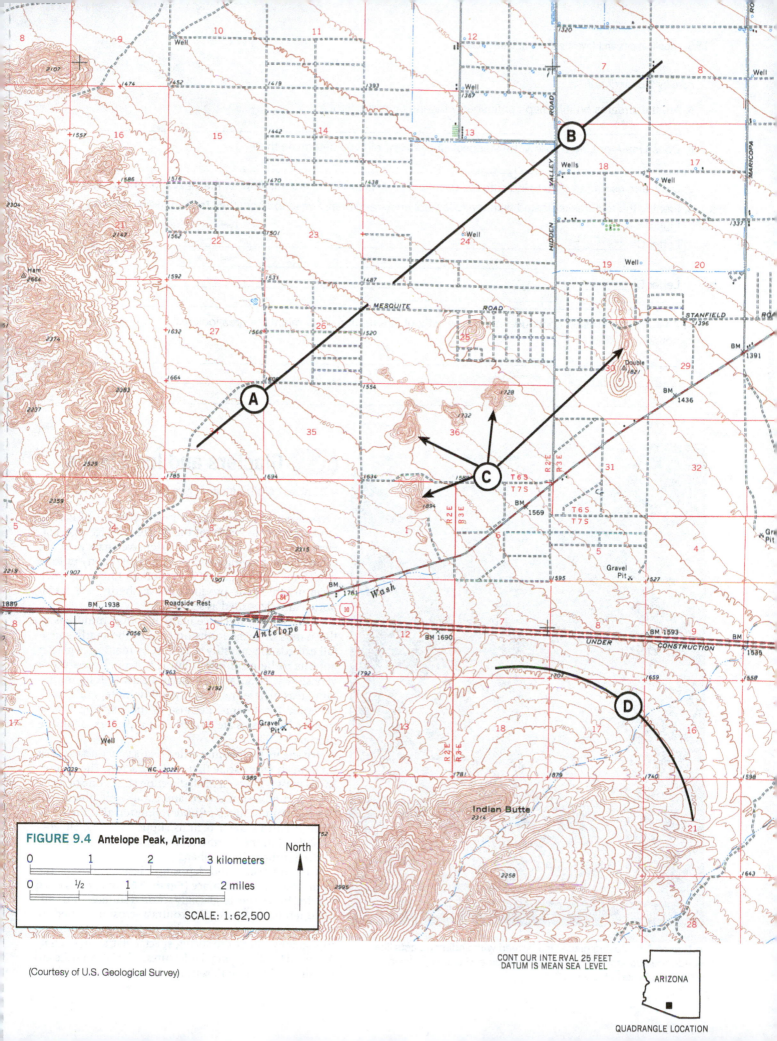

FIGURE 9.4 Antelope Peak, Arizona

North

| 0 | | 1 | | 2 | | 3 kilometers |

| 0 | ½ | | 1 | | 2 miles |

SCALE: 1:62,500

(Courtesy of U.S. Geological Survey)

CONTOUR INTERVAL 25 FEET
DATUM IS MEAN SEA LEVEL

ARIZONA

QUADRANGLE LOCATION

Activity 9.2 continued

5. Are the streams on this map continuously flowing or intermittent?

_____

6. Does Line A or Line B follow the steepest slope? How did you determine this?

_____

7. Draw arrows on the map to indicate the directions that intermittent streams flow as they leave the mountains.

8. Identify the features indicated with the following letters and briefly describe how they formed. (*Hint:* Refer to Figure 9.2.)

Letter C: _____

_____

Letter D: _____

_____

9. Assume that erosion continues in the area without interruption. How might the area look millions of years from now?

_____

_____

_____

Glacial abrasion created the scratches and grooves in this bedrock.

▲ **Figure 9.5** Moving glacial ice, armed with sediment, acts like sandpaper, scratching and polishing rock and creating glacial striations. (Photo by Michael Collier)

9.3 Glaciers and Ice Sheets

■ **Contrast alpine (valley) glaciers and ice sheets.**

Present-day glaciers cover nearly 10 percent of Earth's land area At the height of the Quaternary Ice Age, glaciers were three times more extensive than they are today. These moving masses of ice create many unique landforms and are part of an important link in the rock cycle in which the products of weathering are transported and deposited as sediment.

A **glacier** is a thick ice mass that, over hundreds or thousands of years, forms on land as the yearly snowfall exceeds the quantity of ice lost by melting. A glacier appears to be motionless, but it is not; glaciers move very slowly. Thousands of glaciers exist in lofty mountain areas, where they usually follow valleys originally occupied by streams. Because of their settings, these moving ice masses are termed **valley glaciers**, or **alpine glaciers**.

Ice sheets (sometimes called *continental glaciers*) exist on a much larger scale than valley glaciers. These enormous masses flow out in all directions from one or more snow-accumulation centers and completely obscure all but the highest areas of underlying terrain. Presently each of Earth's polar regions supports an ice sheet—Greenland in the Northern Hemisphere and Antarctica in the Southern Hemisphere.

Glacial erosion and deposition leave unmistakable imprints on Earth's surface (**Figure 9.5**). In regions once covered by ice sheets, glacially scoured surfaces and subdued terrain dominate. By contrast, erosion caused by alpine glaciers accentuates the irregular mountainous topography, often producing spectacular scenery characterized by sharp, angular features. Glacial deposits are usually visible in both settings.

ACTIVITY 9.3
Glaciers and Ice Sheets

1. What percentage of Earth's land surface do glaciers presently cover?

   _____%

2. Identify two locations where ice sheets are currently found.

   _____

3. Briefly compare an ice sheet to a valley glacier.

   _____
   _____

9.4   Depositional Features of Glaciers

■ **Describe the types of glacial deposits.**

The general term **glacial drift** applies to all sediments of glacial origin, regardless of how, where, or in what form they were deposited. There are two types of glacial drift: (1) **till**, which is characteristically unsorted sediment deposited directly by a glacier (**Figure 9.6**), and (2) **stratified drift**, which is material that has been sorted and deposited by glacial meltwater.

The most widespread depositional features of glaciers are **moraines**, which are ridges or layers of till that form along the outer margins of glaciers. An **end moraine** is a ridge of till that forms along the terminus of a glacier whenever the ice front is stationary. The end moraine marking the farthest advance of a glacier is called the *terminal end moraine*. Other end moraines, called *recessional moraines*, form when the ice front becomes stationary during the process of retreat. By contrast, *ground moraine* is a layer of till deposited as the ice front is retreating. The ground moraines deposited during the Quaternary Ice Age were expansive, some covering several states.

Several other depositional features exist, as described below:

- **Outwash plain** A relatively flat, gently sloping plain consisting of materials deposited by meltwater streams (stratified drift) in front of the margin of an ice sheet.
- **Kettle** A depression created when blocks of ice become lodged in glacial deposits and subsequently melt. When these depressions fill with water, they are called **kettle lakes**.
- **Kame** A steep-sided hill composed of stratified drift (sand and gravel) that originates when sediment being carried by streams of meltwater collects in openings in stagnant glacial ice.
- **Drumlin** A streamlined asymmetrical hill composed of glacial till that resembles an inverted spoon. The steep side of the hill faces the direction *from which* the ice advanced.
- **Esker** A sinuous ridge composed largely of sand and gravel deposited by a meltwater stream flowing within, on top of, or beneath a glacier near its terminus.

Glacial till is an unsorted mixture of many different sediment sizes.

A close examination of glacial till often reveals cobbles that have been scratched as they were dragged along by the ice.

▲ **Figure 9.6** Glacial till is an unsorted mixture of many different sediment sizes. (Photos by E. J. Tarbuck)

ACTIVITY 9.4

Depositional Features of Glaciers

Use **Figure 9.7**, which illustrates a hypothetical *retreating glacier*, to complete the following.

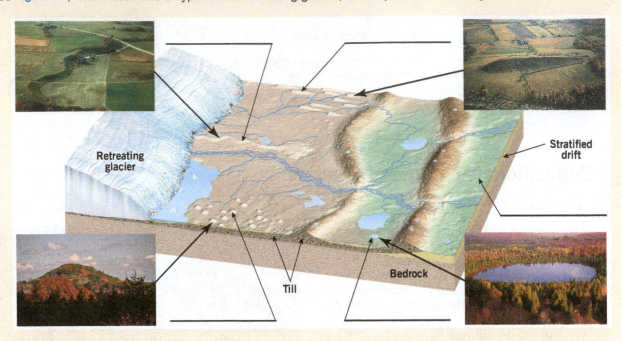

Retreating glacier

Stratified drift

Bedrock

Till

▲ **Figure 9.7** This diagram showing common depositional landforms depicts a hypothetical area affected by ice sheets in the recent geologic past. (Drumlin photo courtesy of Ward's Natural Science Establishment; esker photo by Richard P. Jacobs/ JLM Visuals; kame photo by John Dankwardt; kettle lake photo by Carlyn Iverson/Science Source)

1. As a glacier retreats, it sometimes stalls and deposits a recessional end moraine. Label the recessional end moraine on Figure 9.7.

2. On Figure 9.7, label an area covered by ground moraine.

3. On Figure 9.7, fill in the appropriate blanks with the names of the following depositional features: kettle lake, kame, drumlin, esker, and outwash plain.

4. Which area is composed of stratified drift: the ground moraine or the outwash plain?

_____

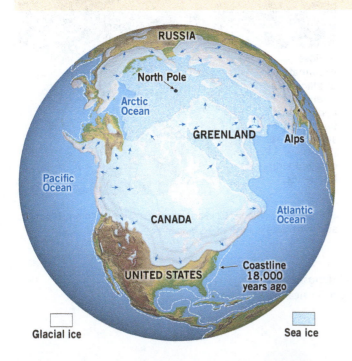

RUSSIA

North Pole

Arctic Ocean

GREENLAND

Alps

Pacific Ocean

CANADA

Atlantic Ocean

UNITED STATES

Coastline 18,000 years ago

☐ Glacial ice

☐ Sea ice

9.5 Depositional Features Associated with Ice Sheets

■ Identify and explain the formation of depositional features commonly found in areas once covered by ice sheets.

During a span known as the *Last Glacial Maximum*, about 18,000 years ago, ice sheets not only covered Greenland and Antarctica but also large portions of North America, Europe, and Siberia (**Figure 9.8**). Such enormous masses flow out in all directions from one or more ice-accumulation centers and completely obscure all but the highest areas of underlying terrain.

Landforms produced by ice sheets, especially those that covered portions of the conterminous United States, are largely depositional. Some of the most extensive glacial deposits are located in the north-central United States. In many places, moraines, outwash plains, kettles, and other depositional features are prominent landscape features.

◄ **Figure 9.8** This map shows the maximum extent of ice sheets in the Northern Hemisphere during the Quaternary Ice Age.

ACTIVITY 9.5 ///
Depositional Features Associated with Ice Sheets

Figure 9.9 is a stereogram showing a portion of the area covered by the Whitewater, Wisconsin, topographic map (**Figure 9.10**, page 160). Use the stereogram and the map to complete the following.

▲ **Figure 9.9** Stereogram of the Whitewater, Wisconsin, area. (Courtesy of U.S. Geological Survey)

1. After examining the map and stereogram, draw a line on the map to outline the area illustrated on the stereogram.

2. What evidence on the map indicates that portions of the area are poorly drained? On what part of the map are these features located?

_____

_____

3. Use **Figure 9.11** to draw a topographic profile of the X–Y line on Figure 9.10.

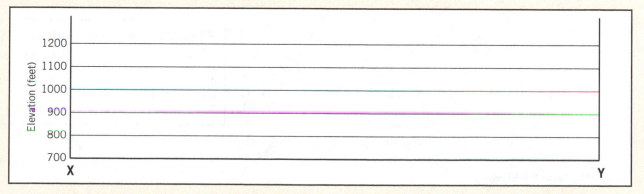

▲ **Figure 9.11** Profile of the Whitewater map along line X–Y.

continued

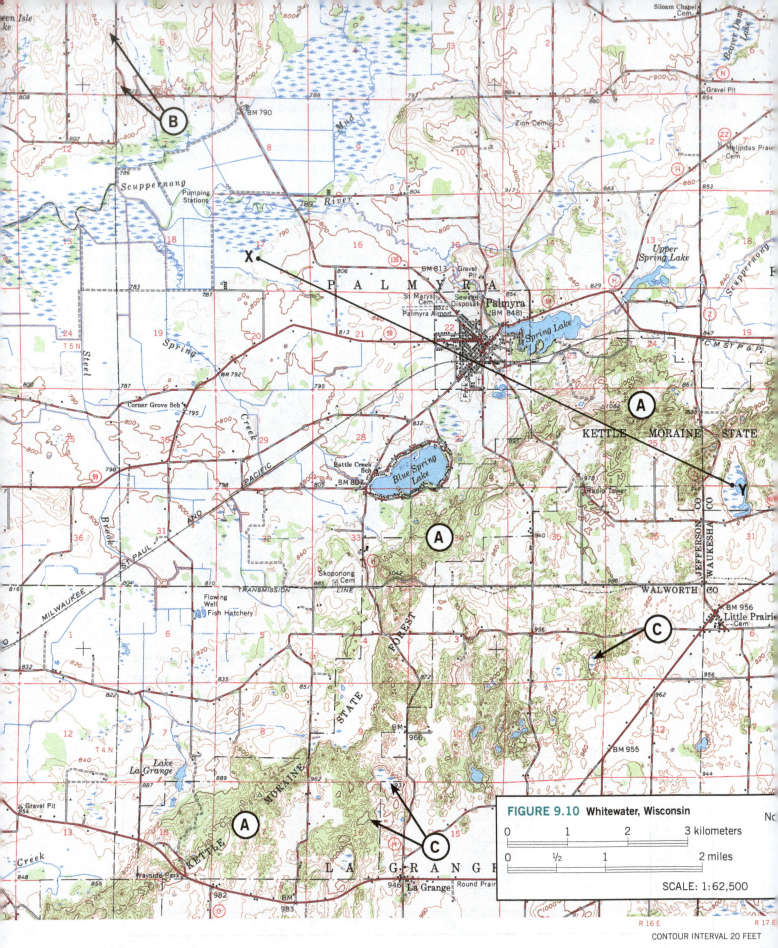

FIGURE 9.10 Whitewater, Wisconsin

| 0 | 1 | 2 | 3 kilometers |

| 0 | ½ | 1 | 2 miles |

SCALE: 1:62,500

CONTOUR INTERVAL 20 FEET

(Courtesy of U.S. Geological Survey)

Activity 9.5 continued

4. Is the general topography of the land in Sections 7 and 8 in the northwest portion of the region higher or lower in elevation than the land around the letter A located near the center of the map? Is it more or less hilly?
The topography in Sections 7 and 8 is _____ in elevation and _____ hilly.

5. Is the area that coincides with Kettle Moraine State Forest higher or lower in elevation than the land to the northwest and southeast?

_____

6. The feature labeled A on the map is a *long* ridge composed of till. Is this ridge an esker, an end moraine, or a drumlin?
This ridge is a(n) _____.

7. The streamlined, asymmetrical hills composed of till, labeled B, are what type of feature?
The streamlined hills are _____.

8. Examine the shape of the features labeled B on the map in Figure 9.10. How can these features be used to determine the direction of ice flow in a glaciated area?

_____

9. Using the features labeled B in Figure 9.10 as a guide, draw an arrow on the map to indicate the direction of ice movement that occurred in this region.

10. What is the likely location of the outwash plain on the map? Identify and label the area "outwash plain." (*Hint:* Refer to Figure 9.7.)

_____

11. Label the area covered by ground moraine.

12. What term is applied to the numerous almost circular depressions designated with the letter C?

_____

9.6 Erosional Features Associated with Valley Glaciers

■ **Describe the evolution of glaciated mountainous areas and the erosional features created by alpine glaciation.**

Alpine glaciers often exaggerate the already mountainous topography of a region by eroding and deepening the valleys they occupy. **Figure 9.12** illustrates changes that an unglaciated mountainous area experiences as a result of alpine glaciation. Some of the landforms produced by glacial erosion include:

- **Arête** A narrow, knifelike ridge that separates two adjacent glaciated valleys.
- **Cirque** An amphitheater-shaped basin at the head of a glaciated valley produced by frost wedging and plucking.
- **Hanging valley** A tributary valley that was not deepened as much as the main valley and that enters a glacial trough at a considerable height above the floor of the trough.
- **Horn** A pyramid-like peak formed by glacial action in three or more cirques surrounding a mountain summit.
- **Tarn** A small lake occupying a cirque.

ACTIVITY 9.6A
How Glaciation Affects Mountainous Topography

Refer to Figure 9.12 to complete the following.

1. How does a glacier change the shape and depth of a mountain valley?

_____

_____

continued

V-shaped valley

A. Unglaciated topography

Arête

Medial moraine · **Arête** · **Horn** · **Cirques**

Main glacier

Cirque

B. Region during period of maximum glaciation

Arête · **Tarn** · **Horn** · **Cirques**

Glacial trough

Hanging valley

Hanging valley

C. Glaciated topography

▲ **SmartFigure 9.12** Erosional landforms created by alpine glaciers. The unglaciated landscape in **A** is modified by valley glaciers in **B**. After the ice recedes, in **C**, the terrain looks very different than it did before glaciation. (Arête photo from James E. Patterson collection; cirque photo by Marli Miller; hanging valley photo by John Warden/SuperStock)

ANIMATION
https://goo.gl/wzstSW

Activity 9.6A continued

2. How is an arête different from a horn?

_____

_____

3. Briefly describe the changes that occur in mountainous areas due to alpine glaciation.

_____

_____

ACTIVITY 9.6B ///

Identifying Glacial Features on a Topographic Map

Refer to **Figure 9.13**, a portion of the Holy Cross, Colorado, topographic map. This is a mountainous area that has been shaped by alpine glaciers.

1. Identify the glacial feature indicated by each of the following letters. Use Figure 9.12C as a reference.

    Letter A: _____

    Letter B: _____

2. A *tarn* is a lake that forms in a cirque. Of the features labeled C, D, E, and F, which indicate(s) a tarn(s)?

    Letter(s): _____

3. The feature marked G on the map is a *depositional feature* composed of glacial till. What type of glacial feature is it? How did it form?

    _____

    _____

4. Explain how Turquoise Lake likely formed.

    _____

    _____

5. Use **Figure 9.14** to draw a topographic profile along the X–Y line from Sugar Loaf Mountain to Bear Lake and mark the position of the Lake Fork stream. (Use only index contours.)

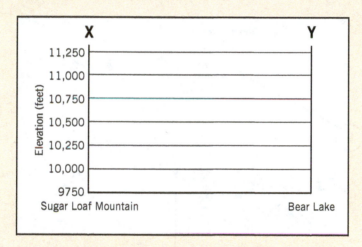

▲ **Figure 9.14** Topographic profile of the valley of Lake Fork on the Holy Cross, Colorado, map.

6. Describe the shape of Lake Fork Valley, based on your profile.

    _____

7. What glacial feature is Lake Fork Valley?

    _____

MasteringGeology™

Looking for additional review and lab prep materials? Go to **www.masteringgeology.com** for Pre-Lab Videos, Geoscience Animations, RSS Feeds, Key Term Study Tools, The Math You Need, an optional Pearson eText, and more.

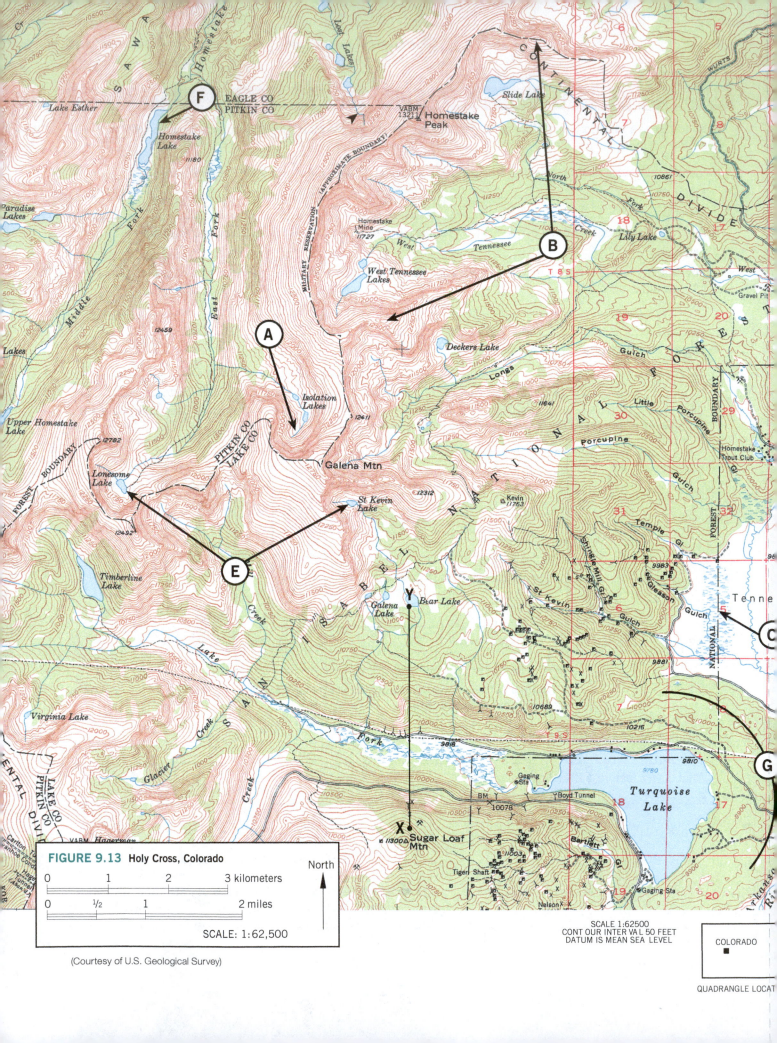

FIGURE 9.13 Holy Cross, Colorado

0   1   2   3 kilometers

0   ½   1   2 miles

SCALE: 1:62,500

North

(Courtesy of U.S. Geological Survey)

SCALE 1:62500
CONTOUR INTERVAL 50 FEET
DATUM IS MEAN SEA LEVEL

COLORADO

QUADRANGLE LOCAT

Shaping Earth's Surface: Arid and Glacial Landscapes

Name _____

Course/Section _____

Date _____

Due Date _____

1. Name the area of the United States that is characterized by fault-block mountains with streams that drain into adjoining basins?

   _____

2. Identify the feature shown by the arrows in **Figure 9.15**.

   _____

▲ **Figure 9.15** Photo to accompany Question 2. (Photo by Michael Collier)

3. List some of the prominent features found in a mountainous desert landscape (see Figure 9.2).

   _____

   _____

4. Toward what general direction does the landscape slope downward on the Antelope Peak topographic map (see Figure 9.4)?

   _____

5. If you were working in the field, explain how you would determine whether a glacial feature is an end moraine or an esker.

   _____

   _____

   _____

6. In the following space, sketch a simplified map view (an area viewed from above) of the Whitewater, Wisconsin, topographic map (Figure 9.10). Include and label the outwash plain, end moraine, drumlins, and a few kettles and kettle lakes.

   [blank box for sketch]

7. What reason did you give for the formation of Turquoise Lake on the Holy Cross, Colorado, topographic map (Figure 9.13)?

   _____

   _____

8. Use **Figure 9.16**, an image
   of Alaska's Bear Glacier, to
   complete the following.

   a. Identify the erosional feature
      labeled A.

   b. Feature B is a ridge composed of
      till. Is it more likely an esker or an
      end moraine? _____

   c. When this photo was taken, is it
      more likely that Bear Glacier was
      advancing or retreating? Provide
      evidence for you answer.

   Bear Glacier was _____
   _____
   _____
   _____
   _____
   _____
   _____.

▲ **Figure 9.16** Photo to accompany Question 8. (Photo by Michael Collier)

9. What type of glacier is shown in **Figure 9.17**?
   _____

▲ **Figure 9.17** Photo to accompany Question 9. (Photo by
Michael Collier)

10. Use **Figure 9.18** to complete the following.

    a. Describe the shape of this classic glacial valley.
       _____

    b. What was the likely shape of the valley prior to
       glaciation?
       _____

▲ **Figure 9.18** Photo to accompany Question 10. (Photo by
Michael Collier)

Geologic Time

LEARNING OBJECTIVES

Each statement represents an important learning objective that relates to one or more sections of this lab. After you complete this exercise you should be able to:

- **Distinguish between relative and numerical dates.**
- **List and explain the principles used to determine the relative ages of geologic events.**
- **Apply appropriate relative dating principles to determine a sequence of geologic events.**
- **Describe several different types of fossilization.**
- **Explain how fossils are used to date rocks and correlate rock layers.**
- **Explain the principle of radiometric dating, including half-life, parent isotope, and daughter product.**
- **Determine numerical dates for sedimentary strata.**
- **List the sequence of major events that have occurred on Earth and compare the span of geologic time to the length of human history.**

MATERIALS

ruler	calculator
meterstick or metric measuring tape	32 nickels
more than 5 meters of cash register paper	32 pennies
red and blue colored pencils	

PRE-LAB VIDEO https://goo.gl/enPyxW

 Prepare for lab! Prior to attending your laboratory session, view the pre-lab video. Each video provides valuable background that will contribute to your understanding and success in lab.

INTRODUCTION

In the 1800s, geologists effectively demonstrated that Earth had experienced many episodes of mountain building and erosion, which must have required great spans of geologic time. Although these pioneering scientists understood

PART 1 **Geology**

167

that Earth was very old, they had no way of determining its age in years. Was it tens of millions, hundreds of millions, or even billions of years old? Long before geologists could establish a geologic time scale that included numerical dates in years, they gradually assembled a time scale using relative dating principles. This exercise examines those relative dating principles, as well as the principles involved in determining numerical dates using radioactivity.

10.1 Numerical and Relative Dates

■ **Distinguish between relative and numerical dates.**

After the fact was established that geologic history was exceedingly long, numerous attempts were made to determine Earth's age in years. Although some of the methods appeared promising at the time, none of those early efforts proved to be reliable. Those scientists were seeking **numerical dates** to specify the number of years that have passed since an event occurred. Today, our understanding of *radioactivity* allows geologists to accurately determine numerical dates for rocks that represent important events in Earth history. We will examine numerical dating later in this exercise.

When we place rocks in their proper *sequence of formation*—which formed first, second, third, and so on—we are establishing **relative dates**. Such dates do not tell us how long ago an event took place—only that it followed one event and preceded another. The relative dating techniques that were developed are valuable and still widely used. Numerical dating methods did not replace these techniques; they simply supplemented them. To establish a relative time scale, a few basic principles or rules had to be discovered and applied. The following sections discuss these basic principles.

ACTIVITY 10.1
Numerical and Relative Dates

1. On the chart in **Figure 10.1** list six events (something as simple as eating lunch or going to Earth science class) that occurred in your life yesterday. Place them in *relative order* (1, 2, 3, etc.) and then assign them a *time of day*, as best you can remember.

2. In your own words, briefly distinguish between relative dates and numerical dates.

_____

_____

▶ **Figure 10.1** Relative versus numerical dating.

RELATIVE DATING	NUMERICAL DATING		
Events in Order of Occurrence	Approximate Times	AM	PM
1.			
2.			
3.			
4.			
5.			
6.			

10.2 Relative Dating Principles: Original Horizontality and Superposition

■ **List and explain the principles used to determine the relative ages of geologic events.**

The **principle of original horizontality** states that most sediments and lava flows are *originally* deposited as *horizontal* layers. Therefore, most sedimentary rocks consist of horizontal layers, called **beds**, or **strata** (**Figure 10.2**).

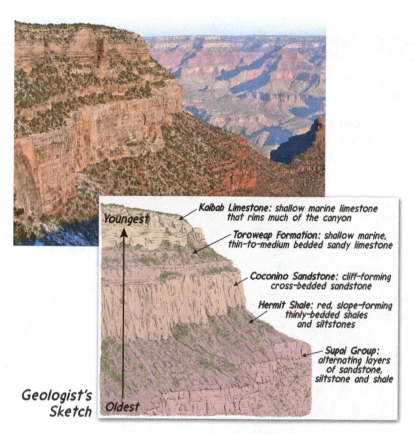

©2019 Pearson Education, Inc.

◄ Figure 10.2 The principle of superposition indicates that among these layers of the Grand Canyon, the Supai Group is oldest and the Kaibab Limestone is the youngest. (Photo by Dennis Tasa)

Youngest

Kaibab Limestone: shallow marine limestone that rims much of the canyon

Toroweap Formation: shallow marine, thin-to-medium bedded sandy limestone

Coconino Sandstone: cliff-forming cross-bedded sandstone

Hermit Shale: red, slope-forming thinly-bedded shales and siltstones

Supai Group: alternating layers of sandstone, siltstone and shale

Geologist's Sketch   *Oldest*

One of the fundamental rules in relative dating is the **principle of superposition**, which states that in any sequence of *undeformed* sedimentary rocks (or lava flows), the oldest rock is always at the bottom, and the youngest is at the top. Therefore, each layer of rock represents an interval of time that is more recent than that of the underlying rocks.

As shown in Figure 10.2, the principle of superposition makes it easy to determine the relative ages of horizontal sedimentary strata. By applying the principle of original horizontality, we can also assume that sedimentary layers that are folded or inclined at steep angles were deformed *after* the layers were deposited (**Figure 10.3**). The principle of superposition must be applied with care when dating deformed rocks because the sequence may have been completely overturned.

A

◄ Figure 10.3 A. This outcrop of sedimentary strata illustrates the characteristic horizontal layering of this group of rocks. (Photo by Adam Burton/Alamy) **B.** When we see strata that are folded or tilted, we can assume that they were moved into that position by crustal disturbances *after* their deposition. (Photo by Marco Simoni/Robert Harding)

B

Principle of Superposition

Assume that the playing cards shown in **Figure 10.4** are layers of sedimentary rocks viewed from above.

1. In the space provided in Figure 10.4, list the cards by number, according to the order in which they appear to have been laid down.

2. Were you able to place all of the cards in sequence? Explain.

_____

_____

3. Apply the law of superposition to determine the relative ages of the sedimentary strata labeled A–D in **Figure 10.5**. On the figure, list the letter of the oldest rock layer first, the next-oldest rock layer next, and so on, in *decreasing* order of age.

▲ **Figure 10.4** Sequence of playing cards illustrating the law of superposition.

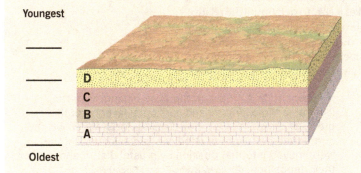

◀ **Figure 10.5** Sedimentary rock layers illustrating the law of superposition.

10.3 Relative Dating Principles: Cross-Cutting and Inclusions

■ **List and explain the principles used to determine the relative ages of geologic events.**

Cross-Cutting Relationships

When a *fault*, or an *intrusive igneous feature*, cuts through an existing rock unit or feature, it is younger than that rock unit or feature. For example, **Figure 10.6** shows a basalt dike cutting through a sandstone layer. The principle of cross-cutting indicates that the sandstone layer must have been there first; therefore, the sandstone is older than the dike.

Inclusions

Inclusions are rock fragments of one rock unit that have been enclosed within another (**Figure 10.7**). The rock mass adjacent to the one containing an inclusion must have been there first in order to provide the fragment. Therefore, the rock mass that contains the inclusion is the younger of the two.

▲ **Figure 10.6** This basalt dike (black) is younger than the sandstone layers it cuts through. (Photo by E. J. Tarbuck)

◀ **Figure 10.7** Inclusions are fragments of an older rock enclosed within a younger rock. In this image, the black-colored inclusion is older than the light igneous rock because it must have been in existence at the time the younger rock formed around it. (Photo by E. J. Tarbuck)

ACTIVITY 10.3

Principles of Cross-Cutting and Inclusions

Figure 10.8 is a geologic cross-section of a hypothetical area. Use it to answer the following questions.

1. Is the igneous intrusion, dike E, older or younger than rock layers A–D?

2. Is fault H older or younger than rock layers A–D?

3. Is fault H older or younger than sedimentary layers F and G?

4. Did fault H occur before or after dike E? Explain how you arrived at your answer.

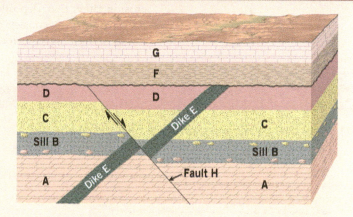

▲ **Figure 10.8** Geologic block diagram of a hypothetical area, illustrating relative dating principles.

5. What evidence supports the conclusion that the igneous intrusion labeled sill B is more recent than the rock layers on either side (A and C)?

10.4 Relative Dating Principles: Unconformities

■ **List and explain the principles used to determine the relative ages of geologic events.**

Layers of rock that have been deposited essentially without interruption are called **conformable**. Numerous areas contain conformable beds that represent long spans of geologic time. However, a *complete set* of conformable strata representing all of geologic time has not

A. Angular unconformity
Represents an extended period during
which deformation and erosion occurred

B. Disconformity
Gap in the rock record represents a
period of nondeposition and erosion

C. Nonconformity
Period of uplift and erosion that
exposed the deep rocks at the surface

▲ **Figure 10.9** Three common types of unconformities. Wavy lines mark each unconformity.

been found anywhere on Earth. Throughout history, the deposition of sediment has been interrupted over and over again. These breaks in the rock record are called *unconformities*. An **unconformity** represents a long period during which deposition ceased, erosion removed previously formed rocks, and deposition resumed. Unconformities are typically shown on a cross-sectional (side view) diagram by a wavy line to illustrate the irregular erosional surface. Three types of unconformities are illustrated in **Figure 10.9** and described below:

- An **angular unconformity** consists of tilted or folded sedimentary rocks that are overlain by younger, more flat-lying strata.
- A **disconformity** is a gap in the rock record that represents a period during which erosion rather than deposition occurred.
- A **nonconformity** consists of younger sedimentary rocks that overlie older metamorphic or igneous rocks.

ACTIVITY 10.4
Unconformities

1. Label the *angular unconformity* and *disconformity* on **Figure 10.10**.

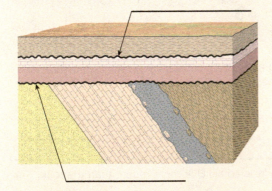

▲ **Figure 10.10** Geologic block diagram of a hypothetical region.

2. Identify the types of unconformities in **Figure 10.11A** and **Figure 10.11B**.

A. _____

B. _____

A.

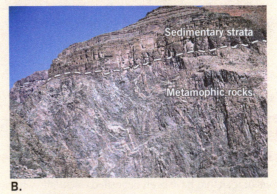

B.

▲ **Figure 10.11** Identify these two unconformities.

10.5 Applying Relative Dating Principles

■ **Apply appropriate relative dating principles to determine a sequence of geologic events.**

Geologists usually must apply several principles of relative dating when investigating the geologic history of an area.

ACTIVITY 10.5
Applying Relative Dating Principles

Use Figure 10.12 to complete the following.

1. Identify and label the unconformities in Figure 10.12.

2. Is rock layer I older or younger than layer H? What relative dating principle did you apply to determine your answer?

   Rock layer I is _____ than layer H.

   Relative dating principle: _____

3. Is fault L older or younger than rock layer D? What principle did you apply to determine your answer?

   Fault L is _____ than rock layer D.

   Relative dating principle: _____

4. Is igneous intrusion J older or younger than layers A and B? What two relative dating principles apply to your answer?

   Intrusion J is _____ than layers A and B.

   Relative dating principles: _____ and _____

5. Is the igneous intrusion labeled dike K older or younger than layers C–F?

   Intrusion K is _____ than layers C–F.

6. List the entire sequence of events, in order from oldest to youngest, by writing the appropriate letters in the spaces provided on Figure 10.12.

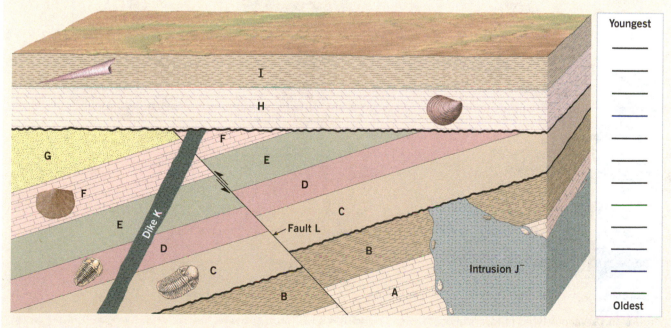

▲ **Figure 10.12** Geologic block diagram of a hypothetical area.

10.6 **Types of Fossils**

■ **Describe several different types of fossilization.**

Fossils, the remains or traces of prehistoric life, are among the most important tools for interpreting the geologic past. Knowing the nature of the life-forms that existed at a particular time helps researchers understand past environmental conditions. Fossils are also valuable *time indicators*.

ACTIVITY 10.6

Types of Fossils

1. Refer to **Figure 10.13**. Which photo(s) (A–I) best illustrate(s) the methods of fossilization or fossil evidence listed below? (Photos/letters may be used more than once.)

Permineralization: The small internal cavities and pores of an original organism that are filled with precipitated mineral matter. Photo(s): _____

Cast: The space once occupied by a dissolved shell or other structure that is subsequently filled with mineral matter. Photo(s): _____

Carbonization: Preservation that occurs when fine sediment encases delicate plant or animal forms and leaves a residue of carbon. Photo(s): _____

Impression: A replica of an organism, such as a leaf, left in fine-grained sedimentary rock. Photo(s): _____

Amber: Hardened resin of ancient trees that preserved delicate organisms such as insects. Photo(s): _____

Indirect evidence: Traces of prehistoric life but not the organism itself. Photo(s): _____

D. Fossil brachiopods

A. Trilobite fossil

B. Fossilized wood

E. Dinosaur track

C. Fossil bee

F. Trilobite fossil

G. Fossil dung

H. Fossil spider

I. Fossil fish

▲ **Figure 10.13** Various types of fossilization. (Photo C by National Park Service; photo F by Dennis Tasa; photo H by Colin Keats/Dorling Kindersley; other photos by E. J. Tarbuck)

10.7 Fossils as Time Indicators

■ **Explain how fossils are used to date rocks and correlate rock layers.**

Earth has been inhabited by different assemblages of plants and animals at different times. As sedimentary rocks form, they often incorporate the preserved remains of these

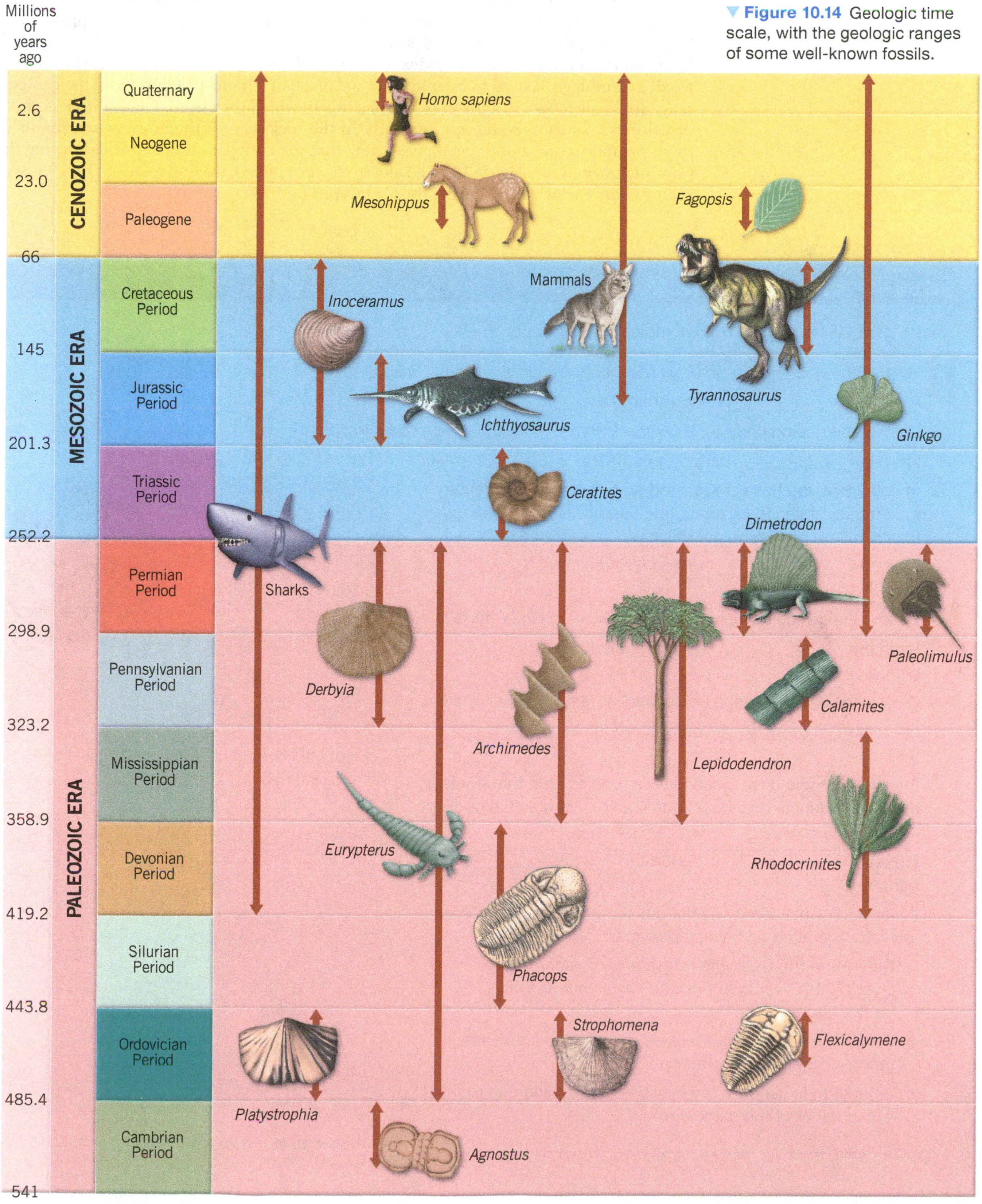

▼ **Figure 10.14** Geologic time scale, with the geologic ranges of some well-known fossils.

organisms as fossils. According to the **principle of fossil succession**, organisms succeed each other in a definite and determinable order. Therefore, the time that sediment was deposited can frequently be determined by identifying fossils found within it. Further, fossils play a key role in matching rocks of similar age found in different places. This task is called **correlation**. For example, if strata in one location (for example, Montana) contain fossils similar to strata in another location (for example, Wyoming), scientists can be quite confident that the rock layers are of the same age.

Some fossils, called **index fossils**, are widespread geographically and are associated with a short span of geologic time. Consequently, the presence of an index fossil provides an important method of matching rocks of the same age in different places. However, rock layers do not always contain an index fossil. In such situations, a group of fossils, called a **fossil assemblage**, is used to establish the age of a bed. **Figure 10.14** provides the geologic range for some fossils and the **eras** and **periods** in which they lived. For example, a genus of trilobites called *Agnostus* appears early in the rock record; therefore, when fossils of these organisms are found, geologists know that the rock unit was deposited during the Cambrian period (see Figure 10.14). Later in the rock record, we find fossils of the dinosaur *Tyrannosaurus*, which lived during the Cretaceous period.

ACTIVITY 10.7
Fossils as Time Indicators

Use Figure 10.14, page 175, to complete the following.

1. What is the geologic range of plants that belong to the group *Ginkgo*?

   From the _____ period through the _____ period.

2. What is the geologic range of *Lepidodendron*, an extinct coal-producing plant?

   From the _____ period through the _____ period.

3. Imagine that you have discovered an outcrop of sedimentary rock that contains fossil shark teeth and fossils of *Archimedes*. In which time periods might this rock have formed?

   From the _____ period through the _____ period.

4. What is the geologic range of the fossil shown in **Figure 10.15**?

   From the _____ period through the _____ period.

5. What is the geologic range of the fossil shown in **Figure 10.16**?

   From the _____ period through the _____ period.

▲ **Figure 10.15** Photo to accompany Question 4.
(Photo by Dennis Tasa)

6. Imagine that you have discovered a rock outcrop that contains the fossils identified in Questions 4 and 5. What is the geologic range of this rock?

   From the _____ period through the _____ period.

7. **Figure 10.17** illustrates two different sequences of sedimentary rock strata located some distance apart.

   a. Determine the geologic range/age of each rock layer by its fossil content and write your answer on the figure.

   b. Draw lines connecting the rock layers of the same age in outcrop 1 to those in outcrop 2.

   c. What term is used for the process of matching one rock unit with another of the same age?

   _____

▲ **Figure 10.16** Photo to accompany Question 5.

d. Based on the ages of the rock layers in outcrop 1, identify an unconformity. Remember that an unconformity is a break in the rock record. Draw a wavy line on the figure to represent the unconformity.

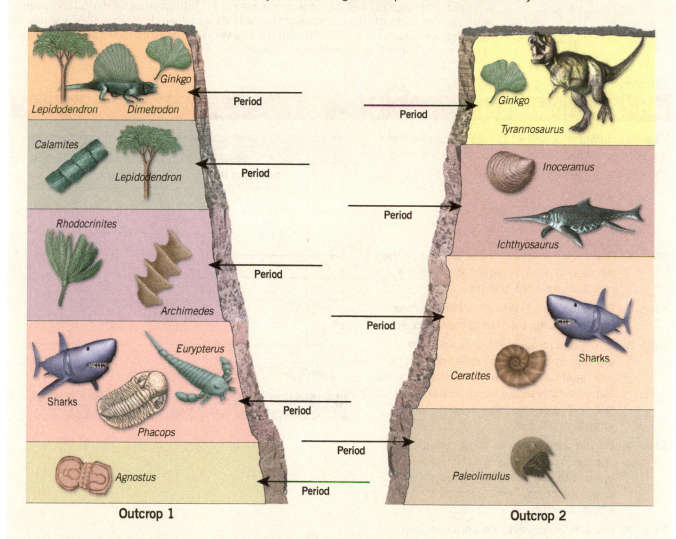

▲ **Figure 10.17** Sequences of sedimentary rock layers to accompany Question 7.

10.8 Numerical Dating with Radioactivity

■ **Explain the principle of radiometric dating, including half-life, parent isotope, and daughter product.**

Radioactivity has provided a reliable means for calculating the **numerical (absolute) age** of rocks, a procedure called **radiometric dating**. Radioactive isotopes, such as uranium-238, emit particles from their nuclei that we call *radiation*. Ultimately, the process of decay produces an isotope that is stable and no longer radioactive. For example, the radioactive decay of uranium-238 produces the stable element lead-206.

An unstable (radioactive) isotope that decays is referred to as the **parent**, while the stable isotope that results is called the **daughter product**. The rate of radioactive disintegration is expressed as the **half-life**—the amount of time it takes for half of the parent atoms to turn into the daughter product. Therefore, after one half-life, 50 percent of the parent isotope will have decayed. After the second half-life, half of the remaining radioactive atoms, or 25 percent $\left(\frac{1}{4}\right)$ of the original amount, will remain, and 75 percent will have decayed to form the daughter product. *With each successive half-life, the remaining parent isotope will be reduced by half.*

Among the many radioactive isotopes that exist in nature, six have proved particularly useful in providing radiometric ages. Rubidum-87, thorium-232, and the two isotopes of uranium have very long half-lives and can be used only for dating rocks that are millions of years old. Although the half-life of potassium-40 is 1.3 billion years, analytical techniques make the detection of tiny amounts of its stable daughter product, argon-40, possible in rocks that are younger than 100,000 years. To date very recent events, the radioactive isotope of carbon, carbon-14, can sometimes be used.

ACTIVITY 10.8
Half-Life Experiment

This experiment is designed to help you understand the concept of half-life. You will need the following materials to complete this activity:

32 nickels
32 pennies
Red and blue colored pencils

Step 1: Place the 32 nickels heads-side up on the lab table. These represent atoms of the radioactive parent isotope.

Step 2: Place the nickels into a container, shake well, and dump them onto the lab table.

Step 3: Count and remove all the nickels that are *tails-side up*. The nickels you remove represent parent atoms that have decayed. Replace each tails-side-up nickel with a penny. The pennies represent daughter atoms.

Step 4: On the chart in **Figure 10.18**, record the number of remaining nickels (heads-side up) and the pennies that represent the daughter atoms. The combined total should equal 32 coins.

Step 5: Place these coins in the container and repeat steps 3 and 4 until only one or two nickels remain.

After you have completed the exercise (Trial 1), use a blue pencil to plot on the graph in **Figure 10.19** the number of nickels that remained at each interval. Connect the points with a solid line, which represents the change in parent atoms over time. Next, use a red pencil to plot the number of pennies at each interval on the same graph. Connect the points with a solid line, which represents the change in the number of daughter atoms over time.

Repeat this experiment (Trial 2), using the table in **Figure 10.20** and the graph in **Figure 10.21**.

1. Compare the graphs you completed with the idealized half-life graph shown in **Figure 10.22**. What differences do you observe?

TRIAL 1		
	Number of Nickels	Number of Pennies
0	32	0
1		
2		
3		
4		
5		
6		

▲ **Figure 10.18** Chart to record data from Trial 1.

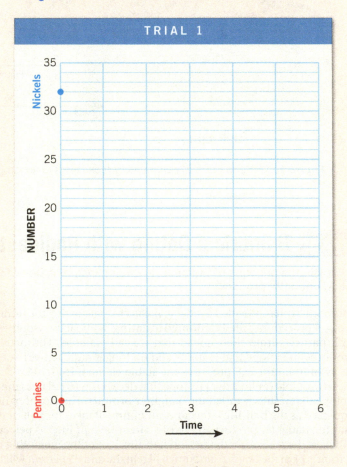

▲ **Figure 10.19** Graph to plot data from Trial 1.

2. If you had conducted this experiment 10 times and averaged your results, how do you think the results would have differed?

_____

Use Figure 10.22 to complete Questions 3–5.

3. What percentage of the original parent isotope remains after each of the following half-lives have elapsed?

PERCENTAGE OF PARENT ISOTOPE REMAINING

One half-life: _____

Two half-lives: _____

Three half-lives: _____

Four half-lives: _____

4. Geologists know that potassium-40 decays to argon-40, with a half-life of 1.3 billion years. If an analysis of a feldspar crystal in a sample of granite shows that 75 percent of the potassium-40 atoms have decayed to form argon-40, what is the age of the granite?

The granite sample is _____ years old.

5. Determine the numerical ages of rock samples that contain a parent isotope with a half-life of 100 million years and have the following percentages of original parent isotope:

50%: Age = _____

25%: Age = _____

6%: Age = _____

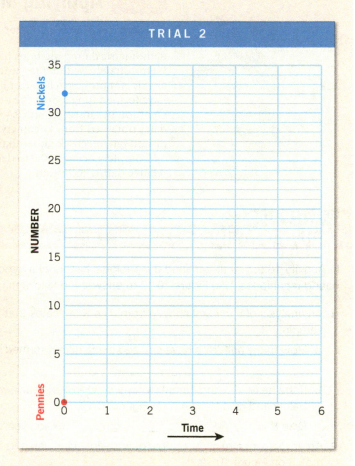

▲ **Figure 10.21** Graph to record data from Trial 2.

TRIAL 2		
	Number of Nickels	**Number of Pennies**
0	32	0
1		
2		
3		
4		
5		
6		

▲ **Figure 10.20** Chart to record data from Trial 2.

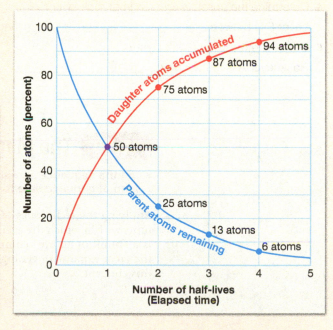

▲ **Figure 10.22** Idealized radioactive decay curves.

10.9   Applying Multiple Dating Techniques

■ **Determine numerical dates for sedimentary strata.**

Radiometric dating methods have produced thousands of dates for events in Earth history. Unfortunately, sedimentary rocks, which contain the vast bulk of the fossil record, generally do not provide reliable radiometric dates because the particles that compose them may have formed long before the sedimentary rock formed. For example, the mineral zircon is common in sandstone and is widely used for radiometric dating. However, the radiometric dates obtained from zircon crystals tell geologists when this mineral crystallized in some distant igneous rock rather than the time when the zircon grains were deposited.

Complete Activity 10.9 to determine how radiometric dates are used in conjunction with relative dating techniques.

ACTIVITY 10.9
Applying Multiple Dating Techniques

In Activity 10.5, you determined the geologic history of a hypothetical region, shown in Figure 10.12, using relative dating techniques. Use Figure 10.12 to answer the following.

1. An analysis of a mineral sample from dike K indicates that 25 percent of the parent isotope is present in the sample.

   a. How many half-lives have elapsed since dike K formed?

   _____

   b. If the half-life of the parent isotope is 50 million years, what is the numerical age of dike K? Write your answer below and next to dike K on Figure 10.12.

   Dike K is _____ million years old.

2. An analysis of a mineral sample of intrusion J indicates its age to be 400 million years. Write the numerical age of rock J on Figure 10.12.

3. Are rock layers H and I younger or older than 100 million years? Explain.

   Layers H and I are _____ than 100 million years.

   _____

4. Determine the possible age range of rock layer E.

   Layer E is between _____ and _____ million years old.

5. Determine the age of rock layer A.

   Rock layer A is greater than _____ million years.

10.10   The Magnitude of Geologic Time

■ **List the sequence of major events that have occurred on Earth and compare the span of geologic time to the length of human history.**

Applying the techniques of geologic dating, the history of Earth has been divided into units within which the events of the geologic past are arranged. Since a human life span is a "blink of an eye" compared to the age of Earth, it is difficult to fully comprehend the great expanse of geologic time. Completing Activity 10.10 will help you understand this concept.

ACTIVITY 10.10
The Magnitude of Geologic Time

Step 1: Obtain a piece of cash register paper slightly longer than 5 meters and a meterstick or metric measuring tape from your instructor. Draw a line at one end of the paper and label it "PRESENT." Using the scale provided, construct a **time line** by completing Steps 2 and 3:

SCALE

1 meter = 1 billion years

10 centimeters = 100 million years

1 centimeter = 10 million years

1 millimeter = 1 million years

Step 2: Using the geologic time scale in **Figure 10.23** as a reference, place lines on your time line indicating the beginning and end of each of these major time divisions: Precambrian, Paleozoic era, Mesozoic era, and Cenozoic era. Label each division and indicate its numerical age.

Step 3: Plot and label the events shown in red in Figure 10.23 on your time line.

Use Figure 10.23, page 182, and your time line to complete the following.

1. What fraction or percentage of geologic time is represented by Precambrian time?

_____%

2. List the major animal groups, in order of their appearance on Earth. (*Note:* These times are approximate and subject to change as more data are obtained.)

_____

_____

_____

_____

_____

3. How many times longer is all geologic time than the time represented by recorded history (about 5000 years)? (Round geologic time to 5 billion years to simplify.)

Geologic time is _____ times longer than recorded history.

MasteringGeology™

Looking for additional review and lab prep materials? Go to **www.masteringgeology.com** for Pre-Lab Videos, Geoscience Animations, RSS Feeds, Key Term Study Tools, The Math You Need, an optional Pearson eText, and more.

Evolution of Life Through Geologic Time

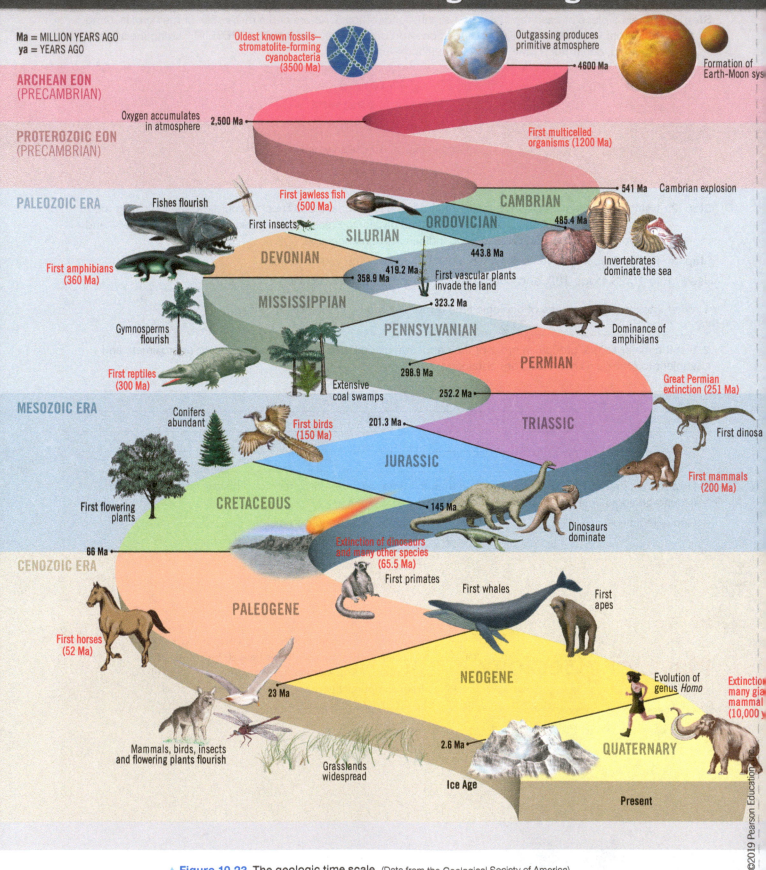

Geologic Time

Name _____

Date _____

Course/Section _____

Due Date _____

Use **Figure 10.24** to complete the Questions 1–8:

1. Place the lettered features in proper sequence, from oldest to youngest, in the space provided on the figure.

2. What type of unconformity separates layer G from layer F?

   _____

3. Which principle of relative dating did you apply to determine whether rock layer H is older or younger than layer I?

   _____

4. Which principle of relative dating did you use to determine whether fault M is older or younger than rock layer F?

   _____

5. Explain how you can determine whether fault N is older or younger than igneous intrusion J.

   _____

6. If rock layer F is 150 million years old, and layer E is 160 million years old, what is the approximate age of fault M?

   Approximate age of fault M: _____
   million years

7. The analysis of samples from layers G and F indicates the following proportions of parent isotope to daughter product. If the half-life of the parent is known to be 75 million years, what are the ages of the two layers?

	PARENT	DAUGHTER	AGE
Layer G	50%	50%	_____
Layer F	25%	75%	_____

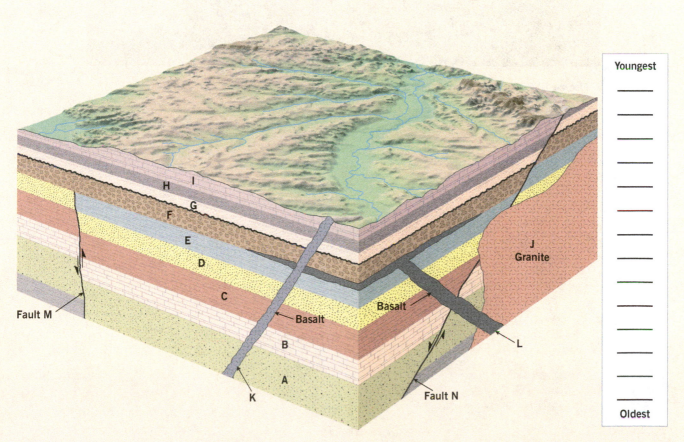

▲ **Figure 10.24** Geologic block diagram of a hypothetical region.

8. What time interval is represented by the disconformity at the base of rock layer G?

Time interval: _____ to _____ million years

9. Examine the photograph in **Figure 10.25**. Describe, as accurately as possible, the relative geologic history of the area, using the principles of relative dating covered in this exercise.

_____

_____

_____

_____

▲ **Figure 10.25** Photo of sedimentary beds to accompany Question 9. (Photo by E. J. Tarbuck)

Introduction to Oceanography

LEARNING OBJECTIVES

Each statement represents an important learning objective that relates to one or more sections of this lab. After you complete this exercise, you should be able to:

- Locate the major water bodies on Earth and contrast the distribution of land and water in the Northern and Southern Hemispheres.
- Use echo sounding data to calculate ocean depths and construct an ocean-floor profile.
- Identify and describe the major features of ocean basins.
- Explain the relationship between salinity and the density of seawater.
- Describe how salinity varies with latitude and depth in the oceans.
- Explain how temperature affects the density of seawater.

MATERIALS

colored pencils	ruler	dye
graduated cylinder	globe, atlas, or wall map	test tubes
salt solutions	rubber band	salt
beaker	ice	

PRE-LAB VIDEO ▶ https://goo.gl/2YBsXC

Prepare for lab! Prior to attending your laboratory session, view the pre-lab video. Each video provides valuable background that will contribute to your understanding and success in lab.

INTRODUCTION

Although the ocean comprises a much greater percentage of Earth's surface than the continents, it has only been in the relatively recent past that the ocean became an important focus of study. This exercise investigates some of the physical characteristics of the oceans.

PART 2 Oceanography

11.1 Extent of the Oceans

■ **Locate the major water bodies on Earth and contrast the distribution of land and water in the Northern and Southern Hemispheres.**

Some scientists aptly refer to Earth as the "water planet" because nearly 71 percent of Earth's surface is covered by the global ocean. **Oceanography** is an interdisciplinary science that draws on the methods and knowledge of geology, chemistry, physics, and biology to study the many complex aspects and interrelationships of the world ocean.

ACTIVITY 11.1A
Extent of the Oceans

1. Refer to a globe, a wall map of the world, or an atlas and locate each of the oceans and major water bodies in the following list. Label each on the world map in **Figure 11.1**.

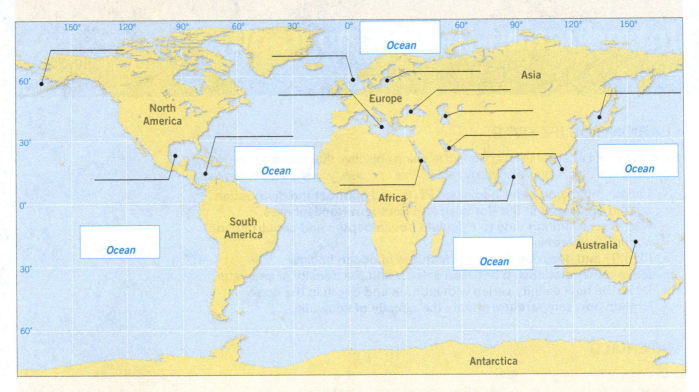

▲ **Figure 11.1** World map.

OCEANS	OTHER MAJOR WATER BODIES		
A. Pacific	1. Caribbean Sea	6. Persian Gulf	11. Red Sea
B. Atlantic	2. Mediterranean Sea	7. North Sea	12. Caspian Sea
C. Indian	3. Coral Sea	8. Black Sea	13. South China Sea
D. Arctic	4. Sea of Japan	9. Baltic Sea	14. Bay of Bengal
	5. Gulf of Mexico	10. Bering Sea	

2. Which ocean covers the greatest area?

_____

3. Which ocean is almost entirely in the Southern Hemisphere?

_____

ACTIVITY 11.1B
Distribution of Land and Water

The total area of Earth is about 510 million square kilometers (197 million square miles). Of this, approximately 360 million square kilometers (140 million square miles) are covered by oceans and adjacent marginal seas. Use this information to complete Questions 1 and 2.

1. What percentage of Earth's surface is covered by oceans and marginal seas?

$$\frac{\text{Area of oceans and marginal seas}}{\text{Area of Earth}} \times 100\% = \text{_____} \% \text{ ocean}$$

2. What percentage of Earth's surface is land?

_____ % land

3. Referring to Figure 11.1, which hemisphere, Northern or Southern, could be called the "water" hemisphere and which the "land" hemisphere?

Water hemisphere: _____

Land hemisphere: _____

4. Use Figure 11.2, which shows the distribution of land and water in 5° latitude belts, to answer the following.

a. How much area is covered by land in the belt between 60° and 65° *north* latitude?

_____ million km²

b. How much area is covered by water in the belt between 60° and 65° *north* latitude?

_____ million km²

c. How much area is covered by land in the belt between 60° and 65° *south* latitude?

_____ million km²

d. How much area is covered by water in the belt between 60° and 65° *south* latitude?

_____ million km²

5. Using the data from Question 4, calculate the percentage of area that is *land* in the belt between 60° and 65° *north* latitude. (To calculate percentage, divide the area covered by land by the total area covered by land and water. Then multiply your answer by 100%.)

_____ % land

6. Using Questions 4 and 5 as a guide, calculate the percentage of the area that is *water* in the belt between 45° and 50° *south* latitude.

_____ % water

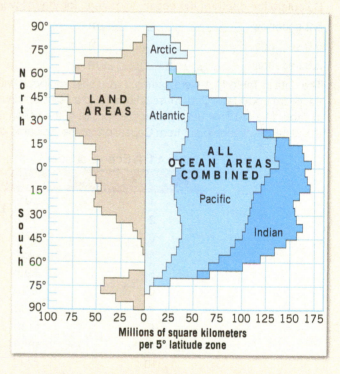

▲ **Figure 11.2** Distribution of land and water in each 5° latitude belt.

11.2 Measuring Ocean Depths

■ **Use echo sounding data to calculate ocean depths and construct an ocean-floor profile.**

Charting the shape of the ocean floor, called **bathymetry**, is a basic task of oceanographers. In the 1920s, a technological breakthrough for determining ocean depths occurred

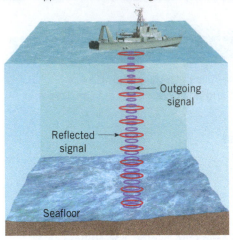

◀ **Figure 11.3** An echo sounder determines water depth by measuring the time required for an acoustic signal to travel from a ship to the seafloor and back. The speed of sound in seawater is 1500 m/sec.

with the invention of electronic depth-sounding equipment. The **echo sounder** (also referred to as *sonar*) measures the precise time that a sound wave, traveling at about *1500 meters per second*, takes to reach the ocean floor and return to the instrument (**Figure 11.3**).

Ships generally don't make single depth soundings. Rather, as a ship moves from one location to another, it continually sends out sound pulses and records the returning echoes. In this way, oceanographers obtain many depth recordings from which a *profile* (cross-sectional view) of the ocean floor can be prepared. Today, in addition to using sophisticated echo sounders such as *multibeam sonar*, oceanographers are also using satellites to map the ocean floor.

ACTIVITY 11.2
Measuring Ocean Depths

1. When using echo-sounding data to determine ocean depths, the following formula is used:

   Ocean depth = 1/2 (1500 m/sec × Echo travel time)

   Calculate the depth of the ocean, in meters, for the following echo soundings:

   5.2 seconds: _____ m

   6.0 seconds: _____ m

   2.8 seconds: _____ m

2. The data in **Table 11.1** were gathered by a ship equipped with an echo sounder as it traveled the North Atlantic Ocean eastward from Cape Cod, Massachusetts, to a point somewhat beyond the center of the Atlantic Ocean. Use the data in Table 11.1 to construct a generalized profile of the ocean floor in the North Atlantic on **Figure 11.4**. Begin by plotting the distance of each point from Cape Cod, at the indicated depth. Complete the profile by connecting the points.

Table 11.1 Echo Sounder Depths Eastward from Cape Cod, Massachusetts

POINT	DISTANCE (KM)	DEPTH (M)
1	0	0
2	180	150
3	270	2700
4	420	3300
5	600	4000
6	830	4800
7	1100	4750
8	1150	2500
9	1200	4800
10	1490	4750
11	1750	4800
12	1800	3100
13	1860	4850
14	2120	4800
15	2320	4000
16	2650	3000
17	2900	1500
18	2950	1000
19	2960	2700
20	3000	2700
21	3050	1000
22	3130	1900

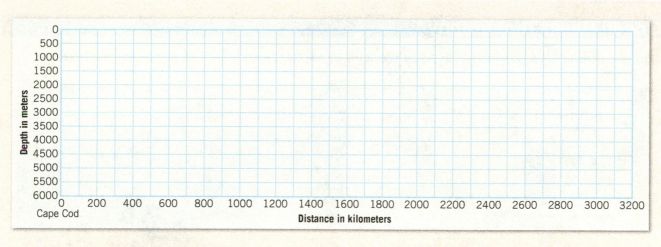

▲ **Figure 11.4** North Atlantic Ocean floor profile.

11.3 Ocean Basin Topography

■ **Identify and describe the major features of ocean basins.**

Oceanographers studying the topography of the ocean floor recognize three major units: the *continental margin*, the *deep-ocean basin*, and the *oceanic* (*mid-ocean*) *ridge*. **Figure 11.5** is a diagram showing the continental margin and the deep-ocean basin off the east coast of the United States. **Continental shelves**, flooded extensions of the continents, are gently sloping submerged surfaces that extend from the shoreline toward the ocean basin The seaward edge of the continental shelf is marked by the **continental slope**, a relatively steep zone (compared to the shelf) that marks the boundary between continental crust and oceanic crust. Deep, steep-sided valleys known as **submarine canyons**, partially eroded by the periodic downslope movements of dense, sediment-laden water called **turbidity currents**, are often cut into the continental slope. The continental slope merges into a more gradual incline known as the **continental rise**. This feature may extend seaward for hundreds of kilometers and consists of a thick accumulation of sediment, much of it deposited by turbidity currents. The deep-ocean basin, which constitutes almost 30 percent of Earth's surface, features remarkably flat areas known as **abyssal plains**, tall volcanic peaks called **seamounts**, **oceanic plateaus** generated by mantle plumes, and **deep-ocean trenches**, which are deep, narrow depressions that are found primarily in the Pacific Ocean basin Near the center of most ocean basins is a rugged elevated zone called the **oceanic ridge**, or **mid-ocean ridge**, which is characterized by extensive faulting and numerous volcanic structures, Along some portions of the oceanic ridge system are deep, down-faulted structures that are called **rift valleys**.

▼ **Figure 11.5** Generalized continental margin similar to what occurs in the Atlantic. Note that the slopes shown for the continental shelf and continental slope are greatly exaggerated.

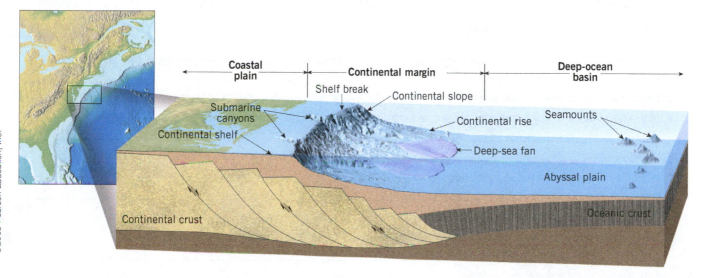

ACTIVITY 11.3
Ocean Basin Topography

The questions in this activity pertain to **Figure 11.6**. It will be helpful to have a world map or an atlas for reference.

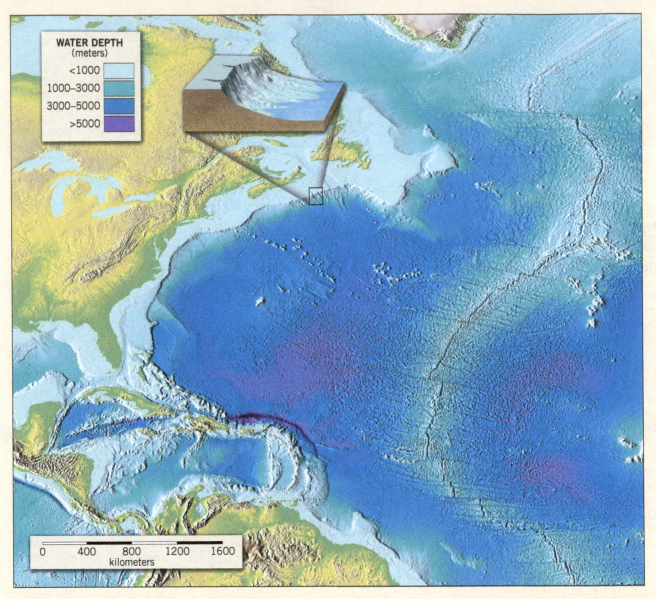

▲ **Figure 11.6** Portion of the North Atlantic basin.

1. Carefully examine the Mid-Atlantic Ridge. At one location label the linear depression, called a *rift valley*, that runs down the axis of the ridge.

2. Draw a line down the middle (axis) of the Mid-Atlantic Ridge by following the rift valleys.

3. Is the Mid-Atlantic Ridge best described as a continuous or discontinuous feature?

_____

4. Label the Puerto Rico trench.

5. Describe the shape of the Puerto Rico trench.

_____

6. Label the continental shelf along the coasts of North and South America.

7. Describe the size of the continental shelf that surrounds the state of Florida.

8. Circle the area along the east coast of the United States where the continental shelf is narrowest.

9. Label two abyssal plains.

10. Label a seamount.

11. Label a submarine canyon.

12. On the profile that you constructed in Figure 11.4, label the continental shelf, the continental slope, an abyssal plain, the Mid-Atlantic Ridge, and a seamount.

11.4   Salinity and Density Currents

▪ **Explain the relationship between salinity and the density of seawater. Describe how salinity varies with latitude and depth in the oceans.**

Ocean circulation has two primary components: surface-ocean currents and deep-ocean circulation. Both are examined in greater detail in Exercise 12. While *surface currents* such as the Gulf Stream are driven primarily by the prevailing global winds, the deep-ocean circulation is largely a result of differences in water *density* (mass per unit volume of a substance). A **density current** is the movement (flow) of one body of water over, under, or through another caused by density differences and gravity. Variations in *salinity* and *temperature* are the two most important factors creating the density differences that result in deep-ocean circulation.

Salinity is the amount of dissolved salts in water, expressed as parts per thousand (parts salt per 1000 parts of water). The symbol for parts per thousand is ‰. Although there are many dissolved salts in seawater, sodium chloride (common table salt) is the most abundant.

In regions where evaporation is high, the concentration of salt increases as the water evaporates. The opposite occurs in areas of heavy precipitation and/or high runoff (where large rivers enter the ocean). The addition of freshwater dilutes the seawater and lowers its salinity. Since the factors that determine the concentration of salts in seawater vary from the equator to the poles, the salinity of seawater also varies with latitude.

ACTIVITY 11.4A
Salinity–Density Experiment

To better understand how salinity affects the density of water, use the lab equipment supplied by your instructor (**Figure 11.7**) to conduct the following experiment.

Step 1: Fill the graduated cylinder with *cool* tap water up to the rubber band or other mark near the top of the cylinder.

Step 2: Fill a test tube about half full of solution A (saltwater) and pour it *slowly* into the graduated cylinder. Record your observations.

Observations: _____

_____

_____

Step 3: Repeat steps 1 and 2 and measure the time required for the front edge of the saltwater to travel from the rubber band to the bottom of the cylinder. Record the times for each trial in Table 11.2. *Make certain* that you drain the cylinder and refill it with cool freshwater and use the same amount of solution each time.

Step 4: Determine the travel time for solution B *exactly* as you did with solution A. Enter your times in Table 11.2.

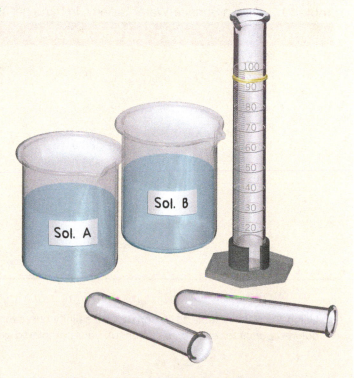

▲ **Figure 11.7** Lab setup for salinity–density experiment.

continued

Step 5: Fill a test tube about half full of solution B and add about $\frac{1}{2}$ teaspoon of salt. Shake the test tube vigorously. Determine the travel time of this solution and enter your results in Table 11.2.

Step 6: Clean all your glassware.

1. Write a brief summary of the results of your salinity–density experiment.

   _____

   _____

2. Is solution A or B most dense?

   Solution _____ is most dense.

3. How did adding salt to solution B affect the travel time?

   _____

Table 11.2 Salinity–Density Experiment Data Table

SOLUTION	TIMED TRIAL #1	TIMED TRIAL #2	AVERAGE OF THE TWO TRIALS
A			
B			
Solution B plus salt		XXXX	XXXX

ACTIVITY 11.4B
Comparing Water Salinity at Various Latitudes

Table 11.3 lists the approximate surface-water salinities at various latitudes in the Atlantic and Pacific Oceans.

Table 11.3 Ocean Surface-Water Salinity, in Parts per Thousand (‰), at Various Latitudes

LATITUDE	ATLANTIC OCEAN	PACIFIC OCEAN
60°N	33.0	31.0
50°N	33.7	32.5
40°N	34.8	33.2
30°N	36.7	34.2
20°N	36.8	34.2
10°N	36.0	34.4
0°	35.0	34.3
10°S	35.9	35.2
20°S	36.7	35.6
30°S	36.2	35.7
40°S	35.3	35.0
50°S	34.3	34.4
60°S	33.9	34.0

1. Using the data in Table 11.3, construct a salinity curve for each ocean on the graph in **Figure 11.8**. *Use a different colored pencil for each ocean.* Consult your completed graph when answering Questions 2–10.

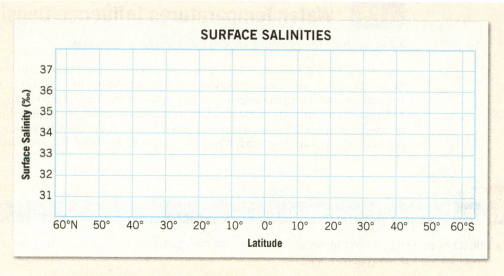

▲ **Figure 11.8** Graph for plotting surface salinities.

2. Which latitudes in the Atlantic Ocean have the highest surface salinities?

_____

3. What are two factors that control the salinity of seawater?

_____ and _____

4. Why is the salinity of the surface waters of the equatorial and subtropical regions (about 30° latitude) in the Atlantic Ocean different?

_____

_____

5. Does the Atlantic Ocean or Pacific Ocean have a higher average surface salinity?

_____

Use **Figure 11.9**, which shows how ocean water salinity varies with depth at different latitudes, to complete the following.

6. In general, does salinity increase or decrease with depth in *equatorial* regions? Does salinity increase or decrease with depth in *polar* regions?

Salinity _____ with depth in equatorial regions.

Salinity _____ with depth in polar regions.

7. In the subtropics, why are the surface salinities much higher than the deep-water salinities?

_____

8. The *halocline* (*halo* = salt, *cline* = slope) is a layer of ocean water where there is a rapid change in salinity with depth. Label the halocline on Figure 11.9.

9. Does the salinity of ocean water below the halocline increase rapidly, remain fairly constant, or decrease rapidly?

The salinity of ocean water _____ below the halocline.

10. In winter, sea ice forms in the polar latitudes. The water in this ice is essentially salt-free freshwater. How does the formation of sea ice influence surface salinity in polar regions? Show this change by sketching a line on Figure 11.9.

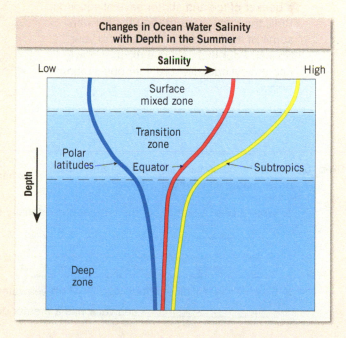

▲ **Figure 11.9** Changes in ocean water salinity with depth for equatorial, subtropical, and polar latitudes.

11.5 Water Temperatures Influence Density

■ **Explain how temperature affects the density of seawater.**

Temperature and density are basic ocean water properties that influence such things as deep-ocean circulation and the distribution and types of marine life. The ocean's surface-water temperature varies with the amount of solar radiation received, which is primarily a function of latitude. The intensity of solar radiation in high latitudes is significantly less than that received in tropical latitudes. Therefore, much lower sea-surface temperatures are found in high-latitude regions, and much higher temperatures are found in low-latitude regions.

ACTIVITY 11.5A
Temperature–Density Experiment

To illustrate the effects of temperature on the density of water, use the lab equipment provided by your instructor (**Figure 11.10**) to conduct the following experiment.

Step 1: Fill a graduated cylinder with *cold* tap water up to the rubber band or other mark near the top of the cylinder.

Step 2: Fill a test tube half full with *hot* tap water and add two or three drops of dye.

Step 3: Pour the contents of the test tube *slowly* into the cylinder. Record your observations below.

_____

_____

_____

Step 4: Empty the graduated cylinder and refill it with *hot* water.

Step 5: Fill a test tube full of cold water and add two or three drops of dye. Then pour it into a beaker of ice and stir for several seconds. Refill the test tube three-fourths full with some of the liquid (but no ice) from your beaker. Pour this cold dye mixture *slowly* into the cylinder. Record your observations below.

_____

_____

_____

Step 6: Clean the glassware and return it and the other materials to your instructor.

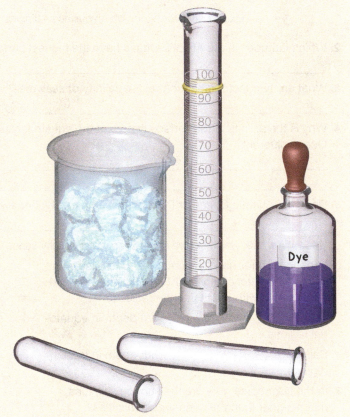

▲ **Figure 11.10** Lab setup for temperature–density experiment.

1. Write a brief summary of the temperature–density experiment.

_____

_____

_____

2. Given equal salinities, does cold or warm seawater have greater density?

_____ seawater has a greater density.

ACTIVITY 11.5B ///
Comparing the Temperature and Density of Seawater

Table 11.4 shows the average surface temperatures and densities of seawater at various latitudes.

1. Using the data in Table 11.4, plot a colored line on the graph in **Figure 11.11** for temperature and a different-colored line for density. Consult your graph when answering Questions 2–5.

2. Do the highest surface densities occur in equatorial regions or at polar latitudes?

_____

3. Describe the relationship between ocean-surface temperatures and density.

_____

4. Ocean-surface temperatures are colder at 60°S latitude than at 60°N latitude. How does this difference affect the surface densities at these locations?

_____

5. Refer to the density curve in Figure 11.11 and the surface salinity graph in Figure 11.8. What evidence supports the fact that the density of seawater is influenced more by temperature than by salinity?

_____

_____

Table 11.4 Idealized Ocean Surface-Water Temperatures and Densities at Various Latitudes

LATITUDE	SURFACE TEMPERATURE (CELSIUS)	SURFACE DENSITY (g/cm^3)
60°N	5°	1.0258
40°N	13°	1.0259
20°N	24°	1.0237
0°	27°	1.0238
20°S	24°	1.0241
40°S	15°	1.0261
60°S	2°	1.0272

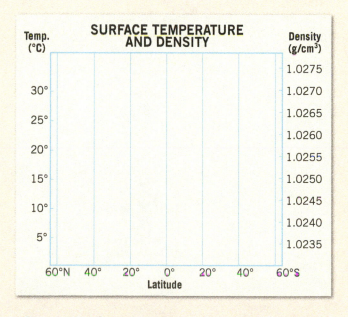

▲ **Figure 11.11** Graph for plotting surface temperatures and densities.

MasteringGeology™

Looking for additional review and lab prep materials? Go to www.masteringgeology.com for Pre-Lab Videos, Geoscience Animations, RSS Feeds, Key Term Study Tools, The Math You Need, an optional Pearson eText, and more.

Notes and calculations:

Introduction to Oceanography

Name _____

Date _____

Course/Section _____

Due Date _____

1. Write a brief statement comparing the distribution of land and water in the Northern Hemisphere to the distribution in the Southern Hemisphere.

   _____

   _____

2. Explain how an echo sounder is used to determine the bathymetry of the ocean floor.

   _____

   _____

   _____

   _____

3. Complete each of the following statements with the correct response.

   a. Does an increase in the salinity of seawater result in an increase or decrease in density?

      An increase in the salinity of seawater results in a(n) _____ in density.

   b. Does a decrease in the temperature of seawater cause its density to increase or decrease?

      A decrease in the temperature of seawater causes a(n) _____ in density.

   c. Are the highest surface salinities found in polar, subtropical, or equatorial regions?

   The highest surface salinities are found in _____ regions.

   d. Does temperature or salinity have the greatest influence on the density of seawater?

      _____ has the greatest influence on the density of seawater.

   e. Does warm or cold seawater, with high or low salinity, have the greatest density?

      _____ seawater with _____ salinity has the greatest density.

4. Summarize the results of the salinity–density experiment (Activity 11.4A) and the temperature–density experiment (Activity 11.5A).

   Salinity–density experiment: _____

   _____

   _____

   Temperature–density experiment: _____

   _____

   _____

5. On the ocean basin profile in **Figure 11.12**, label the following: continental shelf, continental slope, abyssal plain, seamounts, mid-ocean ridge, and deep-ocean trench.

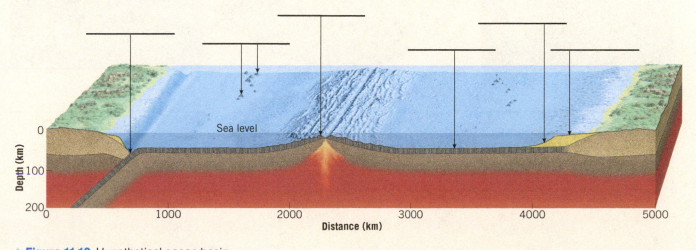

▲ **Figure 11.12** Hypothetical ocean basin.

6. Why is the surface salinity of the ocean higher in the subtropics than in equatorial regions?

_____

_____

_____

7. Given your understanding of the relationship between ocean water temperature, salinity, and density, where would you expect surface water to sink and initiate a deep-water current in the Atlantic Ocean? Explain your choice.

_____

_____

_____

8. Refer to the salinity–density experiment you conducted in Activity 11.4A. Did solution A or B have the greatest density?

Solution _____ had the greatest density.

9. Are the following statements true or false? Circle your response. If the statement is false, correct it to make it true.

T F **a.** The Atlantic Ocean covers the greatest area of all the world's oceans.

T F **b.** Continental shelves are part of the deep-ocean floor.

T F **c.** Deep-ocean trenches are located mostly in the middle of ocean basins.

T F **d.** High evaporation rates in the subtropics (about 30° latitude) cause the surface-ocean water to have a lower-than-average salinity.

10. Processes that _decrease_ seawater salinity include freshwater runoff and melting of icebergs and sea ice, while processes that _increase_ seawater salinity include formation of sea ice and evaporation of seawater. Below each image in **Figure 11.13**, state whether the salinity of the seawater shown is _increasing_ or _decreasing_.

A. _____   B. _____

C. _____   D. _____

▲ **Figure 11.13 Processes that affect seawater salinity.** (Upper left, Bernhard Edmaier/Science Source; upper right, Avalon/Photoshot License/Alamy; lower left, NASA; lower right, Ohotogerson/Shutterstock)

Waves, Currents, and Tides

LEARNING OBJECTIVES

Each statement represents an important learning objective that relates to one or more sections of this lab. After you complete this exercise you should be able to:

- **Explain how ocean waves are generated and describe the motion of water particles as a wave passes.**
- **Calculate the velocity, wavelength, and wave period of a deep-water wave. Explain what causes waves that are moving toward the shore to break and form surf.**
- **Describe the process of wave refraction and how it reshapes the coastline.**
- **Describe the process of longshore transport and the features it produces.**
- **Identify erosional and depositional features along shorelines.**
- **Identify the major surface-ocean currents.**
- **Name and describe the characteristics of the principal deep-water currents in the Atlantic.**
- **Discuss the causes of tides and distinguish among different types of tides.**

MATERIALS

calculator
atlas or world wall map

PRE-LAB VIDEO https://goo.gl/dHnGFo

 Prepare for lab! Prior to attending your laboratory session, view the pre-lab video. Each video provides valuable background that will contribute to your understanding and success in lab.

INTRODUCTION

The restless waters of the ocean are constantly in motion, powered by many differ-ent forces. Winds generate surface currents, which influence climate and provide

PART 2 Oceanography

nutrients that affect marine life in surface waters. Winds also produce waves that carry energy from storms to distant shores, where their impact erodes the land and moves sediment along the shore. Density differences create the deep-ocean circulation, which is important for ocean mixing and nutrient recycling. In addition, the Moon and Sun produce tides, which cause daily changes in sea level.

12.1 Wave Characteristics

- **Explain how ocean waves are generated and describe the motion of water particles as a wave passes.**

Most ocean waves derive their energy and motion from wind. (Tsunamis, which are usually produced by earthquakes that occur on the seafloor, are an exception.) When a breeze is less than 3 kilometers per hour, small wavelets appear. At greater wind speeds, more stable waves form and advance with the wind.

In a fully developed wave, water particles move in circular orbits, as shown in **Figure 12.1**. If you watched a ball floating on the surface of the ocean, you would notice that the wave-form moves forward, while the ball, and hence the water, does not. As water particles reach the highest point in their circular orbits, a **wave crest** forms, while particles at their lowest positions produce **wave troughs**. **Wave height** is the vertical distance between the crest and trough of a wave, and **wavelength** is the horizontal distance separating successive wave crests or successive wave troughs.

Beneath the surface, the circular orbits of water particles become progressively smaller with depth. At a depth equal to about *one-half* the wavelength, the circular motion of water particles becomes negligible. This depth is called the **wave base**.

▼ **SmartFigure 12.1** Features of deep-water and shallow-water waves.

TUTORIAL
https://goo.gl/VgeBTk

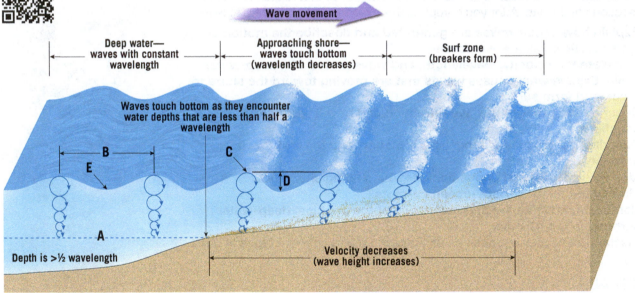

Wave movement

Deep water— waves with constant wavelength

Approaching shore— waves touch bottom (wavelength decreases)

Surf zone (breakers form)

Waves touch bottom as they encounter water depths that are less than half a wavelength

B

E

C

D

A

Depth is >½ wavelength

Velocity decreases (wave height increases)

ACTIVITY 12.1

Wave Characteristics

1. Refer to Figure 12.1 and select the letter that identifies each of the following.

   wave crest _____   wavelength _____

   wave trough _____   wave base _____

   wave height _____

2. Below what depth would a submarine have to submerge so that it would not be swayed by surface waves with a wavelength of 24 meters?

   _____ m

12.2 Deep- Versus Shallow-Water Waves

- **Calculate the velocity, wavelength, and wave period of a deep-water wave. Explain what causes waves that are moving toward the shore to break and form surf.**

For **deep-water waves** traveling where water depths exceed half the wavelength, the wave **velocity (V)** is related to the **wave period (T)** (the time interval between the arrival of successive wave crests) as well as the **wavelength (L)**. The mathematical equation that expresses this relationship is:

$$\text{Velocity} = \frac{\text{Wavelength}}{\text{Wave period}} \text{ or } V = \frac{L}{T}$$

As a wave approaches the shore, the ocean bottom begins to interfere with the orbital motion of the water particles, and the wave begins to "feel bottom" (Figure 12.1). This occurs when the water depth is one-half the wavelength of the incoming wave. At this depth, the seafloor interferes with the water movement at the base of the wave and slows its advance. As the wave slows, wavelength diminishes and wave height increases. At a depth equal to about one-twentieth of the deep-water wavelength (that is, 0.05L), the top of the wave becomes unstable and begins to break. This action creates the **surf zone**, where turbulent water rushes toward the shore.

ACTIVITY 12.2 //

Deep- Versus Shallow-Water Waves

Refer to Figure 12.1 to answer the following questions.

1. Do particles in deep-water waves trace out circular or elliptical paths?

_____

2. Near the shore in shallow water, do water particles trace out circular or elliptical paths?

_____

3. In shallow water, are water particles in the wave crest ahead of or behind those at the bottom of the wave?

_____

4. As waves approach the shore, do their heights increase or decrease? Do wavelengths become longer or shorter?

   Their heights _____, and their wavelengths become _____.

5. In the surf zone, is the water in the crest of a wave falling forward or standing still?

_____

6. What is the velocity of deep-water waves that have a wavelength of 46 meters and a wave period of 6.3 seconds?

$$\text{Velocity} = \frac{\text{Wavelength } (L)}{\text{Wave period } (T)} = \frac{40 \text{ m}}{6.3 \text{ sec}} = \underline{\hspace{2cm}} \text{ m/sec}$$

7. At what water depth will a deep-water wave with a wavelength of 46 meters begin to "feel bottom"?

_____ m

8. At what water depth will the wave described in Question 7 begin to break?

_____ m

9. What two factors determine how far from the shoreline waves will begin to break?

_____

_____

continued

Activity 12.2 continued

10. Imagine that you are standing on a beach but cannot swim. Your friend encourages you to walk into the surf zone created by incoming deep-water waves that have a wavelength of 30 meters. Would it be safe to walk out to where the waves are breaking? Explain how you arrived at your answer.

_____

_____

11. Along some shorelines, incoming waves cause the water to simply rise and fall rather than form a surf zone. What does this tell you about these shorelines?

_____

12. One type of ocean wave, call a *tsunami*, is usually produced by submarine earthquakes. These waves travel at very high velocities in deep water. What is the velocity of a tsunami with a wavelength of 125 miles and a period of 15 minutes?

_____ mph

12.3 Wave Refraction

■ **Describe the process of wave refraction and how it reshapes the coastline.**

As waves approach a shoreline, they are often **refracted** (bent). This occurs because the part of the wave that reaches shallow water first slows down, while the section of the wave in deeper water continues to advance at a greater velocity. The result of refraction is that waves gradually change direction and become roughly parallel to the shoreline.

ACTIVITY 12.3
Wave Refraction

Figure 12.2 is a map view of a **headland** along a coastline. The water depths are shown by blue contour lines. As you complete the following questions, assume that waves with a wavelength of 60 feet are approaching the shoreline from the bottom left of the figure.

1. At approximately what water depth—10, 20, 30, or 40 feet—will the approaching waves begin to touch bottom and slow down? (*Hint:* Recall that this occurs when the water depth is one-half the wavelength.)

_____ ft

2. Using the wave shown in Figure 12.2 as a starting point, sketch a series of lines to illustrate the wave refraction that will occur as the wave approaches the shore by following these steps:

Step 1: Mark the position on the 30-foot contour line where the wave front will first touch bottom.

▶ **Figure 12.2** Coastline with water depth contours (in feet) and an approaching wave.

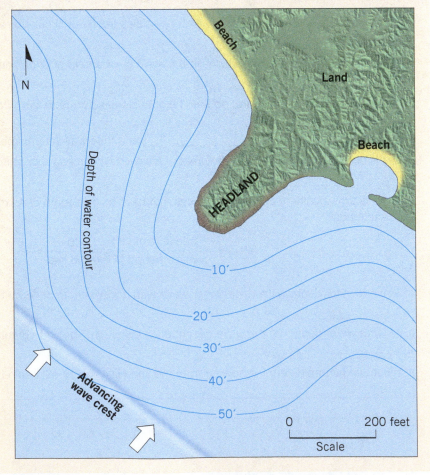

Step 2: Knowing that the section of the wave that touches bottom will slow down first, sketch the shape of the wave front when it reaches the 20-foot contour line.

Step 3: Using the same methodology as in Step 2, sketch the wave front as it approaches the 10-foot contour line.

3. At approximately what water depth—3, 6, 12, or 24 feet—will the waves begin to break?

_____ ft

4. Draw a dashed line on Figure 12.2 to indicate where the waves will begin to break. Write *surf zone* along this line.

5. Will wave erosion be most severe on the headland or in the bays?

_____

6. What effect will the concentrated energy from wave erosion eventually have on the shape of the coastline?

_____

12.4 Longshore Transport

■ Describe the process of longshore transport and the features it produces.

Waves are the dominant agents of erosion and deposition along shorelines. The sediment derived from wave erosion combined with that carried to the oceans by streams is redistributed by waves and wave-generated currents. Although waves are refracted, most still reach the shore at some angle. Consequently, the uprush of water from breaking waves is also at an angle to the shoreline. After the wave energy is spent, the return flow is straight down the slope of the beach. Such water movement causes sediment to move in a zigzag pattern along the beach—a process called **beach drift**. The turbulent water in the surf zone moves in a similar manner, creating a **longshore current** (**Figure 12.3A**). Much of the sediment reworked by waves and carried by longshore currents contributes to **beaches**. When longshore currents enter an adjacent bay, where the water is deeper, they lose energy and deposit their sediment load to form **spits** (**Figure 12.3B**). If a longshore current persists long enough, the spit may grow across the entire bay to form a feature called a **baymouth bar**.

▶ **SmartFigure 12.3** Longshore currents. **A.** Waves that approach the shore at an angle produce longshore currents that parallel the shore. **B.** Formation of a spit by sediment transported by longshore currents.

TUTORIAL

https://goo.gl/2uDCJQ

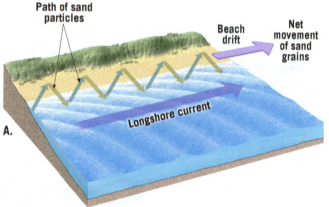

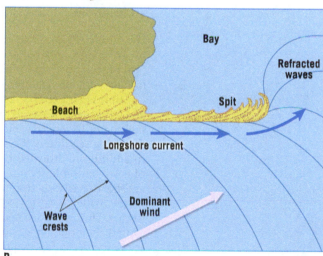

ACTIVITY 12.4
Longshore Transport

1. Describe how beach drift and longshore currents move sediment along the shore.

_____

_____

_____

continued

Activity 12.4 continued

2. Name three shoreline features that are produced, in part, by longshore transport.

_____

_____

3. Use an arrow to indicate the location and direction of the longshore current that would form given the wave pattern in **Figure 12.4A**.

4. Use an arrow to indicate the location and direction of the longshore current that produced the spit shown in **Figure 12.4B**.

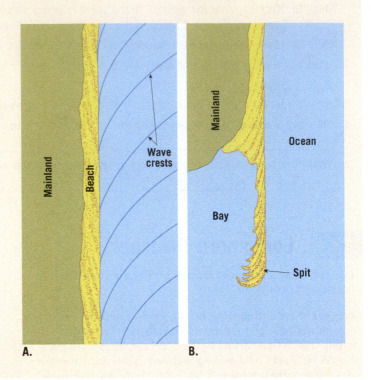

▶ **Figure 12.4** Illustrations to accompany Questions 3 and 4, Activity 12.4.

12.5 Shoreline Features

■ **Identify erosional and depositional features along shorelines.**

The nature of shorelines varies considerably from place to place, depending primarily on the composition of coastal bedrock, the degree of exposure to ocean waves, and the position of the shorelines in relation to sea level. Nevertheless, geologists often classify coasts into two broad categories—*submergent* and *emergent*.

Submergent coasts result from rising sea level or subsiding land. Most U.S. Atlantic and Gulf coastal areas are submergent coasts due to a recent increase in sea level caused by melting of Ice Age glaciers (**Figure 12.5**). Submergent coasts tend to be irregular because the rising sea floods the lower reaches of river valleys, producing bays called **estuaries**. In addition to drowned river mouths, the Atlantic and Gulf coasts, which are composed of easily erodible sediments and sedimentary rocks, are characterized by wide beaches and **barrier islands** that parallel the shorelines (Figure 12.5A). By contrast, the coast of Maine, which is composed of highly resistant igneous and metamorphic rocks, is very rugged. During the Ice Age, glaciers scoured the landscape, and the weight of the ice sheet caused the underlying land to subside. When the ice retreated about 14,000 years ago, the rising ocean flooded the coastal lowlands, giving Maine its distinctive rocky coastline (Figure 12.5B). The coasts of Norway, British Columbia, and Alaska, which are dissected by fiords (long, steep-sided inlets carved by glaciers), represent the most dramatic type of submergent shoreline.

Emergent coasts result from rising land or falling sea level and are often characterized by *wave-cut cliffs* and *wave-cut terraces*. Despite the general rise in sea level over the past several thousand years, tectonic forces have sufficiently uplifted much of the Pacific coast of the United States, producing what is mainly an emergent coastline. Relatively narrow beaches backed by steep cliffs and hilly topography are evident along the California coast, particularly in the area known as Big Sur (**Figure 12.6**).

Despite the differences between submergent and emergent coasts, the effects of wave erosion and sediment deposition by currents cause them to share many features.

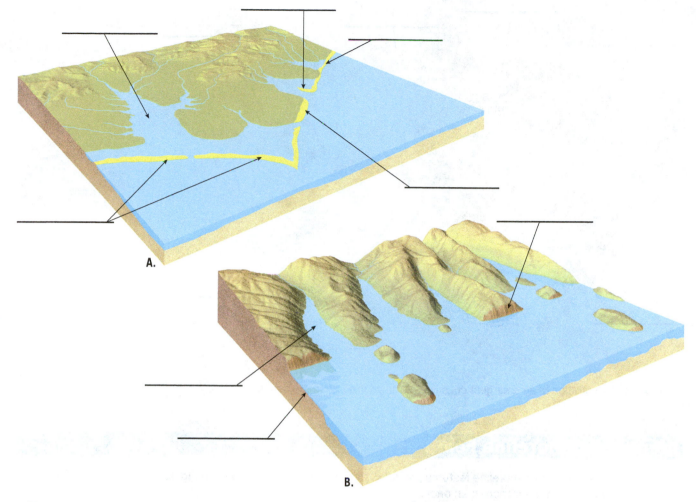

▲ **Figure 12.5** Illustrations showing general features of two different types of submergent coastlines.

Shoreline features associated with submergent and emergent coasts are described below:

- **Headland** Land, possibly containing cliffs, that projects into the ocean.
- **Wave-cut platform** A flat bedrock surface along a shore that is cut by wave erosion.
- **Delta** An accumulation of sediment that was deposited where a stream enters a lake or the ocean.
- **Estuary** A river valley that was flooded either by a rise in sea level or subsidence of the land.
- **Marine terrace** A wave-cut platform that has been uplifted above sea level by tectonic forces.
- **Wave-cut cliff** A cliff cut by the surf into the coastal rocks.
- **Spit** An elongated ridge of sand that projects from the land into the mouth of an adjacent bay.
- **Sea stack** An isolated rock formation that resembles a tiny island left from the collapse and erosion of surrounding rocks.
- **Tombolo** A ridge of sand that connects a sea stack or small island to the mainland or to another island.
- **Beach** An accumulation of sediment on the landward margin of an ocean or a lake that can be thought of as material in transit along the shore.
- **Barrier island** A long, narrow ridge of sand that parallels the coast.
- **Sea arch** A headland that has been cut by erosion into an arch.
- **Baymouth bar** A bar that extends across the mouth of a bay.

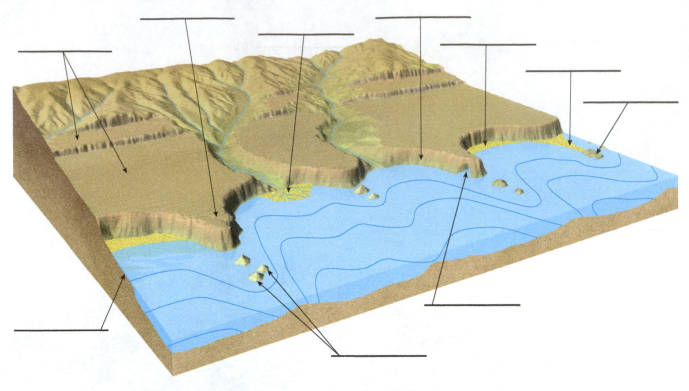

▲ **Figure 12.6** Illustration of an emergent coastline, common in southern California.

ACTIVITY 12.5A
Identifying Shoreline Features

1. Use the descriptions of shoreline features provided above to label the structures in Figures 12.5 and 12.6. The same feature may appear more than once.

2. Next to each of the features listed below, indicate whether it is the result of erosional or depositional processes.

   Sea stack: _____

   Wave-cut cliff: _____

   Spit: _____

   Barrier island: _____

   Baymouth bar: _____

   Marine terrace: _____

3. Label a marine terrace, a wave-cut cliff, and a wave-cut platform in **Figure 12.7**.

   A. _____   B. _____   C. _____

◄ **Figure 12.7** Image to accompany Question 3. (Photo by University of Washington Libraries, Special Collections, John Shelton Collection, KC5902)

4. Label a baymouth bar and a spit in **Figure 12.8**.

A. _____

B. _____

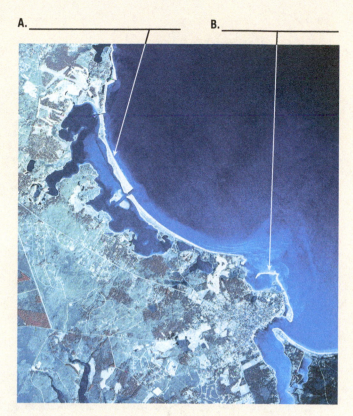

◀ **Figure 12.8** Image to accompany Question 4. (Image courtesy of USDA-ASCS)

5. Label a sea arch, a tombolo, a spit, a wave-cut cliff, and a sea stack on the lines below each image in **Figure 12.9**.

A. _____   B. _____   C. _____

D. _____   E. _____

▲ **Figure 12.9** Images to accompany Question 5. (Part E by Chris Cheadle/All Canada Photos/Getty Images)

▲ Portion of the Point Reyes, California, topographic map. (Courtesy of U.S. Geological Survey)

FIGURE 12.10 Point Reyes, California

North

0   1   2   3 kilometers

0   ½   1   2 miles

CONTOUR INTERVAL 80 FEET   SCALE: 1:62,500

ACTIVITY 12.5B
Identifying Shoreline Features on a Topographic Map

The area shown on the Point Reyes, California, topographic map is located directly to the southwest of the San Andreas Fault, in a very tectonically active region (Figure 12.10). As a result, some of this region has recently been uplifted and exhibits characteristics of an emergent coastline. On the other hand, because of the general rise in sea level over the past several thousand years, other areas exhibit features associated with submergent coastlines. Refer to this topographic map to complete the following.

1. What type of shoreline feature is Drakes Estero (located near the center of the map)?

   _____

2. Point Reyes, located in the bottom-left corner of the map, is a headland undergoing severe wave erosion. What type of feature is Chimney Rock, located off the shore of Point Reyes?

   _____

3. Several depositional features near Drakes Estero are related to the movement of sediment by longshore currents. What type of depositional feature is labeled A on the map?

   _____

4. Follow Limantour Spit from its western tip at the mouth of Drakes Estero eastward to where it joins the mainland. Use an arrow to mark the direction of the longshore current that produced Limantour Spit.

5. What geologic feature is Estero de Limantour?

   _____

6. Limantour Spit may eventually extend across the entire mouth of Estero de Limantour. If the west end of Limantour Spit reaches the mainland, will it be called a tombolo or a baymouth bar?

   _____

7. Locate the bench mark (BM) with an elevation of 552 feet on Point Reyes. If you descended from that bench mark southward to the shore of the Pacific, what landform would you cross?

   _____

8. What feature indicates that the headland called Point Reyes has recently been uplifted?

   _____

9. List at least two features on the map that are characteristic of submergent coasts—similar to the Atlantic coast.

   _____   _____

12.6 Surface-Ocean Currents

■ Identify the major surface-ocean currents.

Ocean currents develop where the prevailing winds cause the surface layer of the ocean to move. Once set in motion, surface-ocean currents are influenced by the **Coriolis effect**, which deflects the path of the moving water to the right in the Northern Hemisphere and to the left in the Southern Hemisphere. *Warm currents* move from the tropics toward the poles, whereas *cold currents* move from higher latitudes toward the equator.

ACTIVITY 12.6
Surface-Ocean Currents

1. On the world map in Figure 12.11, identify each of the major ocean currents listed below. Refer to a wall map, a map in your textbook, or an atlas that depicts surface currents. Warm currents are shown with red arrows and cold currents with blue arrows.

continued

Activity 12.6 continued

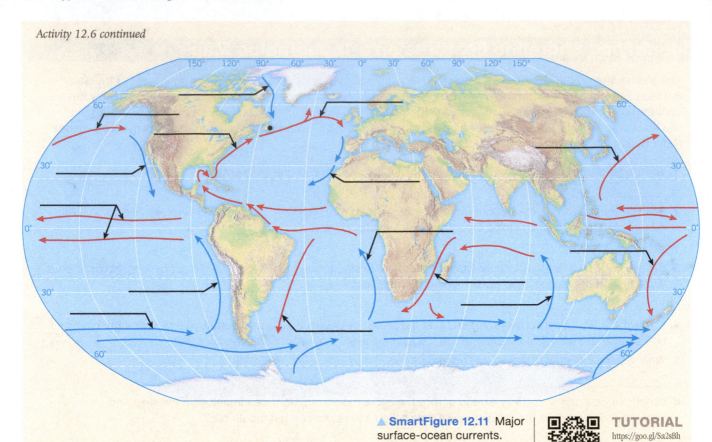

▲ **SmartFigure 12.11** Major surface-ocean currents.

TUTORIAL
https://goo.gl/Sa2sBh

MAJOR SURFACE-OCEAN CURRENTS

1. Equatorial (North and South)
2. Gulf Stream
3. California
4. Canary
5. Brazil
6. Benguela
7. Kuroshio (Japan)
8. Agulhas
9. Labrador
10. North Atlantic Drift
11. North Pacific
12. Peru (Humboldt)
13. West Australian
14. East Australian
15. West Wind Drift (Antarctic Circumpolar)

2. Which surface-ocean current travels completely around the globe, west to east, without interruption?

Name of current: _____

3. Which surface-ocean current flows along the Atlantic coast of the United States? Is it a warm or cold current?

Name of current: _____

This current is a _____ current.

4. What is the name of the surface-ocean current located along the Pacific coast of the United States? Is it a warm or cold current?

Name of current: _____

This current is a _____ current.

5. Is the general circulation of the surface currents in the North Atlantic Ocean clockwise or counterclockwise?

_____

6. In the South Atlantic, is the general circulation clockwise or counterclockwise?

_____

12.7 Deep-Ocean Circulation

■ **Name and describe the characteristics of the principal deep-water currents in the Atlantic.**

Deep currents result when water of greater density sinks beneath water that is less dense. The density of water is determined by its salinity and temperature (covered in Exercise 11). Temperature has a greater influence on density than does salinity. In general, cold seawater with high salinity is densest.

Most deep-ocean circulation begins in high latitudes where the water is cold, and its salinity increases as sea ice forms. When this surface water becomes dense enough, it sinks and then moves laterally through the ocean basins in sluggish currents. Oceanographers estimate that when cold, dense water sinks, it will not reappear at the surface for an average of 500 to 2000 years.

ACTIVITY 12.7
Deep-Ocean Circulation

Figure 12.12 is a cross section of the Atlantic Ocean. Use it to complete the following.

1. Compare Antarctic Bottom Water (ABW) to North Atlantic Deep Water (NADW). Which of these deep currents has the greatest density? Explain how you arrived at your answer.

_____

_____

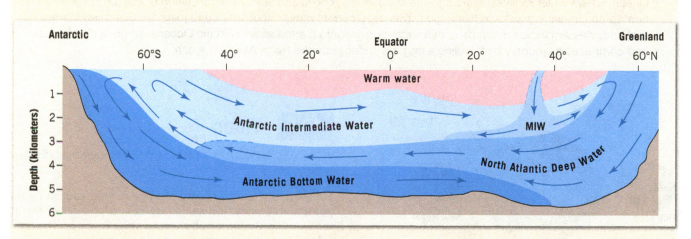

▲ **Figure 12.12** Cross section of the deep circulation in the Atlantic Ocean.

2. What process causes the Mediterranean Intermediate Water (MIW) to become more dense than water in the adjacent Atlantic Ocean? (*Hint:* Water in the Mediterranean Sea is greatly affected by the dry climate of the region.)

_____

_____

3. Figure 12.13 illustrates a simplified model of deep-ocean circulation. What is the name of the deep-water current that begins in the North Atlantic Ocean, adjacent to Greenland?_____

4. Assume that water sinking in the North Atlantic near Greenland takes 1000 years to resurface in the Indian Ocean, a distance of about 18,000 kilometers. What would be the approximate velocity of this deep current, in kilometers per year?

Velocity: _____ km/yr

continued

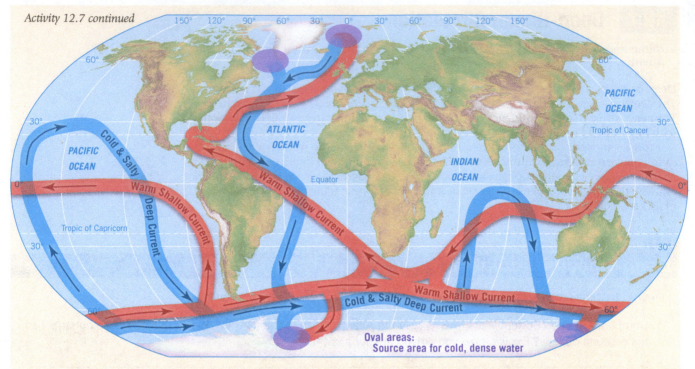

Activity 12.7 continued

▲ **Figure 12.13** In the idealized "conveyor belt" model, deep-ocean circulation is initiated in the North Atlantic Ocean, where water radiates its heat to the atmosphere, cools, and sinks. The formation of sea ice in winter also contributes to sinking water in this region. This deep current moves southward and joins another deep current that encircles Antarctica. From here, this water spreads into the Indian and Pacific Oceans, where it slowly rises and completes its journey by traveling along the surface into the North Atlantic Ocean.

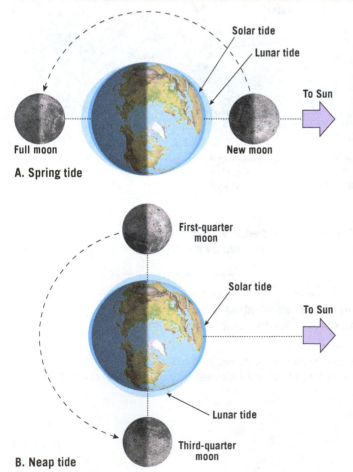

12.8 Tides

- ■ Discuss the causes of tides and distinguish among different types of tides.

Tides are the daily rise and fall of sea level caused by the gravitational attraction of the Moon and, to a lesser extent, the Sun. Gravitational pull creates a bulge in the ocean on the side of Earth nearest the Moon and on the opposite side of Earth from the Moon (**Figure 12.14**). The monthly tidal cycle lasts 29.5 days—the length of time it takes the Moon to complete its orbit around Earth. During this cycle, the position of Earth continually changes relative to the Sun and Moon. During the new-moon and full-moon phases, the Moon and Sun are aligned, and the tide-generating forces combine to produce **spring tides**. The **tidal range** is greatest during this period because of *higher* high tides and relatively *low* low tides (Figure 12.14A). Conversely, when the Moon is in the first- or third-quarter phase, **neap tides** occur. During these times,

◀ **Figure 12.14** Earth–Moon–Sun positions and tides. **A.** When the Moon is in the full- or new-moon phase, the tidal bulges created by the Sun and Moon are aligned, causing a large tidal range called a *spring tide*. **B.** When the Moon is in the first- or third-quarter position, *neap tides* occur. Because the tidal bulges produced by the Moon are at right angles to the bulges created by the Sun, tidal ranges are smaller when neap tides are experienced.

the Sun's tidal force works at right angles to the Moon's tidal force, partly offsetting it. The result is *lower* high tides and *higher* low tides—that is, a smaller tidal range (Figure 12.14B). Ideally, each tidal cycle should consist of two spring tides and two neap tides, each about 1 week in duration. However, many factors, including the shape of the coastline, the configuration of the ocean basin, and water depth, can significantly influence the tidal pattern observed at any particular location.

ACTIVITY 12.8A
Spring and Neap Tides

Use **Figure 12.15**, which illustrates the tidal patterns for three locations, to complete the following.

1. Use lines to divide each tidal pattern into approximately 1-week segments that represent the 2 weeks of spring tides and 2 weeks of neap tides.

2. Write the word *spring* or *neap* above the period of time representing the appropriate type of tide.

3. Which coastal area experiences that largest tidal range (difference in height between the high tide and low tide)?

_____

4. Which coastal area experiences the smallest tidal range?

_____

5. Notice that the phases of the Moon are shown at the top of Figure 12.15. Which coastal location experiences spring and neap tides that align most closely to those predicted by the phases of the Moon?

_____

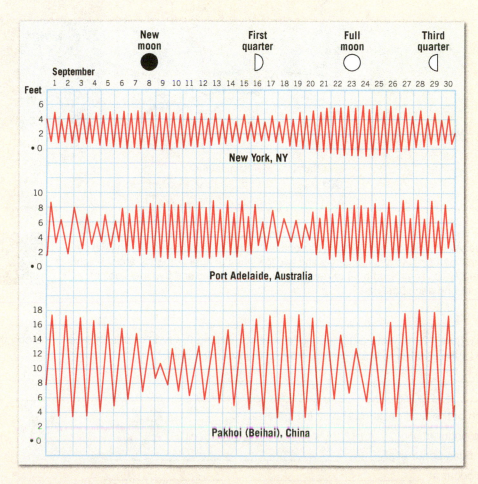

▲ **Figure 12.15** Tidal patterns for the month of September at various locations.
(Data courtesy of U.S. Navy Hydrographic Office)

ACTIVITY 12.8B
Examining Basic Tidal Patterns

Although all coastlines should ideally experience two high tides and two low tides each tidal day, the complexities stated earlier result in three basic tidal patterns:

- **Diurnal (daily) tides** are characterized by a single high tide and a single low tide each tidal day (Figure 12.16A).
- **Semidiurnal (twice daily) tides** exhibit two high tides and two low tides each tidal day (Figure 12.16B).
- **Mixed tides** are similar to semidiurnal tides except that they are characterized by large inequalities in high-water heights, low-water heights, or both (Figure 12.16C). There are typically two high tides and two low tides each day, with both high and low tides of various heights.

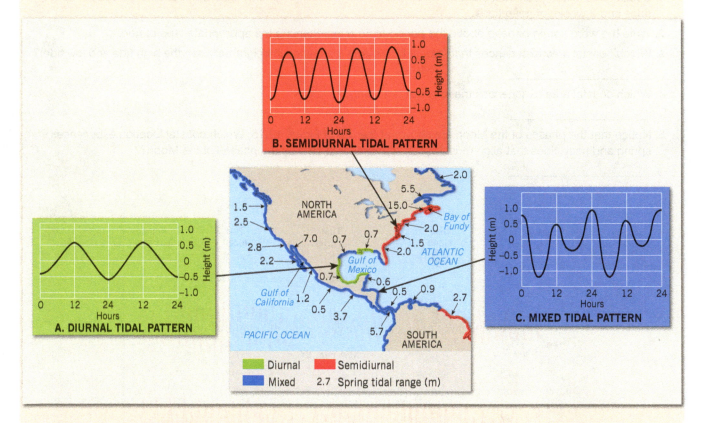

▲ **SmartFigure 12.16** Tidal patterns and their occurrence along North American and Central American coasts. **A.** The diurnal tidal pattern has one high tide and one low tide each tidal day. **B.** The semidiurnal pattern has two high tides and two low tides of approximately equal heights during each tidal day. **C.** The mixed tidal pattern shows two high tides and two low tides of unequal heights during each tidal day.

TUTORIAL
https://goo.gl/5X9Rjh

1. Referring to Figure 12.16, write a general statement comparing the type of tide that occurs along the Pacific coast of the United States to the type found along the Atlantic coast.

   _____

   _____

2. Classify each of the tidal patterns shown in Figure 12.17.

   Diurnal tides occur at _____.

   Semidiurnal tides occur at _____.

   Mixed tides occur at _____.

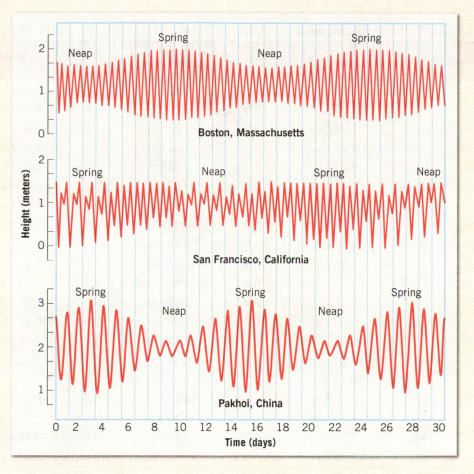

▲ **Figure 12.17** Tidal patterns for three locations, to accompany Question 2, Activity 12.8B.

ACTIVITY 12.8C

Examining Tidal Data for Long Beach, New York

1. Plot the tidal data for a 20-day period in January 2003 at Long Beach, New York, from Table 12.1 on the graph in **Figure 12.18**. (*Note:* The numbers across the top of the graph indicate days 1–20. The number *12* along the bottom of the graph is 12 noon.) After you have plotted the data, complete the following.

2. Is the tide that occurs at Long Beach, New York, diurnal, semidiurnal, or mixed?

   _____

3. On which day does the greatest tidal range occur? On which day does the smallest tidal range occur?

   Greatest tidal range: _____

   Smallest tidal range: _____

4. Using Figure 12.15 as a reference, label the most likely lunar phase (full moon, new moon, first quarter, or third quarter) associated with the tidal pattern above the appropriate days.

5. Assume that at 9:00 A.M. on January 5, a boat was anchored near a beach in 4 feet of water. When the owner returned at 3:30 P.M., the boat was resting on sand. Approximately how long did the owner have to wait before the boat was floating again?

   _____

continued

Activity 12.8C continued

Table 12.1 Tidal Data for Long Beach, New York, January 2003

	TIMES ARE LISTED IN LOCAL STANDARD TIME (LST); HEIGHTS ARE IN FEET							
DAY	TIME	HEIGHT	TIME	HEIGHT	TIME	HEIGHT	TIME	HEIGHT
1	5:45 A.M.	5.5	12:16 P.M.	−0.7	6:12 P.M.	4.4	—	—
2	12:18 A.M.	−0.5	6:35 A.M.	5.6	1:07 P.M.	−0.8	7:03 P.M.	4.4
3	1:10 A.M.	−0.5	7:23 A.M.	5.5	1:56 P.M.	−0.8	7:53 P.M.	4.4
4	1:59 A.M.	−0.4	8:11 A.M.	5.4	2:42 P.M.	−0.7	8:42 P.M.	4.3
5	2:45 A.M.	−0.2	8:59 A.M.	5.1	3:25 P.M.	−0.5	9:32 P.M.	4.2
6	3:30 A.M.	0.0	9:47 A.M.	4.8	4:07 P.M.	−0.3	10:23 P.M.	4.0
7	4:14 A.M.	0.3	10:35 A.M.	4.6	4:49 P.M.	−0.1	11:12 P.M.	3.9
8	5:01 A.M.	0.6	11:22 A.M.	4.3	5:32 P.M.	0.2	11:59 P.M.	3.9
9	5:54 A.M.	0.8	12:09 P.M.	4.0	6:18 P.M.	0.4	—	—
10	12:45 A.M.	3.9	6:56 A.M.	0.9	12:57 P.M.	3.7	7:10 P.M.	0.5
11	1:31 A.M.	3.9	7:59 A.M.	0.9	1:47 P.M.	3.5	8:02 P.M.	0.5
12	2:19 A.M.	4.0	8:57 A.M.	0.8	2:41 P.M.	3.4	8:53 P.M.	0.5
13	3:10 A.M.	4.1	9:50 A.M.	0.6	3:39 P.M.	3.5	9:41 P.M.	0.4
14	4:02 A.M.	4.3	10:38 A.M.	0.3	4:34 P.M.	3.6	10:28 P.M.	0.2
15	4:51 A.M.	4.6	11:26 A.M.	0.1	5:23 P.M.	3.7	11:15 P.M.	0.1
16	5:36 A.M.	4.8	12:12 P.M.	−0.1	6:08 P.M.	3.9	—	—
17	12:02 A.M.	−0.1	6:17 A.M.	5.0	12:57 P.M.	−0.3	6:51 P.M.	4.1
18	12:49 A.M.	−0.2	6:58 A.M.	5.1	1:40 P.M.	−0.5	7:32 P.M.	4.2
19	1:35 A.M.	−0.4	7:38 A.M.	5.2	2:22 P.M.	−0.6	8:15 P.M.	4.3
20	2:20 A.M.	−0.4	8:21 A.M.	5.2	3:30 P.M.	−0.7	9:01 P.M.	4.4

Source: National Ocean Service/NOAA.

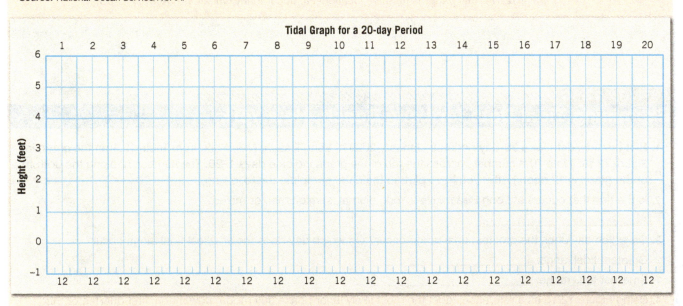

▲ **Figure 12.18** Tidal curve for Long Beach, New York. (*Note:* The numbers across the top of the graph indicate days 1–20. The number *12* along the bottom of the graph is 12 noon.)

Looking for additional review and lab prep materials? Go to
www.masteringgeology.com for Pre-Lab Videos, Geoscience
Animations, RSS Feeds, Key Term Study Tools, The Math You
Need, an optional Pearson eText, and more.

Waves, Currents, and Tides

Name _____   Course/Section _____

Date _____   Due Date _____

1. On **Figure 12.19**, label the surf zone, a wave crest, a wave trough, the wavelength, and a deep-water wave. In addition, sketch the motion of water particles at increasing depths for one deep-water wave and one shallow-water wave.

Depth = ½ wavelength

▲ **Figure 12.19** Features of waves to accompany Question 1.

2. What happens to wavelength and wave height as a wave approaches the shore in shallow water?

_____

_____

3. Refer to Figure 12.16 (page 214). Identify the tidal patterns that occur at each of the following locations.

   Atlantic coast of the United States: _____

   Pacific coast of the United States: _____

   Gulf of Mexico: _____

4. Is the circulation of the surface-ocean currents in the South Atlantic Ocean clockwise or counterclockwise?

_____

5. Describe the difference between a spring tide and a neap tide.

_____

_____

6. Explain what causes spring tides.

_____

_____

7. During which lunar phase(s) are neap tides most likely to occur?

_____

8. Which of the tidal patterns illustrated in Figure 12.17, page 215, exhibits the greatest tidal range?

_____

9. On **Figure 12.20**, page 218, identify the features associated with an evolving shoreline, using the following terms: spit, wave-cut cliff, beach deposits, baymouth bar, wave-cut platform, tombolo, sea arch, longshore current, and sea stack. (*Note:* Terms may be used more than once.)

10. Use **Figure 12.21**, page 218, to complete the following.

   a. Is the shoreline shown in this image a submergent shoreline or an emergent shoreline?

_____

   b. Label the shoreline features on the image.

   c. Place an X on the highest marine terrace shown in this image.

continued

Bay

Island

TIME

▲ **Figure 12.20** Illustrations and images of changes that can take place, through time, along an initially irregular coastline that remains tectonically stable. (Top and bottom photos by E. J. Tarbuck; middle photo by Michael Collier)

◄ **Figure 12.21** Image to accompany Question 10.
(Photo by Michael Collier)

Earth–Sun Relationships

LEARNING OBJECTIVES

Each statement represents an important learning objective that relates to one or more sections of this lab. After you complete this exercise you should be able to:

- Describe the influence of Sun angle on the intensity of solar radiation at Earth's surface.
- Describe the significance of these special parallels of latitude: Tropic of Cancer, Tropic of Capricorn, Arctic Circle, Antarctic Circle, and equator. List the dates and characteristics of the solstices and equinoxes.
- Use the analemma to understand the relationship between latitude and the noon Sun angle on every day of the year.
- Calculate the noon Sun angle for any place on Earth on any day.
- Calculate the latitude of a place by using the noon Sun angle.
- Describe how the average daily solar radiation received at the outer edge of the atmosphere varies with latitude and season.

MATERIALS

metric ruler colored pencils
protractor calculator
globe or world map large rubber band or string

PRE-LAB VIDEO https://goo.gl/DPDwrL

 Prepare for lab! Prior to attending your laboratory session, view the pre-lab video. Each video provides valuable background that will contribute to your understanding and success in lab.

INTRODUCTION

Weather is the state of the atmosphere at a given place and time. In contrast, climate is a description of aggregate weather conditions. Climate is more than the average condition of the atmosphere over a long period of time. Rather, it is

Meteorology

PART 3

(NASA)

the sum of all weather statistics that help describe a place, including variations and extremes. Temperature is a critical part of any description of weather or climate. A primary factor influencing temperatures is the amount of solar energy a place receives, which is strongly influenced by Earth–Sun relationships.

Solar Radiation and the Seasons

■ Describe the influence of Sun angle on the intensity of solar radiation at Earth's surface.

The amount of solar energy (radiation) striking the outer edge of the atmosphere is determined by the Sun's intensity and duration. The **intensity** (strength) of solar radiation depends on the angle at which the Sun's rays strike a surface (**Figure 13.1**). **Duration** refers to the length of daylight. Intensity and duration vary from place to place. In addition, every place on Earth experiences variations throughout the year.

▶ **SmartFigure 13.1**
A. If a flashlight beam strikes a surface at a 90° angle, a small intense spot is produced. However if the flashlight beam strikes at any other angle, the area illuminated is larger—but noticeably dimmer (lower intensity). **B.** Daily paths of the Sun for June and December for an observer in the middle latitudes in the Northern Hemisphere. Notice that the angle of the Sun above the horizon is much greater in the summer than in the winter.

TUTORIAL
https://goo.gl/Hyp8Hd

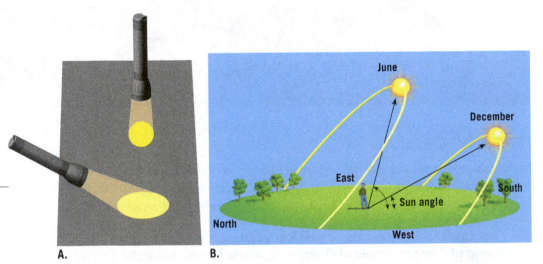

ACTIVITY 13.1
Solar Radiation and Latitude

To illustrate how variations in Sun angle and length of daylight influence the amount of solar radiation reaching Earth's surface, complete the following.

1. On **Figure 13.2**, extend the 1-centimeter-wide beam of sunlight from the Sun to Point A. Extend the second 1-centimeter-wide beam, beginning at the Sun, to the surface at Point B. Notice in Figure 13.2 that the Sun is directly overhead at Point A and that the beam of sunlight strikes the surface at a 90° angle. Refer to Figure 13.2 to answer Questions 2–5.

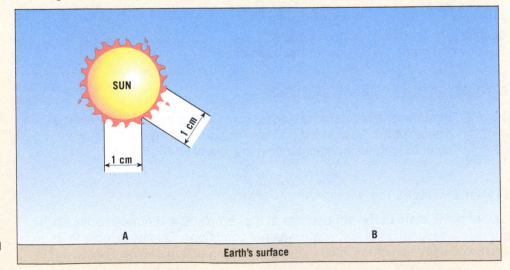

▶ **Figure 13.2** Vertical and oblique Sun beams.

2. Using a protractor, measure the angle between the surface and the beam of sunlight coming from the Sun to Point B.

_____ °

3. What are the lengths of the line segments on the surface covered by the Sun beam at Points A and B? Measure the lengths in millimeters. *Note:* There are 10 millimeters in 1 centimeter.

Point A: _____ mm

Point B: _____ mm

4. Which beam, A or B, is more spread out at the surface and therefore covers a larger area?

Beam: _____

5. Would more solar radiation be received by a square centimeter at Point A or at Point B?

Point: _____

Use **Figure 13.3**, which illustrates the amount of solar radiation intercepted by each 30° segment of latitude, to answer Questions 6–10.

6. With a metric ruler, accurately measure the total width (in millimeters) of incoming rays from Point X to Point Y in Figure 13.3.

_____ mm

7. Assume that the length of line X–Y represents 100% of the solar radiation that is intercepted by Earth's Northern Hemisphere. What percentage of the total incoming radiation is concentrated in each of the following zones? (Recall that you calculate the percentage by dividing the amount in question by the total amount and then multiplying that amount by 100%.)

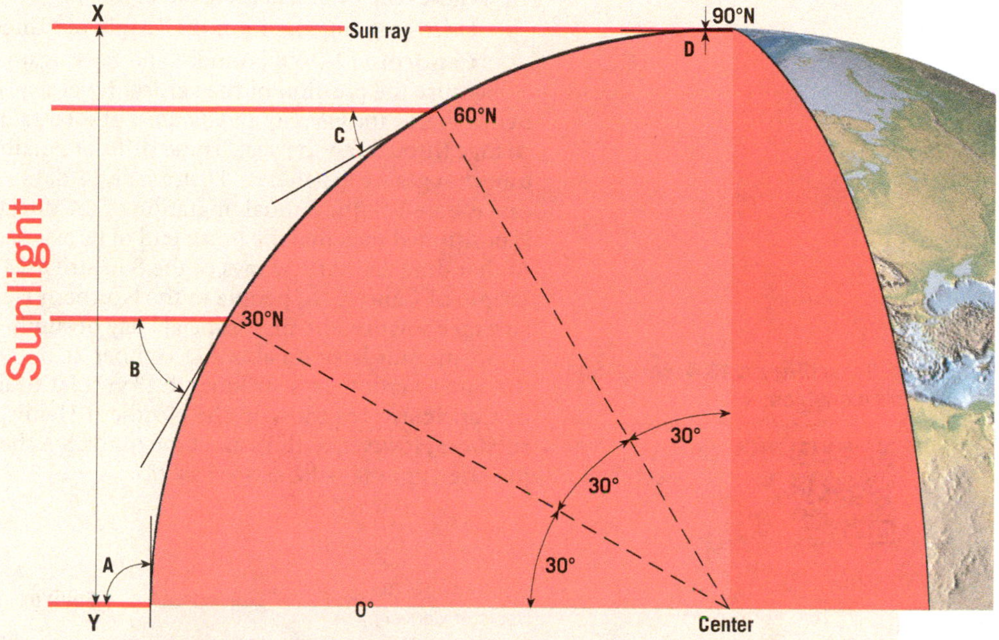

0°–30°N latitude = _____ mm = _____ %

30°–60°N latitude = _____ mm = _____ %

60°–90°N latitude = _____ mm = _____ %

▲ **Figure 13.3** Distribution of solar radiation per 30° segment of latitude in the Northern Hemisphere. The length of line X–Y represents 100% of the solar radiation intercepted by the Northern Hemisphere.

8. Use a protractor to measure the angle between Earth's surface and the Sun ray at each of the following locations. (Angle B has been completed as an example.)

Angle A: _____ °

Angle B: __60°__

Angle C: _____ °

Angle D: _____ °

9. Describe the relationship between the Sun angle and the amount of radiation received in each 30° segment.

_____

_____

10. What fact about Earth creates the unequal distribution of solar energy you noted in Question 9?

_____

13.2 Variations in Solar Energy Throughout the Year

■ **Describe the significance of these special parallels of latitude: Tropic of Cancer, Tropic of Capricorn, Arctic Circle, Antarctic Circle, and equator. List the dates and characteristics of the solstices and equinoxes.**

The amount of solar radiation varies throughout the year at every location on Earth because of the following Earth–Sun relationships:

- Earth rotates on its axis (creating day and night) and also revolves around the Sun (with one revolution equaling 1 year).
- The axis of Earth is inclined (tilted) 23 ½° from the perpendicular to the plane of its orbit.
- Throughout the year, the axis of Earth points to the same place in the sky, which causes the overhead (vertical, or 90°) noon Sun to cross the **equator** (0° latitude) twice as it migrates from the **Tropic of Cancer** (23 ½°N latitude) to the **Tropic of Capricorn** (23 ½°S latitude) and back again.

Because the position of the vertical (overhead) noon Sun gradually shifts between the hemispheres, the *intensity* and *duration* of solar radiation at a particular location continually change throughout the year. These shifts are mainly responsible for the seasonal temperature changes we experience. Historically, 4 days each year have been given special significance, based on the annual migration of the direct rays of the Sun. On June 21 or 22, Earth is positioned such that the north end of its axis is tilted 23 ½° *toward* the Sun (**Figure 13.4**). At that time, the vertical rays of the Sun strike 23 ½°N latitude at a latitude known as the **Tropic of Cancer**. For people in the Northern Hemisphere, June 21 or 22 is known as the **summer solstice**, the first "official" day of summer.

Six months later, on about December 21 or 22, Earth is in the opposite position, with the Sun's vertical rays striking at 23 ½°S latitude. This parallel is known as the **Tropic of Capricorn**. For those in the Northern Hemisphere, December 21 or 22 is the **winter solstice**. However, at the same time in the Southern Hemisphere, people are experiencing just the opposite—the *summer solstice*.

▼ **SmartFigure 13.4** Earth–Sun relationships.

TUTORIAL
https://goo.gl/aN6pBP

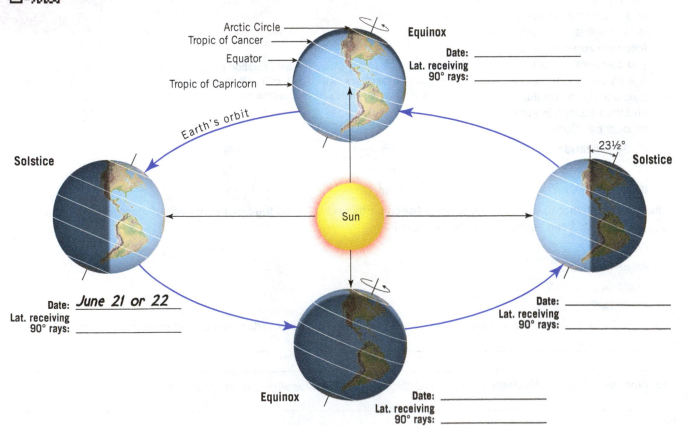

Midway between the solstices are the equinoxes. September 22 or 23 is the date of the **fall equinox** in the Northern Hemisphere, and March 21 or 22 is the date of the **spring equinox**. On these dates, the vertical rays of the Sun strike the equator (0° latitude) because Earth is in such a position in its orbit that the axis is tilted neither toward nor away from the Sun.

ACTIVITY 13.2
Variations in Solar Energy Throughout the Year

To see how the intensity and duration of solar radiation vary throughout the year, answer the following questions.

1. Refer to a globe or world map to locate the important parallels of latitude listed below. On **Figure 13.5**, draw in and label each parallel.

   Equator

   Arctic Circle

   Antarctic Circle

   Tropic of Capricorn

   Tropic of Cancer

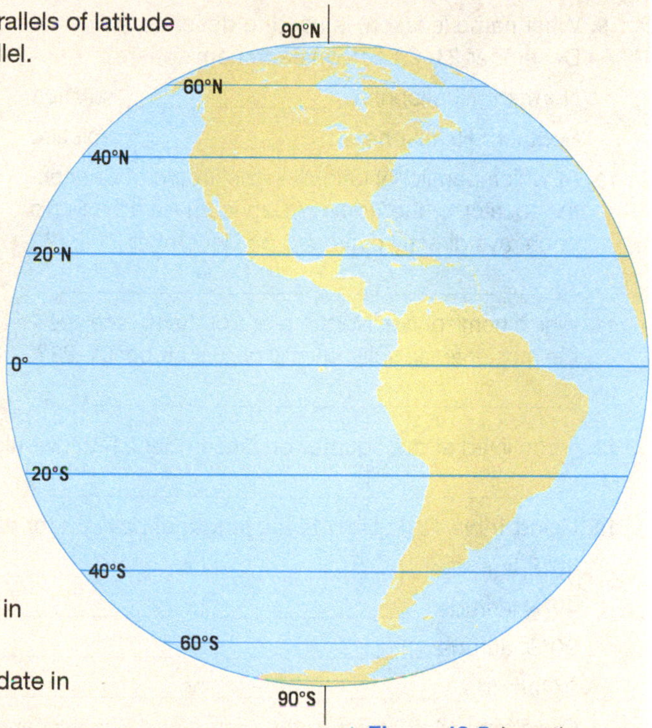

2. Use Figure 13.4, which shows the position of Earth on the traditional first days of summer, fall, winter, and spring, to complete the following. June dates are provided.

   a. Write in the date represented by each of the other three positions.

   b. Label the *circle of illumination*, the line separating the dark half from the lighted half of Earth, on each globe.

   Questions 3–7 refer to the June *solstice* position of Earth in Figure 13.4.

3. What name (season) is used to describe the June 21–22 date in each hemisphere?

   Northern Hemisphere: _____ solstice

   Southern Hemisphere: _____ solstice

4. On June 21–22, are the Sun's noon rays directly overhead at the Tropic of Cancer, the equator or the Tropic of Capricorn?

   _____

5. What latitude receives the most intense solar energy on June 21–22? (Remember to indicate north or south.)

   _____

6. Would you look toward the north or south to see the Sun at noon on June 21–22 if you lived at the following latitudes?

   40°N latitude: _____

   10°N latitude: _____

   10°S latitude: _____

7. Use the *circle of illumination* on the globe representing the June solstice to answer the following.

   a. On June 21–22, what is the length of daylight at latitudes north of the Arctic Circle?

   _____ hr

   b. On June 21–22, what is the length of daylight at latitudes south of the Antarctic Circle?

   _____ hr

▲ **Figure 13.5** Locating important parallels of latitude.

continued

Activity 13.2 continued

8. Using Table 13.1, provide the length of daylight for each of the following latitudes on June 21–22.

90°N latitude: _____ hr, _____ min

40°S latitude: _____ hr, _____ min

40°N latitude: _____ hr, _____ min

0° latitude: _____ hr, _____ min

Questions 9–13 refer to the December *solstice* position of Earth in Figure 13.4.

9. What name (season) is used to describe the December 21–22 date in each hemisphere?

Northern Hemisphere: _____ solstice

Southern Hemisphere: _____ solstice

10. At which parallel of latitude—the Tropic of Cancer, the equator, or the Tropic of Capricorn—are the Sun's noon rays directly overhead on December 21–22?

_____

11. Which hemisphere, Northern or Southern, receives the most intense solar energy on December 21–22?

_____

Table 13.1 Length of Daylight

LATITUDE (DEGREES)	SUMMER SOLSTICE	WINTER SOLSTICE	EQUINOXES
0	12 hr	12 hr	12 hr
10	12 hr 35 min	11 hr 25 min	12 hr
20	13 hr 12 min	10 hr 48 min	12 hr
30	13 hr 56 min	10 hr 04 min	12 hr
40	14 hr 52 min	9 hr 08 min	12 hr
50	16 hr 18 min	7 hr 42 min	12 hr
60	18 hr 27 min	5 hr 33 min	12 hr
66.5	24 hr	0 00	12 hr
70	24 hr (for 2 mo)	0 00	12 hr
80	24 hr (for 4 mo)	0 00	12 hr
90	24 hr (for 6 mo)	0 00	12 h

12. If you lived at the equator on December 21–22, would you look toward the north or south to see the Sun at noon?

_____

13. Using Table 13.1, provide the length of daylight for each of the following latitudes on December 21–22.

40°S latitude: _____ hr, _____ min

40°N latitude: _____ hr, _____ min

90°S latitude: _____ hr, _____ min

0° latitude: _____ hr, _____ min

Questions 14–20 refer to the March and September *equinox* positions of Earth in Figure 13.4.

14. For those living in the Northern Hemisphere, what terms (seasons) are used to describe the following dates?

March 21: _____ equinox

September 22: _____ equinox

15. For those living in the Southern Hemisphere, what terms (seasons) are used to describe the following dates?

March 21: _____ equinox

September 22: _____ equinox

16. On March 21 and September 22, at what latitude are the Sun's noon rays directly overhead?

_____

17. What latitude receives the most intense solar energy on March 21 and September 22?

_____

18. If you lived at 20°S latitude, would you look toward the north or south to see the Sun at noon on March 21 and September 22?

_____

19. What is the relationship between the North Pole and South Pole and the circle of illumination on March 21 and September 22?

_____

20. Briefly describe the length of daylight everywhere on Earth on March 21 and September 22.

_____

13.3 Using an Analemma

■ **Use the analemma to understand the relationship between latitude and the noon Sun angle on every day of the year.**

As you have seen, the latitude where the noon Sun is directly overhead is easily determined for the solstices and equinoxes. **Figure 13.6** is a graph, called an **analemma**, that can be used to determine the latitude where the overhead noon Sun is located for any date. To determine the latitude of the overhead noon Sun from the analemma, find the desired date on the graph and read the coinciding latitude along the left axis. Remember to indicate *north* (N) or *south* (S).

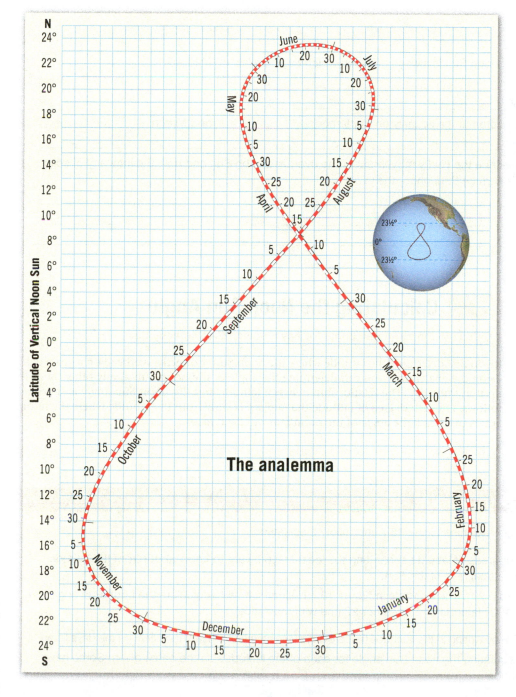

▲ **Figure 13.6** The analemma, a graph illustrating the latitude of the overhead (vertical) noon Sun throughout the year.

ACTIVITY 13.3
Using an Analemma

1. Using a colored pencil, draw and label lines that correspond to the equator, Tropic of Cancer, and Tropic of Capricorn on Figure 13.6.

2. Use the analemma to determine where the Sun is overhead at noon on the following dates. (An example is provided.)

    December 10: _____23°S latitude_____

    March 21: _____

    May 5: _____

    June 22: _____

    August 10: _____

    October 15: _____

3. The position of the overhead noon Sun is always between which two parallels of latitude?

    _____°N (the Tropic of _____)

    _____°S (the Tropic of _____)

4. The overhead noon Sun is located at the equator on September _____ and March _____.

    These two days are called _____.

5. Refer to Figure 13.6 and describe the yearly movement of the overhead noon Sun and how the intensity of solar radiation varies over Earth's surface throughout the year.

    _____

    _____

    _____

13.4   Calculating the Noon Sun Angle

■ Calculate the noon Sun angle for any place on Earth on any day.

If you first determine the latitude where the noon Sun is directly overhead on a given date using the analemma, you can determine the angle of the noon Sun at any other latitude on that same day. The relationship between latitude and the noon Sun angle is stated as follows: For each degree of latitude that a place is *away* from the latitude where the noon Sun is overhead, the angle of the noon Sun will be 1° *lower*.

ACTIVITY 13.4
Calculating Noon Sun Angle

To calculate the noon Sun angle, examine Figure 13.7 and follow these steps:

Step 1: Find the number of degrees of latitude separating the location you want to know about from the latitude that is receiving the vertical rays of the Sun.

Step 2: Subtract that value from 90°. The example in Figure 13.7 illustrates how to calculate the noon Sun angle for a city located at 40°N latitude on December 22 (the winter solstice).

1. Fill in the top row in Table 13.2 by writing in the latitudes that are experiencing the overhead noon Sun on the days listed.

2. Complete Table 13.2 by calculating the noon Sun angle for each of the latitudes on the dates given. Some have been completed for reference.

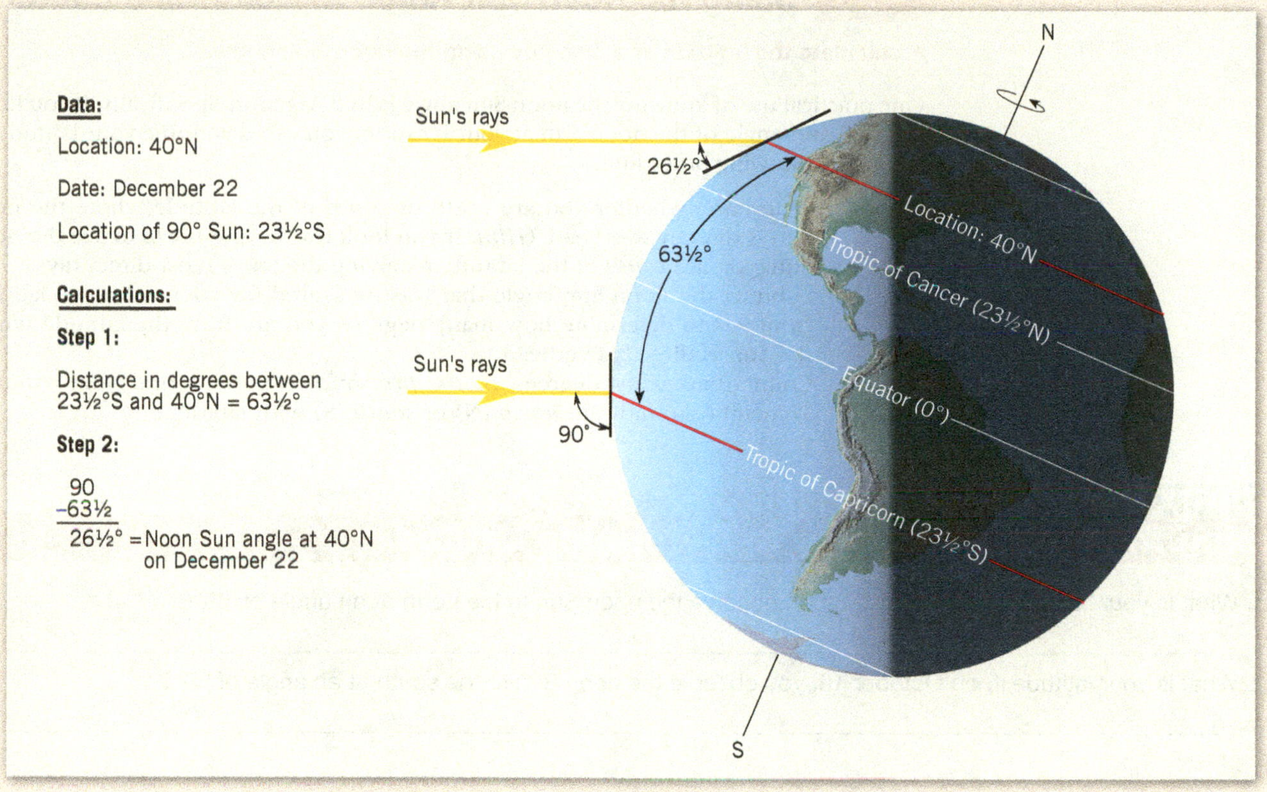

Data:

Location: 40°N

Date: December 22

Location of 90° Sun: 23½°S

Calculations:

Step 1:

Distance in degrees between 23½°S and 40°N = 63½°

Step 2:

```
  90
 −63½
  26½° = Noon Sun angle at 40°N
          on December 22
```

▲ **Figure 13.7** Calculating the noon Sun angle. Remember that on any given day, only one latitude receives vertical (90°) rays of the Sun. A place located 1° away (either north or south) receives an 89° noon Sun angle; a place 2° away, receives an 88° angle, and so forth.

3. Which latitude in Table 13.2 has the highest *average* noon Sun angle?

_____

4. Calculate the noon Sun angle for the latitude where you live for today's date.

Date: _____

Latitude of noon 90° angle Sun: _____

Your latitude: _____

Your noon Sun angle: _____

5. Calculate the yearly maximum and minimum noon Sun angles for your latitude.

MAXIMUM NOON SUN ANGLE

Date: _____

Angle: _____

MINIMUM NOON SUN ANGLE

Date: _____

Angle: _____ °

6. Calculate the yearly average noon Sun angle (maximum plus minimum, divided by 2) and the range of the noon Sun angle (maximum minus minimum) for your location.

Average noon Sun angle: _____ °

Range of the noon Sun angle: ____ °

Table 13.2 Noon Sun Angle Calculations

LATITUDE	NOON SUN ANGLE ON THESE DATES AT THESE LOCATIONS			
	MARCH 21	**APRIL 11**	**JUNE 21**	**DECEMBER 22**
90°N				0°
40°N	50°			26½°
0°			66½°	
20°S		62°		

13.5 Using the Noon Sun Angle to Determine Latitude

■ Calculate the latitude of a place by using the noon Sun angle.

One practical use of knowing the noon Sun angle is in navigation. Specifically, if you know the date and angle of the noon Sun at your location, you can determine your latitude by using the following procedure:

Step 1: Determine whether you are north or south of the latitude where the noon Sun is directly overhead. (*Hint:* If you look toward the *north* to see the noon Sun, you are *south* of the latitude receiving the Sun's most direct rays.)

Step 2: Subtract the noon Sun angle that was measured for your unknown latitude from 90° to determine how many degrees you are from the latitude where the Sun is directly overhead.

Step 3: Count that many degrees *in the proper direction* to find your latitude. Remember to indicate *north* (N) or *south* (S) with latitude.

ACTIVITY 13.5

Using the Noon Sun Angle to Determine Latitude

1. What is your latitude if, on March 21, you observe the noon Sun to the north at an angle of 18°?

_____

2. What is your latitude if, on October 16, you observe the noon Sun to the south at an angle of 39°?

_____

13.6 Solar Radiation at the Outer Edge of the Atmosphere

■ Describe how the average daily solar radiation received at the outer edge of the atmosphere varies with latitude and season.

Table 13.3 shows the average daily radiation received at the outer edge of the atmosphere at select latitudes for different months. These data will be used to help you visualize the pattern of daily radiation received at the outer edge of the atmosphere.

Table 13.3 Solar Radiation at the Outer Edge of the Atmosphere (Langleys/day) at Various Latitudes During Select Months*

LATITUDE	MARCH	JUNE	SEPTEMBER	DECEMBER
90°N	50	1050	50	0
40°N	700	950	720	325
0°	890	780	880	840

*A Langley is a measure of solar intensity. The higher the value, the more intense the solar radiation.

ACTIVITY 13.6

Solar Radiation at the Outer Edge of the Atmosphere

1. Plot the data from Table 13.3 on the graph in **Figure 13.8**. Using a different color for each latitude, draw lines through the monthly values to obtain yearly curves. The graph you prepare will be used to answer Questions 2 through 5.

2. Why are there two periods of maximum solar radiation at the equator?

_____

3. In June, why does the outer edge of the atmosphere at the equator receive less solar radiation than both the North Pole and 40°N latitude?

_____

_____

4. Why does the outer edge of the atmosphere at the North Pole receive no solar radiation in December?

_____

5. What would be the approximate monthly values for solar radiation at the outer edge of the atmosphere at 40°S latitude? Explain how you arrived at the values.

March: _____

June: _____

September: _____

December: _____

Explanation: _____

_____

_____

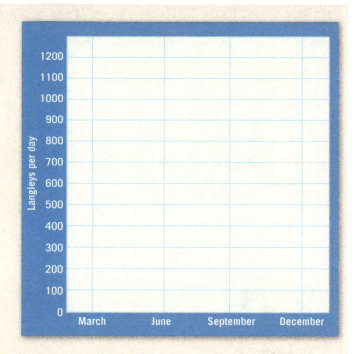

▲ **Figure 13.8** Graph of solar radiation received at the outer edge of the atmosphere.

Notes and calculations:

Earth–Sun Relationships

Name _____

Course/Section _____

Date _____

Due Date _____

1. In Activity 13.1, Question 7, what did you determine was the percentage of solar radiation that is intercepted by each of the following 30° segments of latitude?

   0°–30°N latitude: _____ %

   30°–60°N latitude: _____ %

   60°–90°N latitude: _____ %

2. How many hours of daylight occur at the following locations on the specified dates? (*Hint:* See Table 13.1.)

LATITUDE	MARCH 22	DECEMBER 22
40°N	_____ hr	_____ hr
0°	_____ hr	_____ hr
90°S	_____ hr	_____ hr

3. During the winter solstice in the Northern Hemisphere, do the days get longer or shorter the farther north you travel?

   _____

4. What is the relationship between the angle of the noon Sun and the intensity of solar radiation received at the outer edge of the atmosphere?

   _____

   _____

5. Complete **Figure 13.9**, showing Earth's relationship to the Sun on June 22. Sketch and label the following on Earth:

   Axis

   Equator

   Tropic of Cancer

   Tropic of Capricorn

   Antarctic Circle

   Arctic Circle

   Circle of illumination

   Location of the overhead noon Sun

6. What causes the intensity and duration of solar radiation received at any place to vary throughout the year? That is, what causes the seasons?

   _____

   _____

   _____

7. What are the maximum and minimum noon Sun angles at your latitude, and what are the dates when they occur?

   Maximum noon Sun angle: _____° on _____

   Minimum noon Sun angle: _____° on _____

8. What are the approximate maximum and minimum durations of daylight at your latitude?

   Maximum duration of daylight: _____ hr

   Minimum duration of daylight: _____ hr

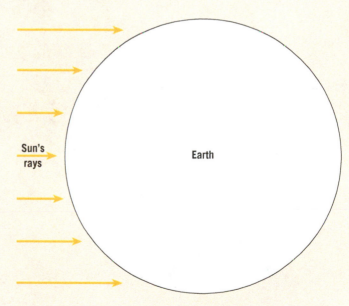

Sun's rays

Earth

▲ Figure 13.9 Earth's relationship to the Sun on June 22.

9. Briefly describe how the intensity and duration of solar radiation change at your location throughout the year.

_____

_____

_____

_____

10. On March 22, you view the noon Sun to the south at an angle of 35°. What is your latitude?

_____

_____

11. What is the date illustrated by the diagram in **Figure 13.10**?

_____

12. Calculate the noon Sun angle at 30°N latitude on the date shown in Figure 13.10 and then write a statement describing the distribution of solar radiation over Earth on this date.

_____

_____

_____

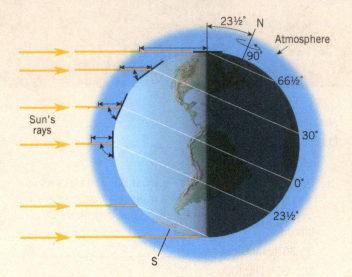

▲ **Figure 13.10** Earth–Sun relationships diagram.

Heating the Atmosphere

LEARNING OBJECTIVES

Each statement represents an important learning objective that relates to one or more sections of this lab. After you complete this exercise you should be able to:

- **Distinguish between short-wave and long-wave radiation.**
- **List the three paths taken by incoming solar radiation and the average percentage of sunlight taking each path.**
- **List the gases that are responsible for absorbing radiation emitted by Earth's surface. Explain how Earth's atmosphere is heated.**
- **Explain how albedo affects the heating of Earth's surface. Describe the differences in the heating of land and water.**
- **Describe daily variations in temperature and how they are influenced by cloud cover.**
- **Contrast the global pattern of surface air temperatures for January and July.**
- **Determine windchill-equivalent temperatures.**

MATERIALS

calculator
light source
black and silver containers
two thermometers

colored pencils
wood splints
beaker of sand
beaker of water

PRE-LAB VIDEO ▶ https://goo.gl/m7Fsst

Prepare for lab! Prior to attending your laboratory session, view the pre-lab video. Each video provides valuable background that will contribute to your understanding and success in lab.

INTRODUCTION

Temperature is a basic element of weather and climate. In this exercise, you will gain a better understanding of how Earth's atmosphere is heated. In addition, you will examine some of the factors that cause temperatures to vary over time and from one place to another.

14.1 # Radiation and the Electromagnetic Spectrum

■ **Distinguish between short-wave and long-wave radiation.**

From everyday experience, we know that the Sun emits light and heat as well as the ultra-violet rays that cause sunburn. Although these forms of energy comprise a major portion of the energy that radiates from the Sun, they are only part of a large array of energy called **radiation**, or **electromagnetic radiation**, shown in **Figure 14.1A**.

Visible light is the only portion of the spectrum we can see. As shown in **Figure 14.1B**, visible light is actually a mixture of wavelengths that each produce one of the colors of the rainbow.

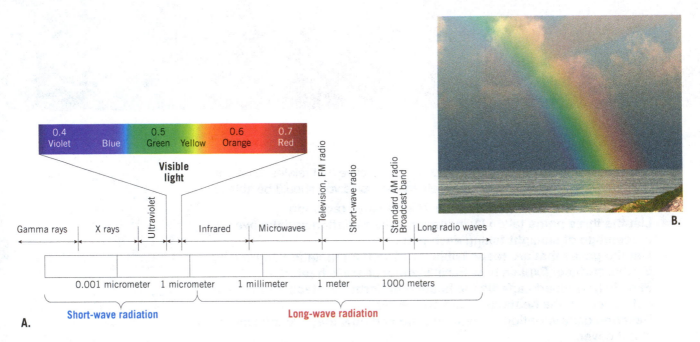

▲ **Figure 14.1** The electromagnetic spectrum. **A.** Names and wavelengths of various types of radiation emitted by the Sun. **B.** Visible light consists of an array of colors we commonly call the "colors of the rainbow." (Image by Dennis Tasa)

ACTIVITY 14.1 ///

Radiation and the Electromagnetic Spectrum

Refer to Figure 14.1 to answer the following questions.

1. Which part (color) of visible light has the longest wavelength?

_____

2. Which part (color) of visible light has the shortest wavelength?

_____

3. Which part of the electromagnetic spectrum—infrared or ultraviolet—has shorter wavelengths?

_____

14.2 | What Happens to Incoming Solar Radiation?

■ **List the three paths taken by incoming solar radiation and the average percentage of sunlight taking each path.**

When radiation strikes an object, there are three possible outcomes. (1) Some energy may be **absorbed**, causing the temperature of the material to rise. (2) Some substances, such as air and water, are transparent to certain wavelengths and simply **transmit** that energy. Radiation that is transmitted *does not* heat the material that transmits it. (3) Some radiation may "bounce off" an object by **reflection** or may be redirected by **scattering**. Dust and cloud droplets in the atmosphere are usually responsible for these latter processes.

ACTIVITY 14.2 //

What Happens to Incoming Solar Radiation?

Figure 14.2 shows what happens to incoming solar radiation, averaged over the entire globe. Refer to this diagram when answering the following questions.

1. What percentage of incoming solar radiation is reflected and scattered back to space?

   _____%

2. Does solar radiation that is reflected and scattered heat the objects it strikes?

   _____

3. What percentage of incoming solar radiation is absorbed by clouds and gases in the atmosphere?

   _____%

4. What percentage of incoming solar radiation passes through the atmosphere and is absorbed at Earth's surface?

   _____%

5. Excluding the radiation that is reflected and scattered back to space, does Earth's atmosphere transmit or absorb most incoming solar radiation?

   The atmosphere _____ most incoming radiation.

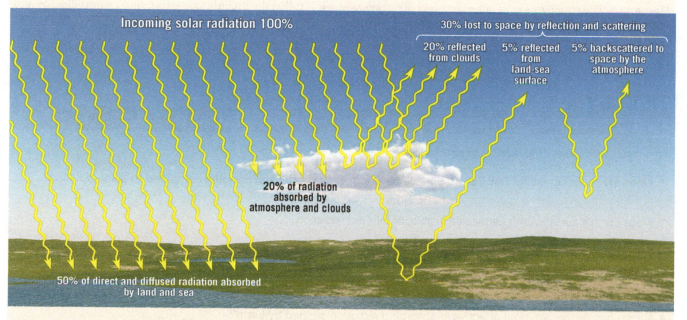

▲ **Figure 14.2** Average distribution of incoming solar radiation.

14.3 Heating Earth's Atmosphere: The Greenhouse Effect

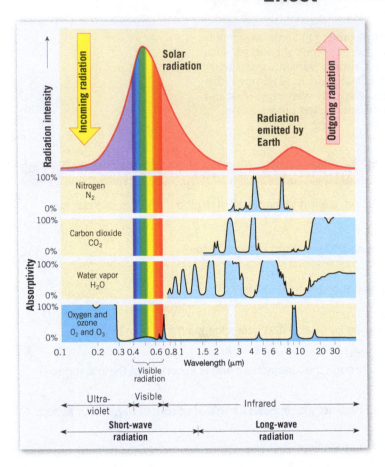

▲ **Figure 14.3** The effectiveness of selected gases of the atmosphere in absorbing incoming short-wave radiation (left side of graph) and outgoing long-wave radiation (right side). The light blue areas represent the percentage of radiation absorbed by the various gases.

■ **List the gases that are responsible for absorbing radiation emitted by Earth's surface. Explain how Earth's atmosphere is heated.**

What happens to the energy absorbed at Earth's surface? In addition to getting hotter, Earth's surface continually emits radiation back to the atmosphere. The radiation emitted by Earth's surface is different from solar radiation because the wavelengths emitted by an object depend on its temperature: *The cooler the object, the longer the wavelengths it emits.* Earth's surface is *much* cooler than the surface of the Sun. Therefore, the energy radiated outward from Earth's surface has longer wavelengths than incoming solar radiation.

Figure 14.3 shows the effectiveness of selected gases in absorbing incoming short-wave radiation (left side of graph) and outgoing long-wave radiation (right side). The light blue areas represent the percentage of radiation absorbed by these gases. The atmosphere as a whole is quite transparent to incoming solar radiation between 0.3 and 0.7 micrometer, which includes the band of visible light. Most solar radiation falls in this range, and this explains why a large amount of solar radiation penetrates the atmosphere and reaches Earth's surface.

Figure 14.3 also shows that carbon dioxide and water vapor are effective absorbers of long-wave radiation emitted by Earth's surface. After long-wave radiation is absorbed by molecules of carbon dioxide and water vapor, these atmospheric gases radiate some of this energy back toward Earth's surface. The process whereby the atmosphere reradiates energy Earthward is called the **greenhouse effect**, and carbon dioxide and water vapor are called **greenhouse gases**. The greenhouse effect is responsible for keeping Earth's surface about 33°C (59°F) warmer than it would otherwise be.

ACTIVITY 14.3
Heating Earth's Atmosphere: The Greenhouse Effect

To answer Questions 1–4, refer to Figure 14.3.

1. Does the radiation emitted by Earth's surface have longer or shorter wavelengths compared to solar radiation?

_____ wavelengths

2. Does Earth's surface emit ultraviolet, visible, or infrared radiation?

_____ radiation

3. Is nitrogen, the most abundant constituent of the atmosphere (78 percent), a good or poor absorber of incoming solar radiation?

_____ absorber

4. Which two gases are most effective in absorbing radiation emitted by Earth?

_____   _____

5. **Figure 14.4** illustrates the *greenhouse effect.* In your own words, briefly describe the greenhouse effect.

_____

_____

Airless bodies like the Moon All incoming solar radiation reaches the surface. Some is reflected back to space. The rest is absorbed by the surface and radiated directly back to space. As a result the lunar surface has a much lower average surface temperature than Earth.

Bodies with modest amounts of greenhouse gases like Earth Greenhouse gases in the atmosphere absorb some of the longwave radiation emitted by the surface. A portion of this energy is radiated back to the surface and is responsible for keeping Earth's surface 33°C (59°F) warmer than it would otherwise be.

Bodies with abundant greenhouse gases like Venus Venus experiences extraordinary greenhouse warming, which is estimated to raise its surface temperature by 523°C (941°F).

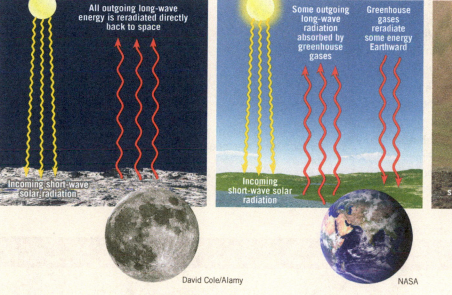
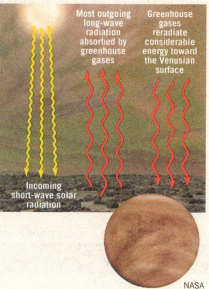

▲ **Figure 14.4** The greenhouse effect.

14.4 Factors Influencing the Heating of Earth's Surface

■ **Explain how albedo affects the heating of Earth's surface. Describe the differences in the heating of land and water.**

Because the atmosphere is heated chiefly by radiation emitted by Earth's surface, differences in the heating properties of various surfaces, such as water, soil, and trees, influences air temperatures.

In addition, albedo influences air temperatures. **Albedo** refers to the reflectivity of a surface and is usually expressed as the percentage of light reflected. Surfaces with high albedos are *not* efficient absorbers of incoming solar radiation. As a result, air above these surfaces is not warmed as much as air above surfaces that are good absorbers (low albedo), all else being equal.

ACTIVITY 14.4A
Experiment: The Influence of Color on Albedo

To better understand the effect of color on albedo, conduct the following experiment:

Step 1: Write a brief hypothesis that relates albedo to the heating and cooling of light-colored and dark-colored surfaces.

_____

_____

continued

Activity 14.4A continued

Step 2: Place the black and silver containers (with lids and thermometers) about 6 inches away from the light source, as shown in Figure 14.5. Place both containers the same distance from the light but ensure that they *do not* touch.

Step 3: Using Table 14.1, record the starting temperature of each container.

Step 4: Turn on the light and record the air temperature in both containers in Table 14.1 at about 30-second intervals for 5 minutes.

Step 5: Turn off the light and continue to record the temperatures at 30-second intervals for another 5 minutes.

Step 6: Plot the temperatures from Table 14.1 on the albedo experiment graph in Figure 14.6. Use a different-color line to connect the points for each container and label each line.

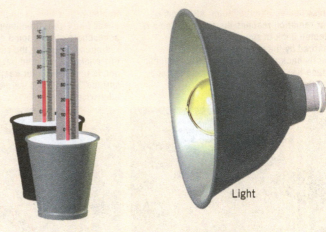

▲ **Figure 14.5** Albedo experiment lab equipment. The thermometers are measuring the air temperature inside the containers.

Table 14.1 Albedo Experiment Data Table

	STARTING TEMPERATURE	30 SEC	1 MIN	1.5 MIN	2 MIN	2.5 MIN	3 MIN	3.5 MIN	4 MIN	4.5 MIN	5 MIN	5.5 MIN	6 MIN	6.5 MIN	7 MIN	7.5 MIN	8 MIN	8.5 MIN	9 MIN	9.5 MIN	10 MIN
Black container																					
Silver container																					

1. For each container, calculate the *rate of heating* (change in temperature divided by the time the light was on) and the *rate of cooling* (change in temperature divided by the time the light was off).

	RATE OF HEATING	RATE OF COOLING
Silver can	_____	_____
Black can	_____	_____

2. Write a statement that summarizes the results of the albedo experiment.

_____

_____

_____

3. List at least two types of Earth surfaces that have a high albedo and two that have a low albedo.

High albedo: _____

Low albedo: _____

4. Given equal amounts of radiation reaching the surface, should air over a snow-covered surface be warmer or colder than air above a dark-colored, barren field? Explain your answer.

_____

_____

5. If you lived in an area with long, cold winters, would a light-colored or dark-colored roof be the best choice for your house?

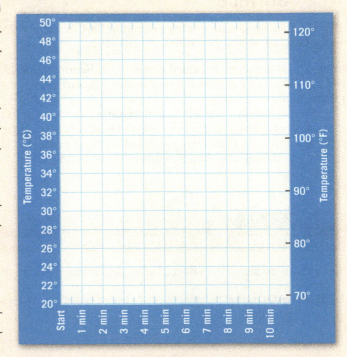

▲ **Figure 14.6** Albedo experiment graph.

ACTIVITY 14.4B ///

Experiment: Differential Heating of Land and Water

Investigate the differential heating of land and water by conducting the following experiment:

Step 1: Fill a beaker three-quarters full with dry sand and fill a second beaker three-quarters full with room-temperature water.

Step 2: Suspend a thermometer in each beaker, as shown in Figure 14.7. *Note:* It is important that the top of each bulb be *barely below* the surfaces of the sand and the water.

Step 3: Suspend the light from a stand so that both beakers receive the same amount of light.

Step 4: Record the starting temperatures for both the dry sand and water in Table 14.2.

Step 5: Turn on the light and record temperatures in Table 14.2 at about 1-minute intervals for 10 minutes.

Step 6: Turn off the light for several minutes. Dampen the sand with water and record the starting temperature of the damp sand in Table 14.2. Turn on the light and record the temperature of the damp sand in Table 14.2 at about 1-minute intervals for 10 minutes.

Step 7: Plot the temperatures for the water, dry sand, and damp sand from Table 14.2 on the graph in Figure 14.8. Use a different-color line to connect and label the points for each material.

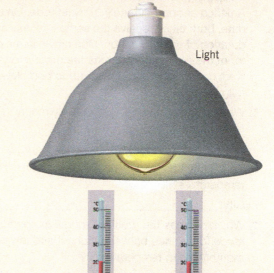
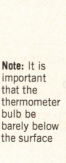

Light

Note: It is important that the thermometer bulb be barely below the surface

▲ **Figure 14.7** Land and water heating experiment lab equipment.

Table 14.2 Land Versus Water Data Table

	STARTING TEMPERATURE	1 MIN	2 MIN	3 MIN	4 MIN	5 MIN	6 MIN	7 MIN	8 MIN	9 MIN	10 MIN
Water											
Dry sand											
Damp sand											

1. How does the temperature differ for *dry sand* and *water* that are exposed to equal amounts of light?

_____

_____

2. How does the temperature differ for *dry sand* and *damp sand* that are exposed to equal amounts of light?

_____

_____

3. Suggest a reason for the differential heating of land and water.

_____

_____

4. Suggest a reason the damp sand heated differently than the dry sand.

_____

_____

continued

Activity 14.4B continued

Figure 14.9 presents the annual temperature curves for two cities, A and B, that are located in North America at approximately 37°N latitude. On any date, both cities receive the same intensity and duration of solar radiation. One city is in the center of the continent, whereas the other is on the west coast. Refer to Figure 14.9 to complete the following.

5. Which city has the highest monthly mean temperature?

   City: _____

6. Which city has the lowest monthly mean temperature?

   City: _____

7. Which city has the greatest *annual temperature range* (difference between highest and lowest monthly mean temperatures)?

   City: _____

8. Which city reaches its maximum monthly mean temperature at an earlier date?

   City: _____

9. Which city maintains more uniform temperatures throughout the year?

   City: _____

10. Which city is along the west coast, and which city is in the center of the continent, far from the influence of the ocean? Explain you answer.

    West coast city: _____

    Center continent city: _____

    Explanation: _____

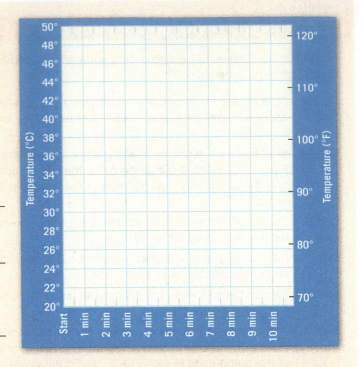

▲ **Figure 14.8** Land and water heating graph.

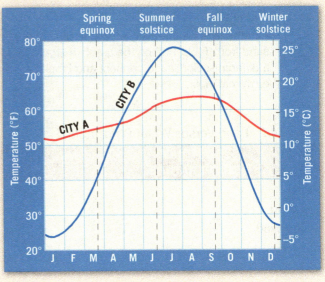

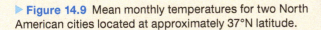

▶ **Figure 14.9** Mean monthly temperatures for two North American cities located at approximately 37°N latitude.

14.5 Air Temperature Data

■ **Describe daily variations in temperature and how they are influenced by cloud cover.**

Temperature is one of the basic elements of weather and climate. The daily **maximum** and **minimum temperatures** are the basis for many of the temperature data compiled by meteorologists. By adding the maximum and minimum temperatures and then dividing the sum by two, the **daily mean temperature** is calculated. The **daily range** of temperature is computed by finding the difference between the maximum and minimum temperatures for a given day.

Cloud cover is one of the factors that influences the daily cycle of temperature. Clouds are important because they reduce the amount of incoming solar radiation, causing daytime temperatures to be lower than if the clouds were not present and the sky were clear. At night, clouds have the opposite effect. They act as a blanket, absorbing radiation emitted by Earth's surface and reradiating a portion of that energy back to the surface.

ACTIVITY 14.5
Daily Temperatures and the Effect of Cloud Cover

The questions below refer to the daily temperature graph in **Figure 14.10**, which is for a midlatitude city on a clear day in early summer.

1. At what time does the coolest temperature occur?

   _____

2. At what time does the warmest temperature occur?

   _____

3. What is the *daily temperature range* (difference between maximum and minimum temperatures for the day)?

   Daily temperature range: _____°F
   (_____°C).

4. What is the *daily mean temperature* (average of the maximum and minimum temperatures)?

   Daily mean temperature: _____°F
   (_____°C).

5. Why does the coolest temperature of the day occur about sunrise?

   _____

   _____

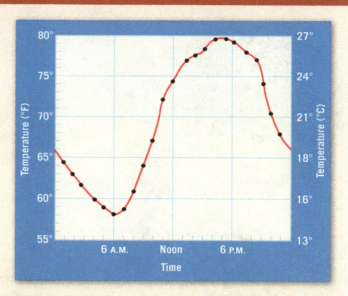

▲ **SmartFigure 14.10** Typical daily temperature graph for a midlatitude city during the summer.

 **TUTORIAL**
https://goo.gl/HsYaD7

6. How would cloud cover influence the daily *maximum temperature*? Explain your answer.

   _____

   _____

7. How would cloud cover influence the daily *minimum temperature*? Explain your answer.

   _____

   _____

8. On Figure 14.10, sketch and label a line that represents a daily temperature graph for a typical cloudy day.

14.6 Global Temperature Patterns

■ **Contrast the global pattern of surface air temperatures for January and July.**

The primary reason for global variations in surface temperatures is the unequal distribution of solar radiation over Earth due to seasonal changes in Sun angle and length of daylight. Among the most important secondary factors affecting the global pattern of temperature are differential heating of land and water and the influence of cold and warm ocean currents.

ACTIVITY 14.6
Global Temperature Patterns

The following questions refer to **Figure 14.11**. The lines on the maps, called **isotherms**, connect places of equal surface temperatures.

1. Is the general trend of the isotherms on the maps north–south or east–west?

   _____

continued

Activity 14.6 continued

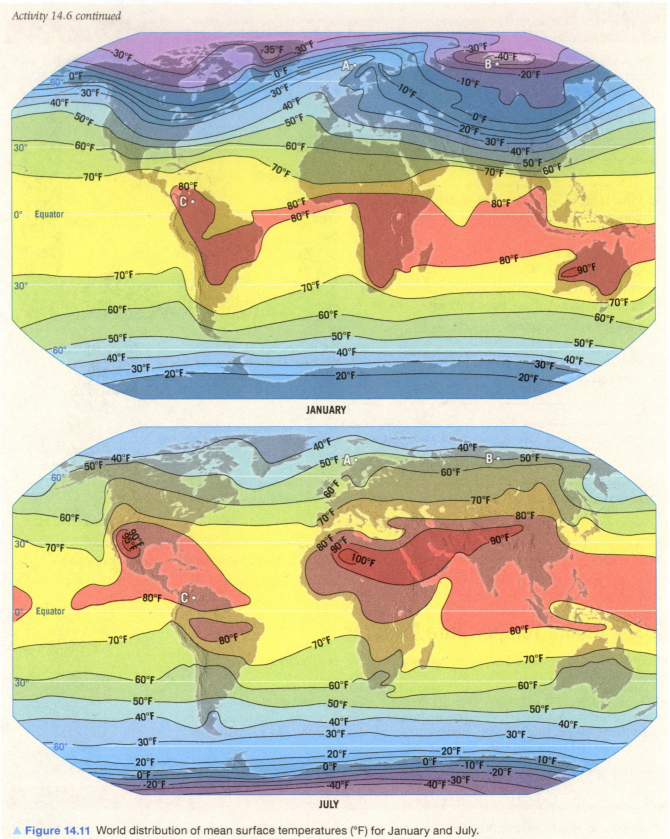

JANUARY

JULY

▲ **Figure 14.11** World distribution of mean surface temperatures (°F) for January and July.

2. In general, how do surface temperatures vary from the equator toward the poles? What is the likely cause of this variation?

_____

_____

_____

3. Do the *warmest* temperatures occur over continents or over oceans? Do the *coldest* temperatures occur over continents or over oceans?

   Warmest global temperatures occur over _____.

   Coldest global temperatures occur over _____.

4. Calculate the *annual temperature range* along the coast of Norway (Point A) and in Siberia at a location having the same latitude (Point B).

   Coastal Norway (Point A): _____°F

   Siberia (Point B): _____°F

5. Explain the wide annual range of temperatures in Siberia.

_____

_____

6. Why is the annual temperature range smaller along the coast of Norway than at the same latitude in Siberia?

_____

_____

7. Compare the average monthly temperatures for January and July for Puerto Ayacucho, Venezuela, marked as Point C on both maps. Explain why temperatures are relatively uniform throughout the year in tropical locations like Puerto Ayacucho.

_____

_____

8. Using the two maps in Figure 14.11, calculate the approximate average annual temperature range for your location. How does your annual temperature range compare with those in the tropics and Siberia?

   Average annual temperature range: _____°F

_____

_____

9. Trace the path of the 40°F isotherm over North America in January. Describe *how* and explain *why* the isotherm deviates from a true east–west trend where it crosses from the Pacific Ocean onto the continent.

_____

_____

10. Trace the path of the 70°F isotherm over North America in July. Describe *how* and explain *why* the isotherm deviates from a true east–west trend where it crosses from the Pacific Ocean onto the continent.

_____

_____

11. Why do the isotherms in the Southern Hemisphere follow a true east–west trend more closely than those in the Northern Hemisphere?

_____

_____

12. Explain why the entire pattern of isotherms has a more northerly location on the July map than on the January map.

_____

_____

14.7 Windchill

- **Determine windchill-equivalent temperatures.**

Windchill is an expression of *apparent temperature*—the temperature that a person perceives, in contrast to the actual temperature of the air as recorded by a thermometer. Wind cools by evaporating perspiration from exposed parts of the body. In addition, wind acts to carry heat away from the body by constantly replacing warm air next to the body with cooler air.

ACTIVITY 14.7
Windchill

1. Refer to the windchill equivalent temperature chart shown in **Figure 14.12**. What is the windchill equivalent temperature sensed by the human body in the following situations?

AIR TEMPERATURE (°F)	WIND SPEED (mph)	WINDCHILL EQUIVALENT TEMPERATURE (°F)
30°	10	_____ °
−5°	20	_____ °
−20°	30	_____ °

2. Examine Figure 14.12 and write a brief summary of how wind speed affects the length of time a person can be exposed to progressively colder temperatures before frostbite develops.

_____

_____

_____

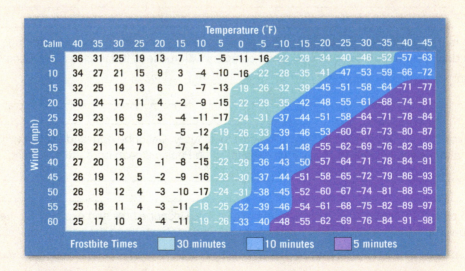

▲ **Figure 14.12** The shaded areas on the windchill chart indicate frostbite danger. Each shaded zone shows how long a person can be exposed before frostbite develops.

Heating the Atmosphere

Name _____

Date _____

Course/Section _____

Due Date _____

1. What percentage of incoming solar radiation is absorbed directly by the atmosphere? What percentage is absorbed by Earth's surface?

   Atmospheric absorption: _____%

   Absorption by Earth's surface: _____%

2. How should an increase in the amount of carbon dioxide in the atmosphere affect atmospheric temperatures?

   _____

   _____

3. Based on your results from the albedo experiment in Activity 14.4A, will the air above a surface that has a high albedo be warmer or cooler than the air above a surface that has a low albedo?

   _____

   _____

4. Briefly explain how Earth's atmosphere is heated.

   _____

   _____

   _____

5. Name the two gases that absorb most of Earth's outgoing long-wave radiation.

   _____   _____

6. What were the starting and 5-minute temperatures you obtained for the black and silver containers in the albedo experiment in Activity 14.4A?

	STARTING TEMPERATURE	5-MINUTE TEMPERATURE
Black container	_____	_____
Silver container	_____	_____

7. Summarize the effect of color on the rate at which an object's temperature rises.

   _____

   _____

8. What were the starting and ending temperatures you obtained for the water and dry sand in the land and water heating experiment in Activity 14.4B?

	STARTING TEMPERATURE	ENDING TEMPERATURE
Water	_____ °	_____ °
Dry sand	_____ °	_____ °

9. In general, do land surfaces heat more or less quickly than water surfaces?

   _____

10. At what locations (continental or marine) do we find the highest and lowest monthly average temperatures?

    Highest average monthly temperature: _____

    Lowest average monthly temperature: _____

11. Why do the middle to high latitudes in the Northern Hemisphere experience a greater annual temperature range than similar latitudes in the Southern Hemisphere?

    _____

    _____

12. Refer to the graph in **Figure 14.13**. It shows monthly mean temperatures for two cities at about 40°N latitude. One city is along the California coast; the other is in central Illinois. Which line on the graph represents the California city, and which represents the city in Illinois? Explain how you determined your answer.

_____

_____

_____

_____

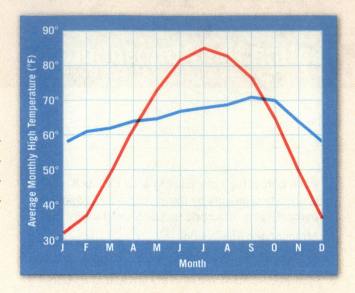

▲ **Figure 14.13** Graph to accompany Question 12.

Atmospheric Moisture, Pressure, and Wind

LEARNING OBJECTIVES

Each statement represents an important learning objective that relates to one or more sections of this lab. After you complete this exercise you should be able to:

- List and describe the processes involved when water changes state.
- Express the relationship between temperature and saturation mixing ratio.
- Define *relative humidity* and explain how changes in mixing ratio and temperature affect relative humidity.
- Relate dew point temperature to relative humidity and saturation mixing ratio.
- Use a psychrometer and appropriate tables to determine the relative humidity and dew-point temperature of air.
- Describe the typical daily pattern of variations in air temperature and relative humidity.
- Explain the process of adiabatic cooling and warming. Calculate the temperature and relative humidity changes that take place in air as a result of adiabatic temperature changes.
- Explain how air pressure is measured and how pressure tendency is related to cloudy or clearing weather.
- Describe the relationship between air pressure and wind.
- Briefly explain how the Coriolis effect influences wind.
- Relate the global distribution of precipitation to global pressure zones. List other factors that influence the distribution of precipitation.

MATERIALS

calculator	cardboard	beaker, ice,
colored pencils	tape	thermometer
hot plate	ruler	world map or atlas
thumbtacks	psychrometer	stopwatch

PRE-LAB VIDEO https://goo.gl/K15mMe

 Prepare for lab! Prior to attending your laboratory session, view the pre-lab video. Each video provides valuable background that will contribute to your understanding and success in lab.

PART 3 Meteorology

INTRODUCTION

This lab focuses on three basic elements of weather: moisture (humidity and precipitation), air pressure, and wind. Although the hour-to-hour and day-to-day changes in air pressure are not perceptible to human beings, they are important in producing changes in our weather. For example, variations in air pressure from one place to another generate wind, which in turn can bring changes in temperature and humidity.

15.1   Changes of State

■ **List and describe the processes involved when water changes state.**

Water vapor is an odorless, colorless gas that mixes freely with the other gases of the atmosphere. Unlike nitrogen and oxygen—the two most abundant components of the atmosphere—water can change from one state of matter to another (solid, liquid, or gas) at the temperatures and pressures experienced at and near Earth's surface. Because of this unique property, water freely leaves Earth's surface as a gas and returns again as a solid or liquid.

When water changes state, heat is exchanged between the water molecules and their surroundings. For example, energy is absorbed by the water molecules in an ice cube as it melts. If you are holding an ice cube tightly in your hand, the energy to melt the ice comes from your hand. Because the heat used to melt ice does not produce a temperature change, it is called **latent heat**. (*Latent* means "hidden," like the latent fingerprints hidden at a crime scene.) Latent heat is energy that is stored in water molecules and is released to its surroundings when water turns to ice.

ACTIVITY 15.1A
Changes of State

1. On **Figure 15.1**, write the name of each process in the appropriate blank on the diagram (choose from the following list).

PROCESSES

Freezing (liquid to solid)   **Evaporation** (liquid to gas)   **Deposition** (gas to solid)

Sublimation (solid to gas)   **Melting** (solid to liquid)   **Condensation** (gas to liquid)

2. When water evaporates, do water molecules absorb or release heat?

_____

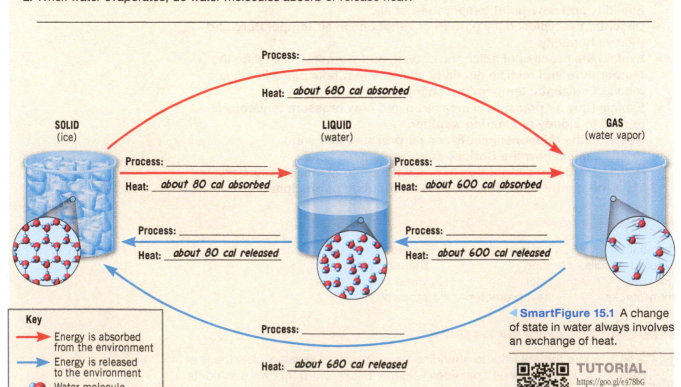

Process: _____

Heat: *about 680 cal absorbed*

SOLID (ice)   **LIQUID** (water)   **GAS** (water vapor)

Process: _____   Process: _____

Heat: *about 80 cal absorbed*   Heat: *about 600 cal absorbed*

Process: _____   Process: _____

Heat: *about 80 cal released*   Heat: *about 600 cal released*

Key

→ Energy is absorbed from the environment

→ Energy is released to the environment

● Water molecule

Process: _____

Heat: *about 680 cal released*

◄ **SmartFigure 15.1** A change of state in water always involves an exchange of heat.

TUTORIAL
https://goo.gl/e478bG

3. During condensation, do water molecules absorb or release heat?

_____

4. Is the amount of latent heat associated with the process of deposition the same as or less than the total energy required to condense water vapor and then freeze the water?

_____

ACTIVITY 15.1B
Latent Heat Experiment

This experiment is intended to help you gain a better understanding of latent heat. You are going to heat a beaker that contains a mixture of ice and water (Figure 15.2). You will record temperature changes *as the ice melts* and continue to record the temperature changes *after the ice melts*. Conduct the experiment by completing the following steps:

Step 1: Write a brief hypothesis about how you expect the temperature of the ice–water mixture to change as it is heated.

_____

_____

_____

Step 2: Turn on the hot plate and set the temperature setting to about three-fourths the maximum (7 on a scale of 1 to 10).

Step 3: Fill a 400-mL or larger beaker approximately half full with ice and add enough *cold* water to cover the ice.

> **CAUTION**
> Do not touch the heating surface of the hot plate.

Step 4: Gently stir the ice–water mixture for about 15 seconds with the thermometer and record the temperature in the "Starting" temperature blank in **Table 15.1**.

Step 5: Place the beaker with the ice–water mixture and thermometer on the hot plate, and while *stirring the mixture frequently*, record the temperature of the mixture at *1-minute intervals* in Table 15.1. Watch the ice closely as it melts. Circle the exact time (minute) in Table 15.1 when all the ice has completely melted.

Step 6: Continue stirring the mixture and recording the temperature for at least 3 or 4 minutes after all the ice has melted.

Step 7: Plot the temperatures from Table 15.1 on the graph in Figure 15.3.

▼ **Figure 15.2** Setup for the latent heat experiment.

Thermometer →

1. How did the temperature of the mixture change before and after the ice melted?

_____

_____

2. Calculate the *average temperature change per minute* of the ice–water mixture *prior to* the ice melting and the average rate *after* the ice melted.

Average rate prior to melting: _____

Average rate after melting: _____

3. With your answers to Questions 1 and 2 in mind, write a statement comparing your results to your hypothesis in Step 1.

_____

_____

continued

Activity 15.1B continued

4. Explain how the absorption of latent (hidden) heat accounts for the temperature changes that occurred before the ice melted compared to after the ice melted.

_____

_____

_____

_____

Table 15.1 Latent Heat Data Table

TIME (MINUTES)	TEMPERATURE
Starting	_____
1	_____
2	_____
3	_____
4	_____
5	_____
6	_____
7	_____
8	_____
9	_____
10	_____
11	_____
12	_____
13	_____
14	_____
15	_____

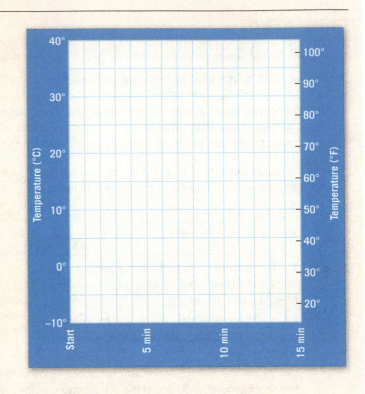

▲ **Figure 15.3** Latent heat experiment graph.

15.2 Humidity: Water Vapor in the Air

■ **Express the relationship between temperature and saturation mixing ratio.**

Water vapor constitutes only a small fraction of the atmosphere, varying from as little as one-tenth of 1 percent up to about 4 percent by volume. Nonetheless, scientists agree that *water vapor* is the most important gas in the atmosphere when it comes to understanding atmospheric processes.

Humidity is the general term for the amount of water vapor in air. Meteorologists use several methods to express the humidity, or water-vapor content, of air. In the following lab activities, we will examine three: mixing ratio, relative humidity, and dew-point temperature.

Before we consider these humidity measurements, it is important to understand the concept of **saturation**. Imagine a closed jar that contains water overlain by dry air (**Figure 15.4**). As water begins to evaporate, a small increase in pressure, called *vapor pressure*, occurs in the air above. As more and more molecules escape from the water surface, the steadily increasing vapor pressure in the air above causes more and more of these molecules to return to the liquid. When the number of water-vapor molecules returning to the surface balances the number leaving, the air is described as being *saturated*.

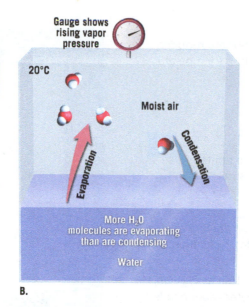
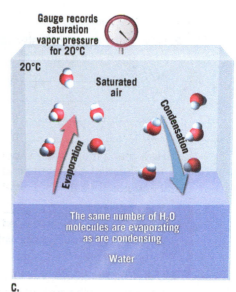

▲ **Figure 15.4** These diagrams illustrate the idea of vapor pressure and saturation. **A.** Initial conditions—dry air at 20°C with no observable vapor pressure. **B.** Evaporation generates measurable vapor pressure. **C.** As more and more molecules escape from the water surface, the steadily increasing vapor pressure forces an increasing number of molecules to return to the liquid. Eventually, the number of water-vapor molecules returning to the surface will balance the number leaving—at which point the air is said to be saturated.

ACTIVITY 15.2
Mixing Ratio

The **mixing ratio** is the mass of water vapor in a unit of air compared to the remaining mass of dry air, commonly expressed as grams/kilogram (g/kg).

Use Table 15.2, which presents the mixing ratios for saturated air at various temperatures, to complete the following.

1. To illustrate the relationship between the amount of water vapor needed for saturation and temperature, prepare a graph by plotting the data from Table 15.2 on Figure 15.5.

Table 15.2 Amount of Water Vapor Needed to Saturate 1 Kilogram of Air at Various Temperatures (the saturation mixing ratio)

TEMPERATURE		SATURATION MIXING RATIO
(°C)	(°F)	WATER-VAPOR CONTENT AT SATURATION (g/kg)
−40	−40	0.1
−30	−22	0.3
−20	−4	0.75
−10	14	2
0	32	3.5
5	41	5
10	50	7
15	59	10
20	68	14
25	77	20
30	86	26.5
35	95	35
40	104	47

continued

Activity 15.2 continued

2. How many grams of water vapor are required to saturate 1 kilogram of air at these temperatures?

40°C: _____ g

20°C: _____ g

0°C: _____ g

−20°C: _____ g

3. How much more water vapor is contained in 1 kilogram of saturated air at 35°C than in 1 kilogram of saturated air at 10°C?

The difference is _____ grams. Or, stated another way, saturated air at 35°C contains _____ times more water vapor than saturated air at 10°C.

4. Using Table 15.2 and/or Figure 15.5 as a guide, describe how the amount of water vapor needed for saturation relates to temperature.

_____

_____

_____

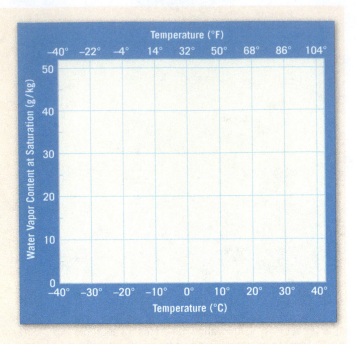

▲ **Figure 15.5** Graph of water-vapor content, at saturation, of 1 kilogram of air at various temperatures.

15.3 Relative Humidity

■ Define *relative humidity* and explain how changes in mixing ratio and temperature affect relative humidity.

Relative humidity is the most common measurement used to describe the amount of water vapor in the air. It expresses how close the air is to being saturated. Relative humidity is a *ratio* of the air's actual water-vapor content (mixing ratio) compared with the amount of water vapor required for saturation at that temperature (saturation mixing ratio), expressed as a percentage. The general formula is:

$$\text{Relative humidity (\%)} = \frac{\text{Mixing ratio}}{\text{Saturation mixing ratio}} \times 100\%$$

For example, from Table 15.2, the saturation mixing ratio of air at 25°C is 20 grams per kilogram. If the mixing ratio (water-vapor content) were 5 grams per kilogram, the relative humidity of the air would be calculated as follows:

$$\text{Relative humidity (\%)} = \frac{5 \text{ g/kg}}{20 \text{ g/kg}} \times 100\% = 25\%$$

ACTIVITY 15.3A
Calculating Relative Humidity

1. Use Table 15.2 and the formula for relative humidity to determine the relative humidity for each of the following situations. Notice that temperature remains unchanged.

AIR TEMPERATURE	WATER-VAPOR CONTENT	RELATIVE HUMIDITY
15°C	2 g/kg	_____ %
15°C	5 g/kg	_____ %
15°C	7 g/kg	_____ %

2. Based on your answers to Question 1, complete the following statements. When air temperature remains constant, will adding water vapor increase or decrease relative humidity? How does relative humidity change when water vapor is removed from the air?

Adding water vapor will _____ relative humidity.

Reducing water vapor will _____ relative humidity.

3. Use Table 15.2 and the formula for relative humidity to determine the relative humidity for each of the following situations in which the *water-vapor content remains unchanged*.

AIR TEMPERATURE	WATER-VAPOR CONTENT	RELATIVE HUMIDITY
25°C	5 g/kg	_____%
15°C	5 g/kg	_____%
5°C	5 g/kg	_____%

4. Based on your answers to Question 3, complete the following statements. If the amount of water vapor in the air remains constant, does cooling the air raise or lower the relative humidity? When air temperature increases, does relative humidity increase or decrease?

Cooling _____ the relative humidity.

Relative humidity _____ when air temperature increases.

5. Many homes are heated in the winter. What effect does heating have on the relative humidity inside the home? What can be done to lessen this effect?

_____

_____

6. Explain why the air in a cool basement is relatively humid (damp) in the summer.

_____

_____

7. Briefly describe the two ways relative humidity can be changed.

a. _____

b. _____

ACTIVITY 15.3B

Relative Humidity Versus Water-Vapor Content

One misconception about relative humidity is that it gives an accurate indication of the actual quantity of water vapor in the air. For example, on a winter day, if you hear on the radio that the relative humidity is 90 percent, can you conclude that the air contains more water vapor than on a summer day that records a relative humidity of 40 percent? Completing the following will help you determine the answer.

1. Refer to Table 15.2 to determine the water-vapor content for each of the following situations. As you do the calculations, keep in mind the definition of relative humidity.

SUMMER	WINTER
Air temperature = 77°F	Air temperature = 41°F
Relative humidity = 40%	Relative humidity = 90%
Content = _____ g/kg	Content = _____ g/kg

2. Explain why relative humidity does *not* give an accurate indication of the actual amount of water vapor in the air.

_____

_____

_____

15.4 Dew-Point Temperature

■ **Relate dew point temperature to relative humidity and saturation mixing ratio.**

The temperature at which saturation occurs is called the **dew-point temperature**. Put another way, the dew point is the temperature at which the relative humidity of the air is 100 percent.

Earlier we determined that 1 kilogram of air at 25°C that contains 5 grams of water vapor has a relative humidity of 25 percent. When the temperature of that same mass of air is lowered to 5°C, the relative humidity increases to 100 percent, and the air is described as being *saturated*. Therefore, 5°C is the dew-point temperature of the air in this example.

ACTIVITY 15.4
Dew-Point Temperature

1. By referring to Table 15.2, determine the dew-point temperature of 1 kilogram of air that contains 7 grams of water vapor.

   Dew-point temperature = _____ °C

2. What are the relative humidity and dew-point temperature of 1 kilogram of 25°C air that contains 10 grams of water vapor?

   Relative humidity = _____ %

   Dew-point temperature = _____ °C

15.5 Using a Psychrometer

■ **Use a psychrometer and appropriate tables to determine the relative humidity and dew-point temperature of air.**

One method of determining the relative humidity and dew-point temperature of air is by using a **psychrometer** (Figure 15.6). A psychrometer consists of two identical thermometers mounted side by side. One of the thermometers, the *dry-bulb thermometer*, measures the air temperature. The other thermometer, the *wet-bulb thermometer*, has a piece of wet cloth wrapped around its bulb. As the psychrometer is spun for approximately 1 minute, water on the wet-bulb thermometer evaporates, and cooling results. In dry air, the rate of evaporation will be high, resulting in a low wet-bulb temperature. After using a psychrometer and recording both the dry- and wet-bulb temperatures, you can determine the relative humidity and dew-point temperatures by using Table 15.3 and Table 15.4 (page 257).

ACTIVITY 15.5
Measuring Relative Humidity and Dew-Point Temperature

1. Use Table 15.3 to determine the relative humidity for each of the following psychrometer readings.

	READING 1	READING 2
Dry-bulb temperature	20°C	32°C
Wet-bulb temperature	18°C	25°C
Difference between dry- and wet-bulb temperatures	_____°C	_____°C
Relative humidity	_____%	_____%

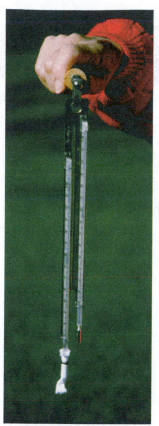

A. The dry-bulb thermometer gives the current air temperature.

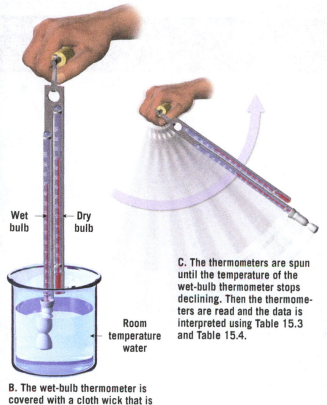

Wet bulb — — Dry bulb

Room temperature water

B. The wet-bulb thermometer is covered with a cloth wick that is dipped in water.

C. The thermometers are spun until the temperature of the wet-bulb thermometer stops declining. Then the thermometers are read and the data is interpreted using Table 15.3 and Table 15.4.

◄ **Figure 15.6** The sling psychrometer, an instrument that is used to determine relative humidity and dew-point temperature. (Photo by E. J. Tarbuck)

2. Based on your answers to Question 1, describe the relationship between the *difference* in the dry-bulb and wet-bulb temperatures compared to the relative humidity of the air.

_____

_____

3. Use Table 15.4 to determine the dew-point temperature for each of these two psychrometer readings.

	READING 1	READING 2
Dry-bulb temperature	8°C	30°C
Wet-bulb temperature	6°C	24°C
Difference between dry- and wet-bulb temperatures	_____ °C	_____ °C
Dew-point temperature	_____ °C	_____ °C

If a psychrometer is available in the laboratory, your instructor will explain the procedure for using it. FOLLOW THE SPECIFIC DIRECTIONS OF YOUR INSTRUCTOR to complete Question 4.

4. Use the psychrometer to determine the relative humidity and dew-point temperature of the air in the room and outside the building. Record your information in the following spaces.

	ROOM	OUTSIDE
Dry-bulb temperature	_____ °C	_____ °C
Wet-bulb temperature	_____ °C	_____ °C
Difference between dry- and wet-bulb temperatures	_____ °C	_____ °C
Relative humidity	_____ %	_____ %
Dew-point temperature	_____ °C	_____ °C

Table 15.3 Relative Humidity (percentage)[*]

DRY-BULB TEMPERATURE (°C)	DEPRESSION OF WET-BULB TEMPERATURE (°C) (DRY-BULB TEMPERATURE – WET-BULB TEMPERATURE = DEPRESSION OF THE WET BULB)																					
	1	2	3	4	5	6	7	8	9	10	11	12	13	14	15	16	17	18	19	20	21	22
−20	28																					
−18	40																					
−16	48	0																				
−14	55	11																				
−12	61	23																				
−10	66	33	0																			
−8	71	41	13																			
−6	73	48	20	0																		
−4	77	54	43	11																		
−2	79	58	37	20	1																	
0	81	63	45	28	11																	
2	83	67	51	36	20	6																
4	85	70	56	42	27	14																
6	86	72	59	46	35	22	10	0														
8	87	74	62	51	39	28	17	6														
10	88	76	65	54	43	33	24	13	4													
12	88	78	67	57	48	38	28	19	10	2												
14	89	79	69	60	50	41	33	25	16	8	1											
16	90	80	71	62	54	45	37	29	21	14	7	1										
18	91	81	72	64	56	48	40	33	26	19	12	6	0									
20	91	82	74	66	58	51	44	36	30	23	17	11	5	0								
22	92	83	75	68	60	53	46	40	33	27	21	15	10	4	0							
24	92	84	76	69	62	55	49	42	36	30	25	20	14	9	4	0						
26	92	85	77	70	64	57	51	45	39	34	28	23	18	13	9	5						
28	93	86	78	71	65	59	53	47	42	36	31	26	21	17	12	8	2					
30	93	86	79	72	66	61	55	49	44	39	34	29	25	20	16	12	8	4				
32	93	86	80	73	68	62	56	51	46	41	36	32	27	22	19	14	11	8	4			
34	93	86	81	74	69	63	58	52	48	43	38	34	30	26	22	18	14	11	8	5		
36	94	87	81	75	69	64	59	54	50	44	40	36	32	28	24	21	17	13	10	7	4	
38	94	87	82	76	70	66	60	55	51	46	42	38	34	30	26	23	20	16	13	10	7	
40	94	89	82	76	71	67	61	57	52	48	44	40	36	33	29	25	22	19	16	13	10	7

Dry-Bulb (Air) Temperature (vertical axis label)

Relative Humidity Values (diagonal label within table)

[*] To determine the relative humidity and dew point, find the air (dry-bulb) temperature on the vertical axis (far left) and the depression of the wet bulb on the horizontal axis (top). Where the two meet, the relative humidity or dew point is found. For example, use a dry-bulb temperature of 20°C and a wet-bulb temperature of 14°C. From Table 15.3, the relative humidity is 51 percent, and from Table 15.4, the dew point is 10°C.

Table 15.4 Dew-Point Temperature (°C)*

DRY-BULB TEMPERATURE (°C)	DEPRESSION OF WET-BULB TEMPERATURE (°C) (DRY-BULB TEMPERATURE – WET-BULB TEMPERATURE = DEPRESSION OF THE WET BULB)																					
	1	2	3	4	5	6	7	8	9	10	11	12	13	14	15	16	17	18	19	20	21	22
−20	−33																					
−18	−28																					
−16	−24																					
−14	−21	−36																				
−12	−14	−22																				
−10	−14	−22																				
−8	−12	−18	−29																			
−6	−10	−14	−22																			
−4	−7	−22	−17	−29																		
−2	−5	−8	−13	−20																		
0	−3	−6	−9	−15	−24																	
2	−1	−3	−6	−11	−17																	
4	1	−1	−4	−7	−11	−19																
6	4	1	−1	−4	−7	−13	−21															
8	6	3	1	−2	−5	−9	−14															
10	8	6	4	1	−2	−5	−9	−14														
12	10	8	6	4	1	−2	−5	−9	−16													
14	12	11	9	6	4	1	−2	−5	−10	−17												
16	14	13	11	9	7	4	1	−1	−6	−10	−17											
18	16	15	13	11	9	7	4	2	−2	−5	−10	−19										
20	19	17	15	14	12	10	7	4	2	−2	−5	−10	−19									
22	21	19	17	16	14	12	10	8	5	3	−1	−5	−10	−19								
24	23	21	20	18	16	14	12	10	8	6	2	−1	−5	−10	−18							
26	25	23	22	20	18	17	15	13	11	9	6	3	0	−4	−9	−18						
28	27	25	24	22	21	19	17	16	14	11	9	7	4	1	−3	−9	−16					
30	29	27	26	24	23	21	19	18	16	14	12	10	8	5	1	−2	−8	−15				
32	31	29	28	27	25	24	22	21	19	17	15	13	11	8	5	2	−2	−7	−14			
34	33	31	30	29	27	26	24	23	21	20	18	16	14	12	9	6	3	−1	−5	−12	−29	
36	35	33	32	31	29	28	27	25	24	22	20	19	17	15	13	10	7	4	0	−4	−10	
38	37	35	34	33	32	30	29	28	26	25	23	21	19	17	15	13	11	8	5	1	−3	−9
40	39	37	36	35	34	32	31	30	28	27	25	24	22	20	18	16	14	12	9	6	2	−2

Dew-Point Temperature Values

* See footnote to Table 15.3.

15.6 Condensation and Dew-Point Temperature

■ Relate dew point temperature to relative humidity and saturation mixing ratio.

If air is cooled below its dew-point temperature, water vapor *condenses* (changes to liquid) if the dew point is above freezing (**Figure 15.7**). *Deposition* (gas to solid) occurs if the dew-point temperature is below freezing. Water vapor needs a surface to condense on. In Figure 15.7, it is the beer mug. In the atmosphere, tiny particles called **condensation nuclei** are required for condensation to occur. Condensation (or deposition) results in the formation of dew or frost on the ground and clouds or fog in the atmosphere.

▲ **Figure 15.7** Condensation, or "dew," occurs when a cold beer mug chills the surrounding layer of air below the dew-point temperature. (Photo by Nitr/Fotolia)

ACTIVITY 15.6

Condensation and Dew-Point Temperature

1. Examine the process of condensation by filling a beaker approximately one-third full of water and then gradually adding ice. As you add the ice, stir the water–ice mixture gently with a thermometer. Note the temperature when water begins to condense on the outside surface of the beaker.

   a. At what temperature did water began to condense on the outside surface of the beaker?
   _____°C

   b. How does the temperature at which water began to condense compare to the dew-point temperature of the air in the room, which you determined earlier, using the psychrometer?
   _____

2. Refer to Table 15.2. How many grams of water vapor will condense on a surface if 1 kilogram of 50°F air with a relative humidity of 100 percent is cooled to 41°F?
   _____ g of water

3. Assume a situation in which 1 kilogram of 25°C air contains 10 grams of water vapor. Refer to Table 15.2 and determine how many grams of water will condense if the air temperature is lowered to each of the following temperatures.
   5°C: _____ g of condensed water
   −10°C: _____ g of condensed water

15.7 Daily Temperature Changes and Humidity

■ **Describe the typical daily pattern of variations in air temperature and relative humidity.**

As we saw earlier, when the water-vapor content of air remains constant, a decrease in air temperature results in an *increase* in relative humidity. Conversely, an increase in temperature causes a *decrease* in relative humidity. **Figure 15.8** shows a graph of typical daily variations in air temperature, relative humidity, and dew-point temperature during two consecutive spring days in a middle-latitude city.

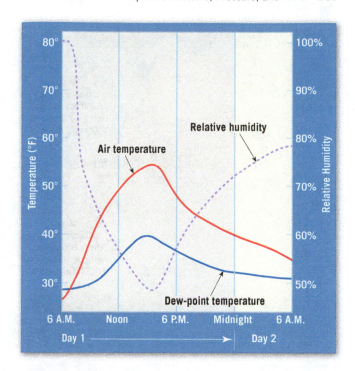

▶ **Figure 15.8** Typical variations in air temperature, relative humidity, and dew-point temperature during a 24-hour period in a middle-latitude city.

ACTIVITY 15.7

Daily Temperature Changes and Humidity

Use the graph in Figure 15.8 to complete the following.

1. On which day and at what time is relative humidity highest?
   Day: _____, time: _____

2. On which day and at what time does the lowest temperature occur?
   Day: _____, time: _____

3. On which day and at what time does the lowest relative humidity occur?
   Day: _____, time: _____

4. Write a general statement describing the relationship between temperature and relative humidity throughout the time period shown in the figure.
   _____
   _____

5. Did dew or frost occur on either of the two days represented in the graph? If so, indicate the day and time and explain how you arrived at your answer.
   _____
   _____

15.8 Adiabatic Processes and Cloud Formation

■ **Explain the process of adiabatic cooling and warming. Calculate the temperature and relative humidity changes that take place in air as a result of adiabatic temperature changes.**

Cloud formation occurs above Earth's surface when air is cooled below its dew-point temperature. The process responsible for most cloud formation is easy to visualize. You have probably experienced the cooling effect of a propellant gas as you applied hair spray or spray deodorant. As the compressed gas in the can is released, it quickly expands and cools. This drop in temperature occurs even though heat is neither added nor subtracted. Such changes, known as **adiabatic temperature changes**, result when air expands or when it is compressed. *When air is allowed to expand, it cools, and when it is compressed, it warms.*

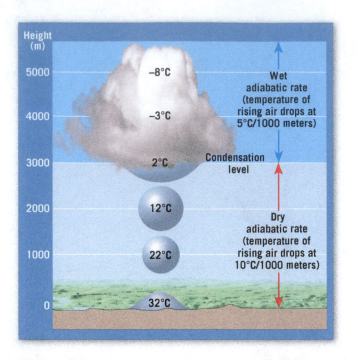

Because air pressure drops rapidly with increasing altitude above Earth's surface, rising air passes through regions of successively lower pressure. Therefore, ascending air expands and cools adiabatically. Rising *unsaturated* air cools at the constant rate of 10°C per 1000 meters (1°C per 100 m), called the **dry adiabatic rate**. The dry adiabatic rate also applies to descending air, which is compressed and warmed.

Once rising air reaches the dew-point temperature and water vapor condenses to form cloud droplets, latent heat that was stored in the water vapor is liberated. The heat released during condensation slows the rate of cooling of the air. Rising *saturated* air continues to cool by expansion, but at a lesser rate of about 5°C per 1000 meters (0.5°C per 100 m)—the **wet adiabatic rate**. **Figure 15.9** summarizes this cloud-forming process. Take a minute to carefully examine this figure before beginning Activity 15.8.

◄ **Figure 15.9** Adiabatic cooling and cloud formation. Rising air cools at the dry adiabatic rate of 10°C per 1000 meters, until the air reaches the dew-point temperature and condensation (cloud formation) begins. As air continues to rise, the latent heat released by condensation reduces the rate of cooling. The wet adiabatic rate is therefore always less than the dry adiabatic rate.

ACTIVITY 15.8
Adiabatic Processes and Cloud Formation

Figure 15.10 shows air that begins with a temperature of 25°C and a relative humidity of 50 percent flowing from the ocean over a coastal mountain range. Assume that the dew-point temperature remains constant in dry air (relative humidity less than 100 percent). When the air parcel becomes saturated, the temperature and dew-point temperature become equal. As the air continues to ascend, the temperature and dew-point temperature will decrease at the wet adiabatic rate, but the dew-point temperature does not change as the air parcel descends. (Completing Figure 15.10 may be helpful.)

1. Use Table 15.2 to determine the saturation mixing ratio, water-vapor content, and dew-point temperature of the air at sea level in Figure 15.10.

   Saturation mixing ratio: _____ g/kg of air

   Water-vapor content: _____ g/kg of air

   Dew-point temperature: _____ °C

2. Is the air at sea level saturated or unsaturated?

   _____

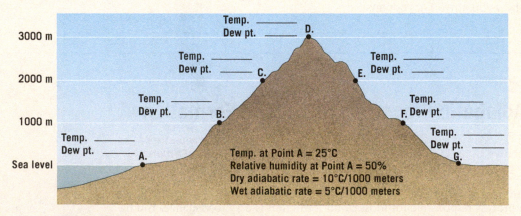

▲ **Figure 15.10** Adiabatic processes and cloud formation associated with air being lifted over a mountain barrier.

3. As the air moves up the windward side of the mountain, will it get cooler or warmer?

   _____

4. At what rate will the temperature of the rising air change?

   _____°C/1000 m

5. At what height will the rising air reach its dew-point temperature, causing the water vapor to begin to condense?

   _____ m

6. From the altitude where condensation begins until the air reaches the top of the mountain, the rising air will continue to expand and cool at a rate of about _____°C per 1000 meters. What term is applied to this rate of cooling?

   _____

7. Determine the air temperature and dew-point temperature of the rising air at the summit of the mountain.

   Temperature: _____°C

   Dew-point temperature: _____°C

8. If the air descends the leeward side of the mountain, will it expand or be compressed?

   _____

9. Will the temperature of the descending air increase or decrease?

   _____

10. Assume that the relative humidity of the air is below 100 percent during its descent to the valley. At what rate will the temperature change?

    _____°C/1000 m

11. As the air descends the leeward side of the mountain, will its relative humidity increase or decrease?

    _____

12. What will the air temperature and dew-point temperature be when the air reaches the valley on the leeward side of the mountain?

    Temperature: _____°C

    Dew-point temperature: _____°C

13. Explain why mountains often exhibit dry conditions on their leeward sides.

    _____

    _____

14. Explain why mountains often have abundant precipitation on their windward slopes.

    _____

    _____

15.9   Atmospheric Pressure

■ **Explain how air pressure is measured and how pressure tendency is related to cloudy or clearing weather.**

Atmospheric pressure is the force exerted by the weight of the atmosphere. The first instrument used for measuring air pressure was introduced in 1643, when Torricelli, a student of the famous Italian scientist Galileo, invented the **mercury barometer**. Torricelli correctly described the atmosphere as a vast ocean of air that exerts pressure on us and on all objects around us. To measure this force, he filled a glass tube, which was closed at one end, with mercury. The tube was then inverted into a dish of mercury (**Figure 15.11**). As Torricelli expected, the mercury flowed out of the tube until the weight of the column was balanced by the pressure exerted by the atmosphere on the surface of the mercury in the dish. Torricelli also discovered that when air pressure increases, the mercury in the tube rises. Conversely, when air pressure decreases, so does the height of the mercury column.

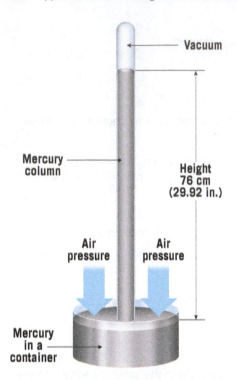

Figure 15.11 Diagram showing the principle of a mercury barometer. The weight of the column of mercury is balanced by the pressure exerted on the dish of mercury by the air above.

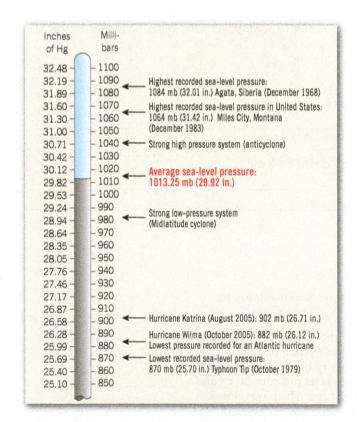

Figure 15.13 Scale for comparing pressure readings in millibars and inches of mercury.

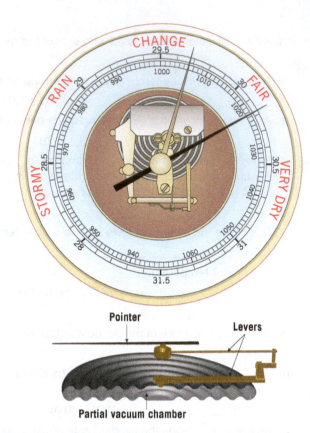

Figure 15.12 Aneroid barometer. The black pointer shows the current air pressure. When the barometer is read, the observer moves the other pointer to coincide with the current air pressure. Later, when the barometer is checked, the observer can see whether the air pressure has been rising, falling, or steady. The bottom diagram is a cross section. An aneroid barometer has a partially evacuated chamber that changes shape, compressing as outside air pressure increases and expanding as pressure decreases.

Figure 15.12 is a sketch of a smaller, more portable, instrument for measuring air pressure—the **aneroid barometer**. The face of an aneroid barometer is inscribed with words such as *fair, change, rain,* and *stormy.* Notice that "fair weather" corresponds with high-pressure readings, and "rain" is associated with low pressures. Although current barometer readings may indicate the present weather, this is not always the case. If you want to "predict" the weather in a local area, *the change in air pressure* over the past few hours (known as the **pressure tendency**) is much more useful than the current pressure reading. Most aneroid barometers have a dial that you can move (shown in gold on Figure 15.12) to coincide with the current reading. Later, when you check the instrument, you can see if the pressure has gone up or down. Falling pressure is usually associated with increasing cloudiness and the possibility of precipitation, whereas rising air pressure generally indicates clearing conditions.

Two units that are frequently used to express air pressure are *inches of mercury* and *millibars.* Inches of mercury refers to the height to which a column of mercury will rise. The millibar is a unit that expresses the actual force of the atmosphere pushing down on a surface. Standard pressure at sea level is 29.92 inches of mercury, or 1013.2 millibars (**Figure 15.13**). Since surface elevations vary, barometric readings are usually adjusted to indicate what the pressure would be if the barometer were located at sea level. This provides a common standard for mapping pressure distribution and comparing places, regardless of elevation.

Atmospheric Pressure

1. What is standard sea-level pressure, in millibars (mb) and inches of mercury (in Hg)?

_____ mb

_____ in Hg

2. When calculating the barometric pressure for a city located 200 meters *above* sea level, would you add or subtract units to its barometric reading in order to correct its pressure to sea level?

_____

3. What two instruments are used to measure air pressure?

_____   _____

4. Refer to Figure 15.12 and Figure 15.13 to answer the following.

   a. What is the current reading (black dial) on the instrument in Figure 15.12?

   _____ in Hg, or _____ mb

   b. Is the pressure tendency rising or falling? (*Hint:* Compare the black dial (current pressure) with the gold dial (previous pressure reading).)

   _____

5. Is falling pressure usually associated with increasing cloudiness or clearing conditions? Does rising air pressure generally indicate increasing cloudiness or clearing conditions?

Falling air pressure: _____

Rising air pressure: _____

15.10 The Driving Force of Wind: The Pressure Gradient Force

■ Describe the relationship between air pressure and wind.

Wind is the horizontal movement of air that results from horizontal differences in air pressure. *Air flows from areas of higher pressure toward areas of lower pressure. The greater the pressure differences, the stronger the wind speed.*

Over Earth's surface, variations in air pressure are determined from measurements taken at hundreds of weather stations. These data are plotted on maps using **isobars**—lines that connect places of equal air pressure (**Figure 15.14**). The spacing of isobars indicates the amount of pressure change occurring over a given distance and is termed the **pressure gradient**. *Closely spaced isobars indicate a steep pressure gradient and strong winds, whereas widely spaced isobars indicate a weak pressure gradient and light winds.*

Figure 15.14 is a simplified weather map that shows only the distribution of air pressure. Notice that isobars are seldom straight but instead form broad curves and elongated circles. If each succeeding isobar *decreases* in value going toward the center, the isobars are depicting a *low-pressure center*, or **cyclone**, usually labeled at the center with a large letter **L**. If the value of isobars *increases* toward the center, the concentric isobars are depicting a *high-pressure center*, or **anticyclone**, labeled at the center with a large letter **H**.

The Driving Force of Wind: The Pressure Gradient

Refer to Figure 15.14 to complete the following.

1. At what interval are isobars drawn on this map? That is, what is the difference, in millibars, between each succeeding isobar?

_____ mb

continued

Activity 15.10 continued

2. The map has two large areas of concentric isobars. Label the center of each as either a low-pressure center (*L*) or a high-pressure center (*H*).

3. Low-pressure centers are referred to as _____, while high-pressure centers are called _____.

4. Assume for a moment that the pressure gradient force is the *only* factor affecting winds. Draw arrows through Points A, B, C, and D that represent the wind direction. An arrow has already been drawn at Point E.

5. Which of the five stations (A–E) likely has the strongest winds? How were you able to figure this out?

_____

6. Identify a location on the map where winds should be weak. Mark the place with an *X*. Explain why you selected that location.

_____

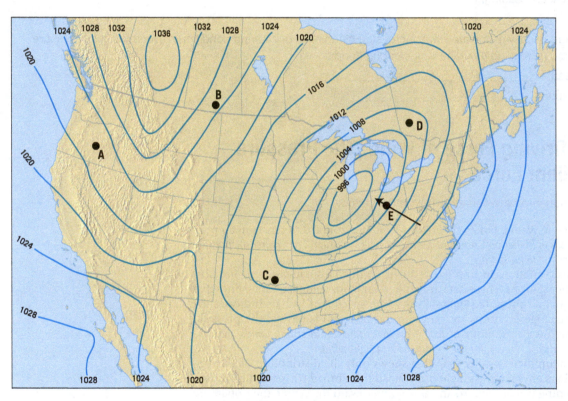

◀ **Figure 15.14**
Simplified weather map with isobars (labeled in millibars).

15.11 Factor Affecting Wind Direction: The Coriolis Effect

■ **Briefly explain how the Coriolis effect influences wind.**

In the preceding section, you learned that the pressure gradient force creates winds and strongly affects wind speed and direction. Another factor affecting wind is the **Coriolis effect**—the deflective effect of Earth's rotation. *All free-moving objects, including wind, are deflected to the right of their path of motion in the Northern Hemisphere and to the left in the Southern Hemisphere* (**Figure 15.15**). This deflecting effect influences only wind direction, not wind speed.

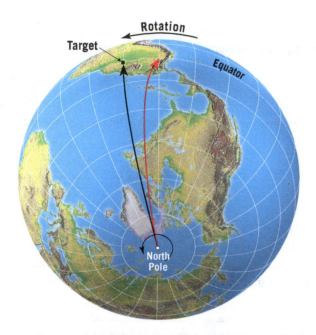

Rotation

Target

Equator

North
Pole

◀**SmartFigure 15.15** The deflection caused by Earth's rotation is illustrated by a rocket traveling for 1 hour from the North Pole to a location on the equator. On a nonrotating Earth, the rocket would fly straight to its target as shown by the black arrow. However, Earth rotates 15° each hour. Thus, although the rocket travels in a straight line, when we plot the path of the rocket on Earth's surface, it follows a curved path that veers to the right of the target, as shown by the red arrow. The video illustrates the Coriolis effect by using a playground merry-go-round.

VIDEO
https://goo.gl/RYDZKZ

ACTIVITY 15.11

Factor Affecting Wind Direction: The Coriolis Effect

To better understand how the Coriolis effect influences the motion of objects as they move across Earth's surface, conduct the following experiment:

Step 1: Working in groups of two or more, construct a rotating "table" that represents Earth's surface by first taping a thumbtack upside down on the table top (or inserting it through a piece of cardboard). Next, center a sheet of heavy-weight paper or thin cardboard over the point of the tack and push the sharp end of the tack through the paper (Figure 15.16). Turning the paper about the thumbtack represents the rotating Earth viewed from above the pole.

Step 2: Place a ruler (or other straight edge) across the paper, resting it on the sharp point of the thumbtack.

Step 3: Using the ruler as a guide and beginning at the edge nearest you, draw a straight line across the entire piece of paper. Mark the beginning of the line you drew with an arrow pointing in the direction the pencil moved. Label the line "no rotation."

Step 4: Have one person hold the ruler and another person turn the paper *counterclockwise* (the direction the Earth rotates when viewed from above the North Pole) about the thumbtack. Next, while spinning the paper at a slow, constant rate, draw a second line along the ruler, using a different-color pencil. Mark the beginning of the line you drew with an arrow pointing in the direction the pencil moved. Label the resulting line "Northern Hemisphere."

Step 5: Repeat Step 4 but this time rotate the paper *clockwise* (representing the Southern Hemisphere). Label the resulting line "Southern Hemisphere."

Step 6: Repeat Step 4 three more times, varying the speed of rotation of the paper with each trial. Identify each new line by labeling it either "slow," "fast," or "very fast."

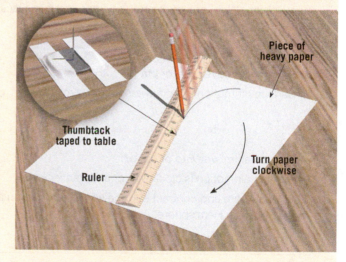

Piece of heavy paper

Thumbtack taped to table

Ruler

Turn paper clockwise

▲ **Figure 15.16** Coriolis effect experiment setup—Southern Hemisphere.

continued

Activity 15.11 continued

1. In the Northern Hemisphere, is the apparent path of a free-moving object deflected to the right or left? In the Southern Hemisphere, is the deflection to the right or left?

   Northern Hemisphere: _____

   Southern Hemisphere: _____

2. Summarize your observations regarding the rate of rotation and the magnitude of the Coriolis effect you observed in Step 6.

   _____

   _____

   _____

   _____

3. Considering what you have learned about the Coriolis effect, briefly describe the Coriolis effect on the atmosphere of the planet Venus, which is about the same size as Earth but has a period of rotation of 244 Earth days.

   _____

4. Compare the strength of the Coriolis effect on Jupiter with that on Earth. (*Note:* Jupiter is a planet much larger than Earth, with a 10-hour period of rotation.)

   _____

5. To summarize, examine surface wind directions associated with high- and low-pressure cells in both hemispheres by combining your understanding of pressure gradient and Coriolis effect. Refer to **Figure 15.17**. The concentric circles represent isobars. Add arrows to show the winds. The diagram for a low-pressure cell in the Northern Hemisphere is already completed. Complete the other three pressure cells with arrows that show the wind patterns.

6. In the following spaces, indicate the movement of air associated with high- and low-pressure cells in each hemisphere. Write one of the two choices given in italics for each blank.

   NORTHERN HEMISPHERE

	HIGH	LOW
Surface air moves *into* or *out of*	_____	_____
Surface air motion is *clockwise* or *counterclockwise*	_____	_____

   SOUTHERN HEMISPHERE

	HIGH	LOW
Surface air moves *into* or *out of*	_____	_____
Surface air motion is *clockwise* or *counterclockwise*	_____	_____

7. Briefly compare and contrast surface winds between an anticyclone in the Northern Hemisphere and an anticyclone in the Southern Hemisphere.

   _____

   _____

   _____

Northern Hemisphere

High (Anticyclone) Low (Cyclone)

Southern Hemisphere

High (Anticyclone) Low (Cyclone)

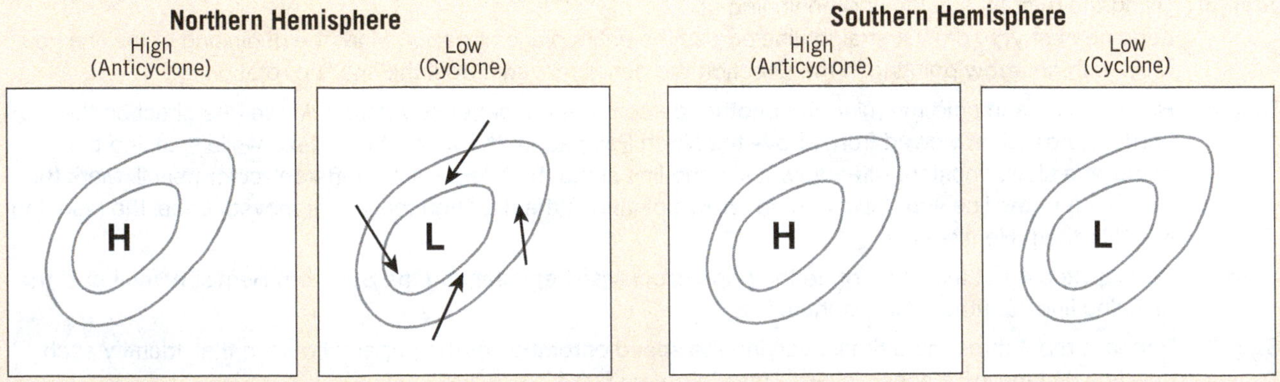

▲ **Figure 15.17** Northern and Southern Hemisphere high- and low-pressure cells.

15.12 Global Patterns of Precipitation

■ Relate the global distribution of precipitation to global pressure zones. List other factors that influence the distribution of precipitation.

Figure 15.18 shows the distribution pattern for average annual precipitation around the globe. Although this map may appear complicated, the general features of the pattern can be explained using knowledge of global winds and pressure systems.

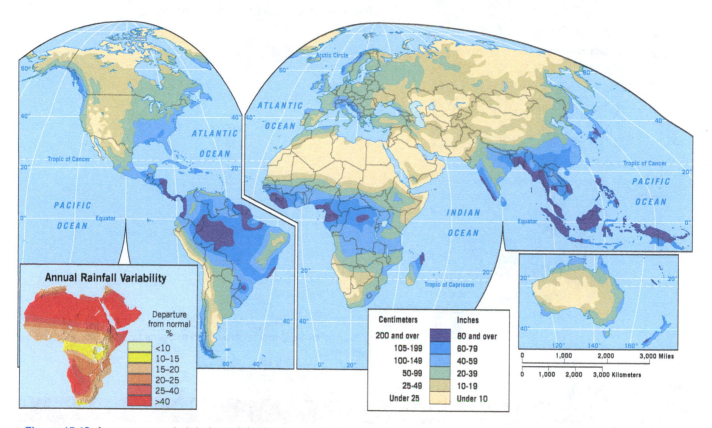

▲ Figure 15.18 Average annual global precipitation, in centimeters and inches.

We will first examine the zonal distribution of precipitation that we would expect on a uniform Earth covered entirely by water and then add the variations caused by the distribution of land and water. As shown in Figure 15.19, four major pressure zones emerge in each hemisphere. These zones are the *equatorial low*, the *subtropical high*, the *subpolar low*, and the *polar high*. In general, regions under the influence of low pressure, with their converging winds and ascending air, such as the equatorial low and subpolar low, receive ample precipitation. Conversely, regions influenced by high pressure, like the subtropical high and polar high, with their associated subsidence and divergent winds, experience dry conditions. Consequently, abundant precipitation occurs in the equatorial and midlatitude regions, whereas substantial portions of the subtropical and polar realms are relatively dry. As we will see in Exercise 16, much of the United States receives the bulk of its precipitation from fronts and associated cyclonic storms that travel eastward though the midlatitudes.

If the wind-pressure regimes were the only control of precipitation, the pattern shown in Figure 15.18 would be much simpler. Air temperature is also important in determining precipitation potential. Because cold air tends to contain less moisture than warm air, we would expect a latitudinal variation in precipitation, with low latitudes (tropical regions) receiving the greatest amounts of precipitation and high latitudes (cold regions) receiving the least.

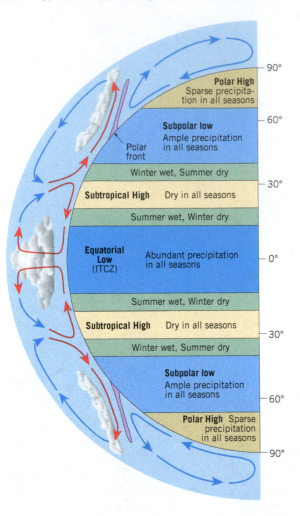

The distribution of land and water also complicates the precipitation pattern. Large landmasses in the middle latitudes commonly experience decreased precipitation toward their interiors. For example, North Platte, Nebraska, receives less than half the precipitation that falls on the coastal community of Bridgeport, Connecticut, despite being located at the same latitude. Furthermore, mountain barriers alter precipitation patterns. Windward mountain slopes tend to receive abundant precipitation, whereas the leeward side of a mountain range, such as Nevada's Great Basin, are usually deficient in moisture.

◀ **Figure 15.19** Zonal precipitation patterns for a uniform Earth.

ACTIVITY 15.12

Global Patterns of Precipitation

1. Are regions that are dry throughout the year dominated by high or low pressure?

_____

2. List two reasons that explain why polar regions experience meager precipitation.

_____

_____

3. Earth's major subtropical deserts are located between about 20° and 35° latitude.

   a. Referring to Figure 9.1, page 152, name five subtropical deserts.

   1. _____   **2.** _____   **3.** _____
   4. _____   **5.** _____

   b. Are the subtropical deserts you named in Question 3a located mainly on the western or eastern sides of continents?

   _____

Use Figure 15.18 to complete the following.

4. Place an X on four areas of the world that receive the greatest average annual (over 200 centimeters per year) precipitation. In what latitudinal zone are these locations found?

_____

5. What is the average annual precipitation at your location?

_____ cm per year, which is equivalent to _____ in per year

6. Describe the precipitation pattern in North America as you progress *westward* from the east coast along the 40th parallel (40°N latitude).

_____

_____

7. Describe the precipitation pattern in North America as you move northward from roughly New Orleans to northern Canada.

_____

_____

8. Describe three factors, in addition to global wind and pressure systems, that influence the global distribution of precipitation.

_____

_____

_____

9. The inset (left) in Figure 15.18 illustrates the annual variability, or reliability, of annual precipitation in Africa. After you compare the inset to the annual precipitation map for Africa, summarize the relationship between the amount of precipitation an area receives and the variability or reliability of that precipitation.

_____

_____

_____

MasteringGeology™

Looking for additional review and lab prep materials? Go to www.masteringgeology.com for Pre-Lab Videos, Geoscience Animations, RSS Feeds, Key Term Study Tools, The Math You Need, an optional Pearson eText, and more.

Notes and calculations:

Atmospheric Moisture, Pressure, and Wind

Name _____   Course/Section _____

Date _____   Due Date _____

1. Label the diagram in **Figure 15.20** with the appropriate terms for the changes of state shown.

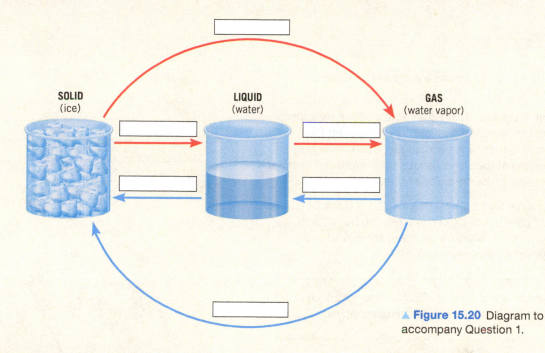

▲ **Figure 15.20** Diagram to accompany Question 1.

2. Complete the following.

    a. By what process does liquid water change to water vapor?

    _____

    b. Does warm or cold air have the highest saturation mixing ratio?

    _____

    c. Does lowering the air temperature increase or decrease relative humidity?

    _____

    d. What is the relative humidity of air that has reached its dew-point temperature?

    _____

    e. When condensation occurs, does water vapor absorb or release heat?

    _____

    f. Does rising air warm or cool, and does it do so by expansion or compression?

    _____ by _____

    g. In the early morning hours, when the daily air temperature is often coolest, is relative humidity generally at its lowest or highest?

    _____

3. Explain how a sling psychrometer works for determining relative humidity.

    _____

    _____

    _____

4. Describe the adiabatic process and explain how it is responsible for causing clouds to form.

    _____

    _____

    _____

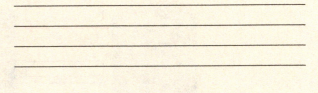

▲ **Figure 15.21** Diagram to accompany Question 5.

5. Label the sketch in **Figure 15.21**, using the appropriate terms from the following list: leeward (dry); warming at the dry adiabatic rate; condensation begins; cooling at the wet adiabatic rate; windward (wet); and cooling at the dry adiabatic rate.

_____

_____

_____

_____

6. Assume that a parcel of air at the surface has a temperature of 29°C and a relative humidity of 50 percent. If the parcel rises, at what altitude should clouds begin to form?

_____

7. Describe the Coriolis effect and its influence in both the Northern and Southern Hemispheres.

_____

_____

_____

8. Low-pressure centers are also referred to as _____, while high-pressure centers are called _____.

9. Assume that you are an observer checking the aneroid barometer shown in **Figure 15.22** several hours after it was last checked. What is the pressure tendency? How did you figure this out? What does the tendency shown on the barometer indicate about forthcoming weather?

▲ **Figure 15.22** Diagram to accompany Question 9.

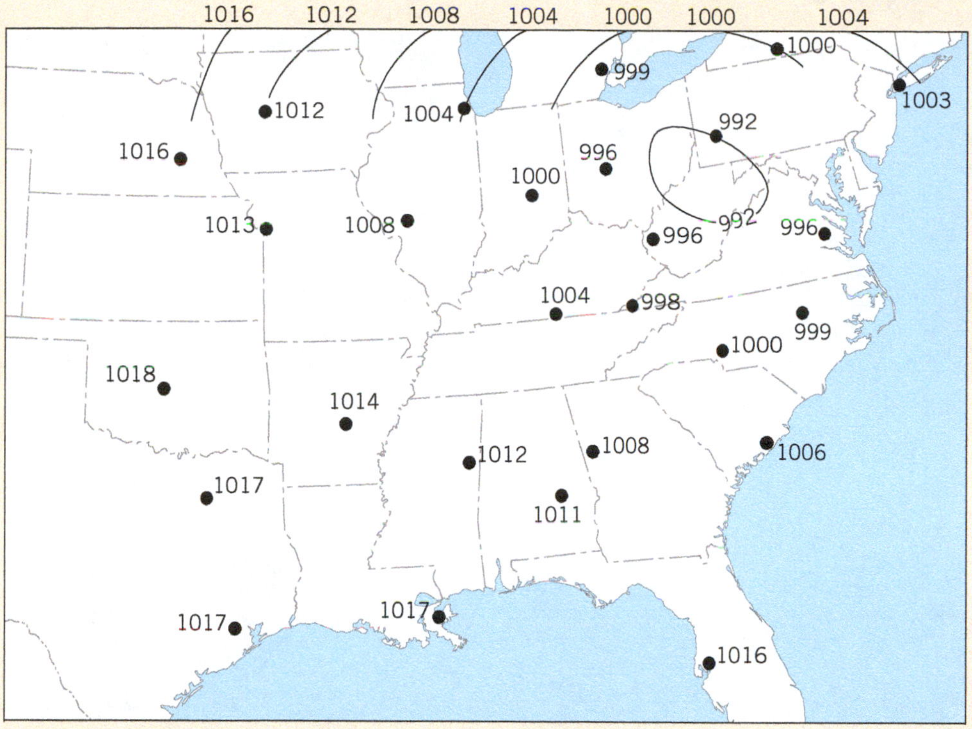

1016   1012   1008   1004   1000   1000   1004
1000
999
1003
1012   1004
1016
992
996
1000
1013   1008
996   992   996
1004   998
1000   999
1018
1014
1012   1008   1006
1017
1011
1017   1017
1016

▲ **Figure 15.23** Atmospheric pressures for select cities.

10. Complete **Figure 15.23** by using the indicated surface barometric pressures to assist in drawing appropriate isobars. Begin with the 996 mb isobar and use a 4 mb interval between successive isobars. (Refer to Activity 7.7, page 122, if you need help drawing isobars. Isobars are a type of isoline, like contour lines, that are used to connect points of equal value.)

11. Are low-pressure systems (cyclones) usually associated with fair weather or rainy conditions? Are high-pressure systems (anticyclones) generally associated with fair weather or rainy conditions?

Low pressure: _____ conditions

High pressure: _____ conditions

12. The sketch in **Figure 15.24** shows a cross section of the pressure belts in the Northern Hemisphere. Match the appropriate number from the sketch to each of the following features.

a. Equatorial low: _____

b. Subpolar low: _____

c. Subtropical high: _____

d. Polar high: _____

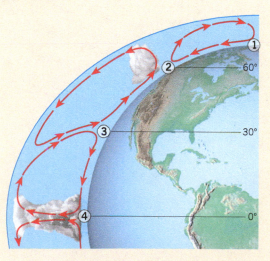

▲ **Figure 15.24** Diagram to accompany Question 12.

Air Masses, Midlatitude Cyclones, and Weather Maps

LEARNING OBJECTIVES

Each statement represents an important learning objective that relates to one or more sections of this lab. After you complete this exercise you should be able to:

- List the characteristics and source regions of the air masses that influence North America.
- Define and sketch a profile of a typical warm front and a typical cold front.
- Diagram and label all parts of an idealized, mature midlatitude cyclone.
- Interpret the data presented on a surface weather map.
- Prepare a simple surface weather map and use it to make a short-range forecast.

MATERIALS

colored pencils   U.S. map or atlas

PRE-LAB VIDEO https://goo.gl/7BjBYs

 Prepare for lab! Prior to attending your laboratory session, view the pre-lab video. Each video provides valuable background that will contribute to your understanding and success in lab.

INTRODUCTION

In this exercise you will study the atmospheric phenomena that most often affect our day-to-day weather: air masses, fronts, and traveling midlatitude cyclones. You will see the interplay of the elements of weather encountered in Exercises 13, 14, and 15 and examine how these elements are displayed on a weather map.

PART 3 Meteorology

16.1 **Air Masses**

■ **List the characteristics and source regions of the air masses that influence North America.**

An **air mass** is an immense body of air, typically 1600 kilometers (1000 miles) or more wide and perhaps several kilometers thick, that is characterized by its temperature and moisture content. The area where an air mass acquires its characteristic properties is called its **source region**. An air mass moving from its source region may affect a large portion of a continent.

Air masses are classified according to their source region (**Figure 16.1**). **Polar (P) air masses** originate in high latitudes and are composed of cool or cold air, whereas **tropical (T) air masses** originate in low latitudes and are warm by comparison. If its source region is over land, an air mass tends to be dry and is referred to as being **continental (c)**. If an air mass forms over the ocean, it is more humid and is classified as **maritime (m)**. Using this classification, four basic types of air masses emerge: **continental polar (cP)**, **continental tropical (cT)**, **maritime polar (mP)**, and **maritime tropical (mT)**.

▶ **Figure 16.1** Source regions for air masses that influence North America.

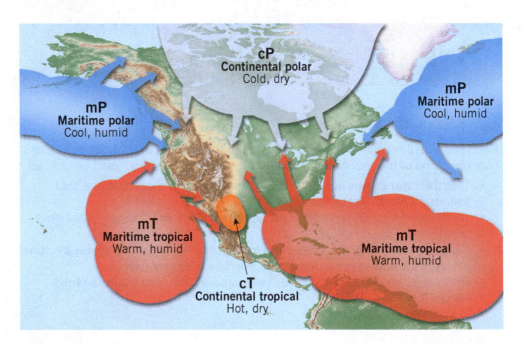

ACTIVITY 16.1 ///

Air Masses

Refer to Figure 16.1 to answer the following questions.

1. What temperature—warm, cool, hot, or cold—and what moisture conditions—dry or humid—are associated with the four basic types of air masses?

   cP: _____ and _____

   mP: _____ and _____

   cT: _____ and _____

   mT: _____ and _____

2. Which two air masses have the greatest impact on the Pacific coast of North America?

   _____ and _____

3. The southwestern United States and northwestern Mexico is the source region for what type of air mass?

   _____

4. What type of air mass originates in the northern reaches of the North Atlantic Ocean?

_____

5. What type of air mass originates mainly in northern Canada?

_____

6. Which two air masses appear to have the greatest impact on the midlatitudes east of the Rocky Mountains?

_____ and _____

16.2 Fronts

■ Define and sketch a profile of a typical warm front and a typical cold front.

Fronts are boundaries that separate different air masses, one warmer and often having a higher moisture content than the other. Generally, an air mass on one side of a front moves faster relative to the frontal boundary than the air mass on the other side. As a result, one air mass actively advances into another. Limited mixing occurs along the frontal boundary. Air masses retain their distinct identities as one is displaced upward over the other, in a process called *overrunning*. No matter which air mass is advancing, *it is always the warmer (less dense) air that is forced aloft*, whereas *the cooler (denser) air acts as the wedge on which lifting takes place*. When air is forced aloft, adiabatic cooling and cloud formation are to be expected.

When the surface position of a front moves so that warm air occupies territory formerly covered by cooler air, it is called a **warm front** (**Figure 16.2A**). By contrast, when dense cold air is actively advancing into a region occupied by warmer air, the boundary is called a **cold front** (**Figure 16.2B**). On average, cold fronts are about twice as steep as warm fronts and travel at speeds up to 50 percent faster than warm fronts. These two differences—faster rate of movement and a steeper boundary—largely account for the fact that the weather along a cold front is more violent than the weather that generally accompanies a warm front. Also, because warm fronts advance more slowly, the precipitation generated is light to moderate but may last for a day or longer.

A third type of front, called an **occluded front**, is generated when a cold front overtakes a warm front (**Figure 16.3**). As the advancing cold air wedges the warm front upward, a new front emerges between the advancing cold air and the cool air over which the warm front is gliding. Most of the clouds and precipitation associated with occluded fronts are generated within the layer of warm air that is being forced aloft by the advancing cold air (Figure 16.3).

▼ **SmartFigure 16.2** Profiles of **A.** a warm front and **B.** a cold front.

ANIMATION https://goo.gl/CSP9wC

ANIMATION https://goo.gl/Qi7ZJS

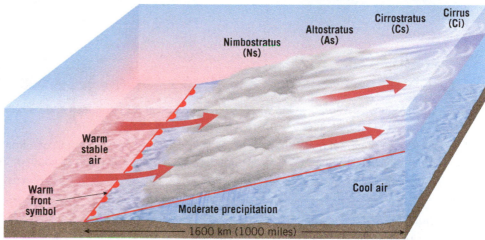
A. Warm front

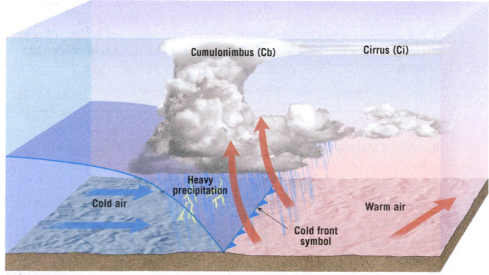
B. Cold front

◄ **Figure 16.3** Stages in the formation of an occluded front. **A.** The air behind the cold front is colder (denser) than the cool air ahead of the warm front. **B.** The cold front moves faster than the warm front and overtakes it to form an occluded front. **C.** The denser cold air lifts both the warm air and the cool air that the warm air was overrunning.

ACTIVITY 16.2

Fronts

1. On Figure 16.1, draw a line where continental polar (cP) air mass and maritime tropical (mT) air masses are likely to collide and, as a result, where a front is likely to develop.

2. Will a continental polar air mass most likely be found north or south of a front east of the Rocky Mountains in the central United States? Will a maritime tropical air mass lie north or south of the front?

   Continental polar air masses usually lie _____ of a front.

   Maritime tropical air masses usually lie _____ of a front.

The following questions refer to Figure 16.2 and Figure 16.3, which illustrate a typical warm front, cold front, and occluded front.

3. Does warm air rise at the steepest angle along a cold or warm front?

   _____ front

4. Are extensive areas of nimbostratus clouds and periods of prolonged precipitation most likely to occur along a cold or warm front? Explain why you expect longer periods of precipitation to be associated with this type of front.

   _____

   _____

   _____

5. Assume that the fronts are moving from left to right in Figure 16.2A,B. Is a drop in temperature most likely to occur with the passage of a cold or warm front?

   _____

6. Is a cold or warm front most likely to produce clouds of vertical development and perhaps thunderstorms?

   _____

7. Do clouds gradually become lower, thicker, and cover more of the sky as a cold front approaches or as warm front approaches?

   _____

©2019 Pearson Education, Inc.

8. During the development of an occluded front, is the cold or warm front advancing most rapidly?

_____

9. What happens to the warm (mT) air during the development of an occluded front? (*Hint:* See Figure 16.3.)

_____

10. Sketch the symbols for a cold front, a warm front, and an occluded front in the space below. (*Hint:* See Figures 16.2 and 16.3.)

Cold front:

Warm front:

Occluded front:

16.3 | Midlatitude Cyclones

■ **Diagram and label all parts of an idealized, mature midlatitude cyclone.**

To understand day-to-day weather patterns in the continental United States and large parts of Canada, it is useful to have an understanding of **middle-latitude cyclones**, also called **midlatitude cyclones**. These large centers of low pressure generally travel from west to east and last from a few days to more than a week. Most midlatitude cyclones contain a cold front and a warm front, and when they reach maturity, they may develop an occluded front that extends from the central area of low pressure. Convergence, which causes air to rise, and forceful lifting along fronts initiate cloud development associated with these pressure systems. In addition, abundant precipitation typically occurs at locations along the paths of these storm systems.

ACTIVITY 16.3
Midlatitude Cyclones

Figure 16.4 depicts an idealized, mature middle-latitude cyclone. Refer to this figure to complete the following questions.

1. Are the surface winds in a cyclone converging or diverging?

_____

2. Is the air in the center of the cyclone subsiding or rising? What effect will this have on the potential for condensation and precipitation? Explain your answer.

_____

_____

3. On Figure 16.4 label the cold front, warm front, and occluded front.

continued

Activity 16.3 continued

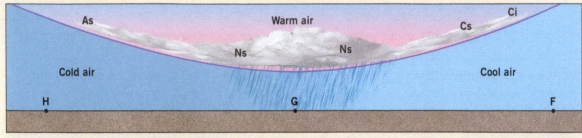

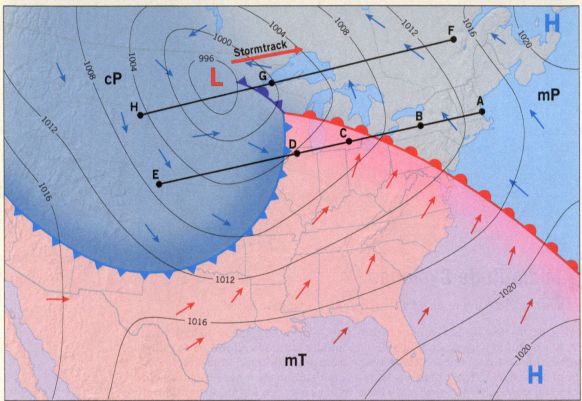

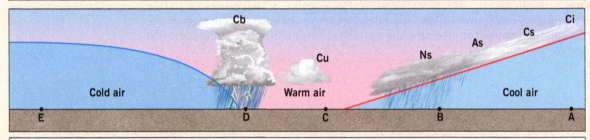

KEY: **As** (Altostratus), **Cb** (Cumulonimbus), **Ci** (Cirrus), **Cs** (Cirrostratus), **Cu** (Cumulus), **Ns** (Nimbostratus)

▲ **SmartFigure 16.4** Idealized drawing of a mature midlatitude cyclone, showing isobars, wind direction, and fronts.

ANIMATION
https://goo.gl/74HJB9

4. Which location (A–H) is most likely receiving precipitation associated with the cold front? Warm front? Occluded front?

Cold front: _____

Warm front: _____

Occluded front: _____

5. Considering the air mass types and their locations in a midlatitude cyclone, describe the likely weather at each of the locations listed below, using two of the following words: *cold, cool, warm, dry,* and *humid.*

Location A: _____ and _____

Location C: _____ and _____

Location E: _____ and _____

Location F: _____ and _____

Location H: _____ and _____

6. Describe how the barometric pressure will change at Point F as the midlatitude cyclone moves eastward.

_____

7. Describe how the wind will shift (change direction) as the cold front passes Point D.

The wind will shift from _____ to _____.

8. As the cold front passes Point D, will the temperature rise or fall?

_____

9. Will the water vapor content of the air at Point D most likely increase or decrease after the cold front passes?

_____

10. As the warm front passes Point B, will the temperature rise or fall?

_____

11. Will the amount of water vapor in the air most likely increase or decrease at Point B after the warm front passes?

_____

12. Describe the change in wind direction that will likely occur at Point B as the warm front passes.

_____

13. Complete **Table 16.1** by providing the current conditions for each of the stations listed (A–E).

14. Refer to your completed version of Table 16.1 and briefly describe the temperature changes that occurred at Point A as the middle-latitude cyclone passed by.

_____

_____

Table 16.1 Sequence of Weather

STATION	TEMPERATURE (WARM, COOL, OR COLD)	PRESSURE TENDENCY (RISING, FALLING, STEADY)	WIND DIRECTION	CLOUD TYPE	PRECIPITATION (YES OR NO)
A	_____	_____	_____	_____	_____
B	_____	_____	_____	_____	_____
C	_____	_____	_____	_____	_____
D	_____	_____	_____	_____	_____
E	_____	_____	_____	_____	_____

16.4 Weather Station Analysis and Forecasting

■ Interpret the data presented on a surface weather map.

To manage the great quantity of data needed for accurate weather maps, meteorologists developed a system for coding weather data **Figure 16.5** (page 282) shows that system and many of the symbols that are used to record data for a weather station.

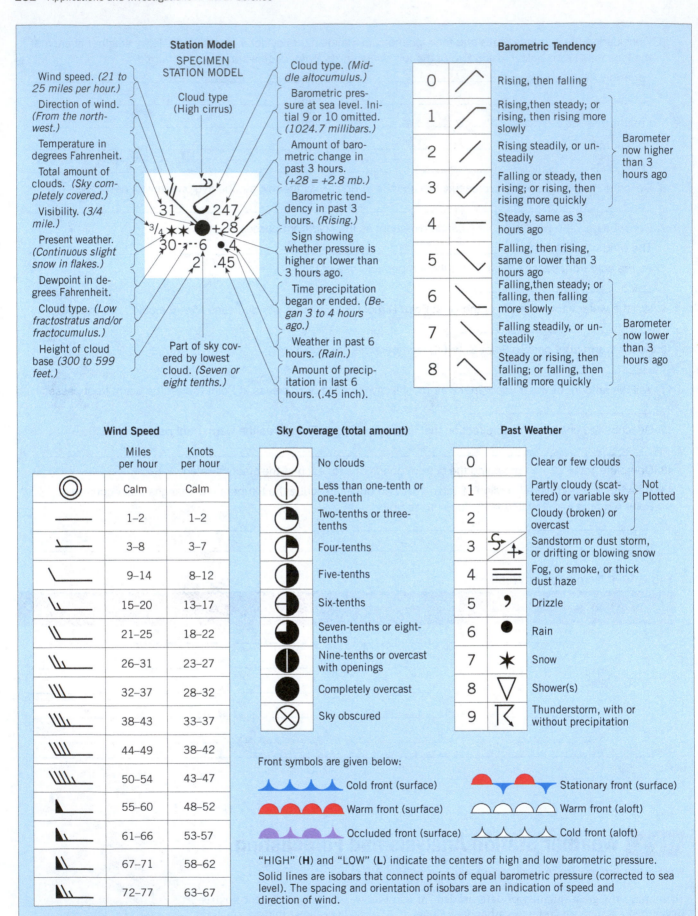

▲ **Figure 16.5** Weather station model and standard symbols. (National Weather Service)

When plotting barometric pressure in millibars, the initial number 9 or 10 is omitted (to conserve space), and the last digit refers to tenths of a millibar. For example, on a map, a barometric pressure of 216 would be read as 1021.6 mb. For numbers that begin with 0–4, add 10, and for 5–9, add 9.

ACTIVITY 16.4
Weather Station Analysis

Figure 16.6 is a coded weather station, shown as it would appear on a simplified surface weather map.

1. Using the station model and explanations shown in Figure 16.5 as a guide, interpret the weather conditions reported at the station illustrated in Figure 16.6.

   Percentage of sky cover: _____%

   Wind direction: _____

   Wind speed: _____ mph

   Temperature: _____ °F

   Dew-point temperature: _____ °F

   Barometric pressure: _____ mb

   Barometric tendency in the past 3 hours: _____ (rising, falling, or steady)

   Weather during the past 6 hours: _____

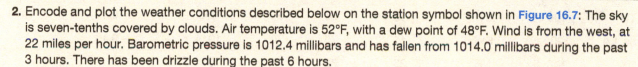

▲ **Figure 16.6** Coded weather station (abbreviated).

2. Encode and plot the weather conditions described below on the station symbol shown in Figure 16.7: The sky is seven-tenths covered by clouds. Air temperature is 52°F, with a dew point of 48°F. Wind is from the west, at 22 miles per hour. Barometric pressure is 1012.4 millibars and has fallen from 1014.0 millibars during the past 3 hours. There has been drizzle during the past 6 hours.

   Use Figures 16.5 and 16.6 as guides.

▲ **Figure 16.7** Station symbol for plotting weather conditions provided in Question 2 of Activity 16.4.

16.5   Preparing a Weather Map and Forecast

■ **Prepare a simple surface weather map and use it to make a short-range forecast.**

Table 16.2 displays weather data for several cities in the central and eastern United States on a December day. Data for several of the cities have been plotted on the map in Figure 16.8 (page 285). (*Note:* Complete barometric pressure readings have been supplied to simplify the process.)

Table 16.2 December Surface Weather Data for Selected Cities in the Central and Eastern United States

STATION	% CLOUD COVER	WIND DIRECTION	WIND SPEED (MPH)	TEMP. (°F)	DEW PT. TEMP.	PRESSURE* (MB)	PRESSURE PAST 3 HOURS (MB)	PRECIP.
Atlanta, GA	20	S	18	61	56	1004	−2.0	
Birmingham, AL	**80**	**SW**	**15**	**70**	**64**	**1004**	**−1.4**	
Charleston, SC	10	S	12	63	58	1008	−2.0	
Charlotte, NC	**70**	**SW**	**14**	**60**	**54**	**1002**	**−4.4**	
Chattanooga, TN	60	SW	12	66	60	1000	−2.5	Drizzle
Chicago, IL	100	NE	13	34	21	1004	−2.2	
Columbus, OH	100	E	10	34	28	996	−5.8	Snow
Evansville, IN	100	NW	7	45	43	997	−2.7	Snow
Fort Worth, TX	0	NW	5	46	43	1011	+1.4	
Indianapolis, IN	**100**	**NE**	**30**	**34**	**32**	**996**	**−5.6**	**Snow**
Jackson, MS	40	SW	10	72	67	1008	+1.2	Thunderstorm
Kansas City, MO	30	N	18	30	27	1010	+1.7	
Little Rock, AR	0	NW	10	46	43	1006	+2.7	
Louisville, KY	100	E	12	34	34	992	−4.2	Snow
Memphis, TN	**80**	**NW**	**12**	**50**	**45**	**1003**	**+5.8**	
Mobile, AL	60	SW	10	72	68	1011	−0.3	
Nashville, TN	**100**	**SW**	**18**	**56**	**55**	**996**	**−0.1**	**Rain**
New Orleans, LA	**20**	**SW**	**11**	**75**	**70**	**1014**	**−0.1**	
New York, NY	100	NE	23	36	18	1014	−2.1	
Oklahoma City, OK	10	NW	13	41	37	1012	+1.5	
Raleigh, NC	100	SW	14	54	52	1005	−0.6	Rain
Richmond, VA	100	E	10	45	45	1005	−2.4	Rain
Roanoke, VA	**100**	**SE**	**10**	**39**	**39**	**1000**	**−3.6**	**Snow**
Savannah, GA	30	SE	7	61	55	1010	−2.0	
Shreveport, LA	**0**	**NW**	**8**	**46**	**43**	**1009**	**+1.4**	
Springfield, MO	10	NW	9	36	20	1008	+3.2	
St. Louis, MO	**100**	**NW**	**10**	**32**	**32**	**1004**	**+1.7**	**Showers**

* *Note:* Barometric pressure readings have been rounded to the nearest whole number for simplification.

ACTIVITY 16.5
Preparing a Weather Map and Forecast

1. Plot the data in Table 16.2 for the stations shown in boldface on the map in Figure 16.8. Refer to the station model in Figure 16.5 if needed. (*Note:* Plot the barometric pressure readings as written.)

2. Use the map in Figure 16.8 to complete the following steps:

   Step 1: Beginning with 996 millibars, draw and label the appropriate isobars as accurately as possible at 4-millibar intervals (996 mb, 1000 mb, 1004 mb, etc.). The 992-millibar isobar has been completed, as have partial isobars for 1008 and 1012 millibars. (*Note:* You will have to estimate pressures between cities to determine the locations of isobars. Also, it is a good idea to first lightly sketch the isobars in pencil.)

continued

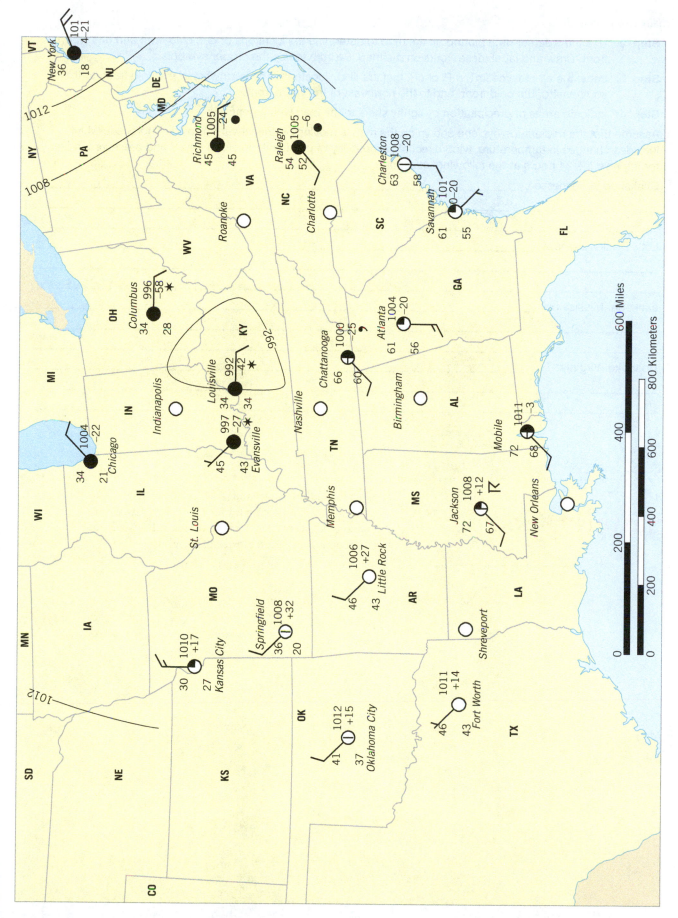

▲ **Figure 16.8** Weather map for a December day for selected cities in the central and eastern United States.

Activity 16.5 continued

Step 2: Use the weather data plotted on the map to determine the locations of one cold front and one warm front. Draw and label these fronts on the map. Be sure to use the proper symbols.

Step 3: Label the air masses (mT, mP, or cP) that are likely located to the northwest of the cold front, to the southeast of the cold front, and to the northeast of the warm front.

Step 4: Indicate areas of precipitation by lightly shading the map with a pencil.

3. Assume that the midlatitude cyclone shown on the map is moving northeastward. Provide a forecast, which includes changes in temperature, wind direction, probability of precipitation, cloud cover, and pressure tendency, for the next 12–24 hours at the following locations:

Chattanooga, Tennessee: _____

_____

_____

Little Rock, Arkansas: _____

_____

_____

Jackson, Mississippi: _____

_____

_____

Roanoke, Virginia: _____

_____

_____

MasteringGeology™

Looking for additional review and lab prep materials? Go to www.masteringgeology.com for Pre-Lab Videos, Geoscience Animations, RSS Feeds, Key Term Study Tools, The Math You Need, an optional Pearson eText, and more.

Air Masses, Midlatitude Cyclones, and Weather Maps

Name _____

Date _____

Course/Section _____

Due Date _____

1. List the source region(s) and winter temperature/ moisture characteristics of each of the following North American air masses:

   cP: _____

   mT: _____

   mP: _____

2. In the following space, sketch a profile (side view) of a cold front. Label the cold air and warm air and sketch the probable cloud type at the appropriate location. Draw an arrow on the diagram to show the direction the front is moving.

   [ ]

3. Indicate the type of front, cold or warm, that is associated with each of the following statements.

   a. Advancing wall of cold air: _____

   b. Warm air replaces cool air: _____

   c. Thunderstorms: _____

   d. Drop in temperature: _____

   e. After the front passes, winds switch to a southwesterly direction: _____

   f. Narrow belt of heavy precipitation: _____

   g. Gradual lifting of warm air over cool air: _____

4. Describe the sequence of weather events that a city would experience with the passage of an idealized, mature midlatitude cyclone that has not developed an occluded front. Assume that the center of the cyclone is moving west to east and is 150 miles to the north of the city. Sketching a diagram may be helpful.

   _____

   _____

   _____

   _____

5. Based on the weather map that you constructed in Activity 16.5, how will the weather at Jackson, Mississippi, change over the next 12–24 hours?

   _____

   _____

continued

6. Using **Figure 16.9**, briefly describe the current weather conditions in each of the following areas:

Southern Minnesota: _____

_____

_____

Pennsylvania: _____

_____

_____

Southern Georgia: _____

_____

_____

_____

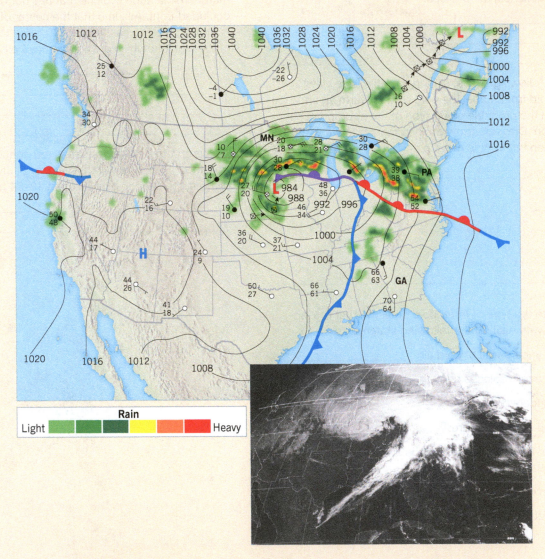

▲ **Figure 16.9** Weather map for a March day. Inset is a satellite image showing the cloud patterns that day. (Courtesy of NOAA)

Global Climates

LEARNING OBJECTIVES

Each statement represents an important learning objective that relates to one or more sections of this lab. After you complete this exercise you should be able to:

- List the six major controls of climate and their effects.
- Read climographs to interpret data on temperature and precipitation.
- Understand the nature of classification systems.
- List the criteria used to define the five principal climate groups in the Köppen system of classification.
- Describe the general locations and characteristics of the principal climate groups.
- Use climate data to classify climates according to the Köppen system.

MATERIALS

ruler   calculator
colored pencils   world map or atlas

PRE-LAB VIDEO https://goo.gl/fKmiFk

 Prepare for lab! Prior to attending your laboratory session, view the pre-lab video. Each video provides valuable background that will contribute to your understanding and success in lab.

INTRODUCTION

The focus of this exercise is **climate**, the long-term aggregate of weather. Climate is more than just an expression of average atmospheric conditions. In order to accurately portray the character of a place or an area, variations and extremes must also be included. We find such an incredible variety of climates around the world that it is hard to believe they could all occur on the same planet.

PART 3 Meteorology

17.1 Controls of Climate

■ List the six major controls of climate and their effects.

Before you examine Earth's major climates, it will be worthwhile to review the primary factors that control climate:

- **Latitude** Because latitude determines the average amount of solar radiation received at a location, it is the most influential control of temperature. Places near the equator receive more solar radiation on average than do places near the poles. In addition, variations in the amount of solar radiation received at Earth's surface, which result from seasonal changes in Sun angle and length of daylight, are also latitude dependent. As a result, temperatures in the tropical realm are consistently high because the Sun's vertical rays are never far away, and the length of daylight is relatively constant throughout the year. As one moves poleward, greater seasonal fluctuations in the receipt of solar energy are reflected in greater annual temperature ranges.

- **Land and water** The distribution of land and water is the second most important control of temperature. Land heats more rapidly and to higher temperatures than does water, and it cools more rapidly and to lower temperatures than does water. Consequently, variations in air temperatures are much greater over land than over water. This differential heating of Earth's surface has led to climates being divided into two broad classes: marine and continental. **Marine climates** are considered relatively mild for their latitude because of the moderating effect of large water bodies; summers are cooler and winters are warmer in marine locations than in comparable continental locations. In contrast, **continental climates** tend to be much more extreme. Although a marine station and a continental station along the same parallel in the middle latitudes may have similar annual mean temperatures, the annual temperature *range* will be far greater at the continental station.

- **Geographic position and prevailing winds** To fully understand the influence of land and water on the climate of an area, the position of that area on the continent and its relationship to the prevailing winds must be considered. The moderating influence of water is much more pronounced along the windward side of a continent, where the prevailing winds may carry the maritime air masses far inland. Places on the leeward side of a continent, where prevailing winds move from land toward the ocean, are likely to have a more continental temperature regime.

- **Mountains and highlands** Mountains and highlands play an important part in the distribution of climates. This impact can be illustrated by examining western North America, where prevailing winds are from the west and where the mountain chains that trend north–south are major barriers. These topographic barriers prevent the moderating influence of maritime air masses from reaching far inland. In addition, they trigger rainfall on their windward slopes, often leaving a dry rain shadow on the leeward side.

    Extensive highlands create their own climatic regions. Because of the drop in temperature with increasing altitude, areas such as the Tibetan Plateau and the highlands of Bolivia are cooler and drier than their latitudinal locations alone would indicate.

- **Ocean currents** The effect of ocean currents on the temperatures of adjacent land areas can be significant. Warm, poleward-moving currents, such as the Gulf Stream, cause winter air temperatures to be warmer than would be expected, particularly in coastal areas at high latitudes. Conversely, cold currents such as the California Current along the west coast of the United States moderate summer temperatures.

- **Pressure and wind systems** The world distribution of precipitation shows a close relationship to the distribution of Earth's major pressure and wind systems. In the realm of the equatorial low, the convergence of warm, moist, unstable air makes this zone one of heavy rainfall. In the regions dominated by the subtropical highs, general aridity prevails, creating major deserts. Farther poleward, in the zone dominated by the irregular subpolar low, the influence of frequent traveling midlatitude cyclones increases precipitation. In polar regions, where temperatures are low and the air can hold only small quantities of moisture, precipitation totals decline.

ACTIVITY 17.1A
Controls of Climate

1. List the six major climate controls.

   _____   _____

   _____   _____

   _____   _____

2. Which climate control has the greatest influence on global temperatures?

   _____

3. Contrast continental and marine climates.

   _____

4. Describe the connection between pressure systems and the world distribution of precipitation.

   _____

5. Describe how a mountain system affects the distribution of precipitation on its windward side as compared to its leeward side.

   _____

6. The contiguous 48 states are located in the midlatitudes, where the westerlies are the dominant wind system. Should we expect a more pronounced marine climate along the west or east side of the continent?

   _____

7. Do low-latitude locations experience larger or smaller annual temperature ranges than high-latitude locations?

   _____

8. Would you expect a warming or cooling effect from a poleward-moving (northward-moving in the Northern Hemisphere) ocean current in winter?

   _____

9. Briefly describe the amount of precipitation that would be expected for locations under the influence of each of the following pressure and wind systems. (Refer to Figure 15.19, page 268.)

   Equatorial low: _____

   Subtropical high: _____

   Subpolar low: _____

   Polar high: _____

10. Which one of the pressure zones listed in Question 9 is most influential in the middle latitudes?

    _____

ACTIVITY 17.1B
Global Temperatures

Figure 17.1 is a graph that shows temperature data for New York City. Refer to it to answer Questions 1 and 2.

1. a. What is the record high maximum temperature for the entire year?

   _____°C (_____°F)

   b. How does the record high compare to the average high temperature for the same month?

   _____°C (_____°F) higher

2. What are the values of the record high temperature and the record low temperature for the month of January?

   Record January high: _____ °C (_____ °F)

   Record January low: _____ °C (_____ °F)

   Complete Questions 3–5 using the temperature data in **Table 17.1**.

continued

Activity 17.1B continued

3. Stations 1–3 in Table 17.1 are representative of three North American cities at approximately 40°N latitude. Choose which station represents each of the following locations and give the reason for your selection.

Interior of the continent: _____

_____

West coast of the continent: _____

_____

East coast of the continent: _____

_____

4. Selecting from Stations 4–10, complete the following.

a. Which of these stations is in the Southern Hemisphere? How did you determine your answer?

_____

_____

b. Identify two stations that must be very near the equator. How did you determine your answer?

_____

_____

c. Which station must be a high-altitude location? Explain your answer.

_____

_____

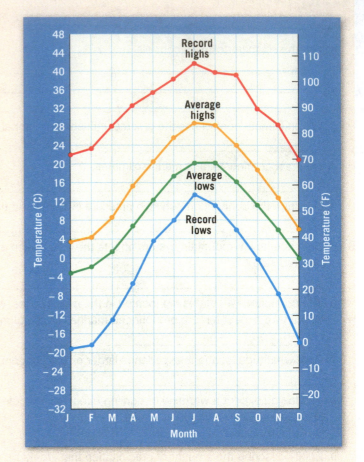

▲ **Figure 17.1** Graph showing temperature data for New York City. In addition to showing the average maximum and minimum temperatures for each month, the graph also shows extremes.

Table 17.1 Monthly and Annual Temperature Data (°F)

STATION	J	F	M	A	M	J	J	A	S	O	N	D	ANNUAL MEAN
1	25	28	36	48	61	72	75	74	66	55	39	28	50
2	48	48	49	50	54	55	57	57	57	54	52	49	52
3	31	31	38	49	60	69	74	73	69	59	44	35	52
4	25	26	35	45	58	70	75	72	65	55	45	35	50
5	20	34	38	46	51	54	60	53	49	40	32	25	40
6	42	48	50	55	56	58	59	60	59	56	48	46	51
7	76	76	76	76	78	78	77	77	76	75	74	74	76
8	90	88	84	79	77	73	70	71	76	79	82	87	80
9	−2	11	25	36	48	54	70	60	50	36	18	10	34
10	59	58	59	59	60	58	60	59	60	58	59	60	59
11	70	69	69	66	61	57	56	60	66	70	71	70	66
12	−46	−35	−10	16	41	59	66	60	42	16	−21	−41	12
13	22	25	37	51	62	72	77	75	66	54	39	27	50
14	82	82	82	83	83	82	82	83	83	83	82	83	83
15	27	30	34	41	48	54	57	55	50	43	36	31	42
16	40	44	48	54	61	69	77	76	67	57	47	41	57

5. Selecting from Stations 11–16, complete the following.

   a. Which of these stations is located near the Arctic Circle in the interior of a large continent? Explain your choice.

_____

   b. Which of these stations is located about 40°S latitude near the coast? How did you figure this out?

_____

17.2   Using a Climograph

■ **Read climographs to interpret data on temperature and precipitation.**

Temperature and precipitation are presented on a **climograph** such as the one shown in **Figure 17.2**. Average monthly temperatures are connected with a single line and read from the temperature scale on the left side of the graph. Average precipitation for each month is usually presented as a bar graph or as a different-colored line and read from the precipitation scale on the right side of the graph.

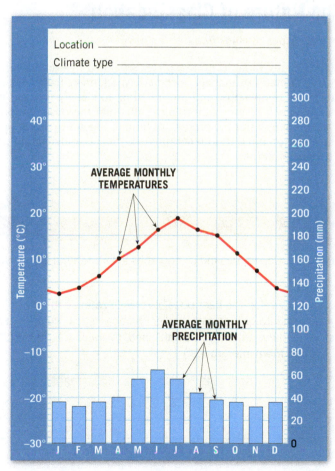

▲ **Figure 17.2** Typical climograph. Letters along the bottom represent the months. Average monthly temperatures (°C) are plotted with points, using the scale on the left axis. Precipitation for each month (in mm) is plotted as a bar, using the right axis.

ACTIVITY 17.2
Using a Climograph

Refer to the climograph in Figure 17.2 to answer the following questions.

1. What month has the greatest amount of precipitation?

_____

2. What month has the lowest average monthly temperature?

_____

3. What month has the highest average temperature?

_____

4. What is the *total* annual precipitation for this location?

_____ mm

5. Is the place represented by the climograph in the Northern Hemisphere or Southern Hemisphere?

_____ Hemisphere

6. During what 3-month span does this location receive the greatest amount of precipitation?

_____

17.3 The Nature of Classification

■ **Understand the nature of classification systems.**

Because weather is highly variable, no two places separated by even a short distance have precisely the same climate. To manage such variety, scientists have devised ways, based mainly on patterns of temperature and precipitation, to classify the vast array of meteorological data available. It is important to remember that the classification of climates is not a natural phenomenon but the product of human ingenuity. Bringing order to large quantities of information not only aids comprehension and understanding but also facilitates analysis and explanation.

ACTIVITY 17.3
The Nature of Classification

To help understand how a classification scheme is developed, use the climographs in **Figure 17.3**, found on page 307, to answer the following questions.

1. Working in groups of three to five students each, develop the best possible classification scheme for the stations represented by the climographs by arranging them into groups with similar characteristics. When you have finished, describe your classification system, listing the criteria you established. (*Note:* Cut Figure 17.3 into individual stations so you can more easily arrange them into groups.)

_____

_____

_____

_____

_____

_____

2. Why is the classification your group devised better than other possible systems you considered?

_____

_____

_____

3. Compare the classification your group devised with those developed by two other groups. Which is the best classification scheme? Why?

_____

_____

_____

_____

17.4 Köppen System of Climate Classification

■ **List the criteria used to define the five principal climate groups in the Köppen system of classification.**

Figure 17.4 presents the climate classification system devised by Wladimir Köppen. Since its introduction, the **Köppen system**, with some modifications, has become the best-known and most-used classification system for presenting the general world pattern of climates (**Figure 17.5**).

Köppen established five principal climate groups. Four groups are defined on the basis of temperature characteristics, and the fifth has precipitation as its primary criterion. Further division of the groups into climate types allows for more detailed climate descriptions. Köppen believed that the distribution of natural vegetation is the best expression of the totality of climate. Therefore, he chose boundaries based largely on the distribution of certain plant associations.

ACTIVITY 17.4 //
Köppen System of Climate Classification

On **Table 17.2**, list the names and general characteristics of each principal climate group next to its designated classification letter. Use Figure 17.4 as a reference.

Table 17.2 Characteristics of the Principal Climate Groups

CLIMATE GROUP	NAME	TEMPERATURE AND/OR PRECIPITATION CHARACTERISTICS
A	_____	_____
B	_____	_____
C	_____	_____
D	_____	_____
E	_____	_____

continued

Letter Symbol 1st 2nd 3rd				
A			Average temperature of the coldest month is 18°C or higher.	**A** Humid Tropical Climates
	f		Every month has 6 cm of precipitation or more.	
	m		Short dry season; precipitation in driest month less than 6 cm but equal to or greater than 10 – $R/25$ (R is annual rainfall in cm).	
	w		Well-defined winter dry season; precipitation in driest month less than 10 – $R/25$.	
	s		Well-defined summer dry season (rare).	
B			Potential evaporation exceeds precipitation. The dry–humid boundary is defined by the following formulas: (Note: R is the average annual precipitation in cm, and T is the average annual temperature in °C.) When R is less than the calculated value the climate is dry. $R < 2T + 28$ when 70% or more of rain falls in warmer 6 months. $R < 2T$ when 70% or more of rain falls in cooler 6 months. $R < 2T + 14$ when neither half year has 70% or more of rain.	**B** Dry Climates
	S		Steppe ⟶ The BS–BW boundary is 1/2 the dry–humid boundary.	
	W		Desert Desert (BW) climates are dryer then steppes (BS).	
		h	Average annual temperature is 18°C or greater.	
		k	Average annual temperature is less than 18°C.	
C			Average temperature of the coldest month is under 18°C and above –3°C.	**C** Humid Middle-Latitude Climates (Mild Winters)
	w		At least 10 times as much precipitation in a summer month as in the driest winter month.	
	s		At least three times as much precipitation in a winter month as in the driest summer month; precipitation in driest summer month less than 4 cm.	
	f		Criteria for w and s cannot be met.	
		a	Warmest month is over 22°C; at least 4 months over 10°C.	
		b	No month above 22°C; at least 4 months over 10°C.	
		c	One to 3 months above 10°C.	
D			Average temperature of coldest month is –3°C or below; average temperature of warmest month is greater than 10°C.	**D** Humid Middle-Latitude Climates (Severe Winters)
	w		Same as under C.	
	s		Same as under C.	
	f		Same as under C.	
		a	Same as under C.	
		b	Same as under C.	
		c	Same as under C.	
		d	Average temperature of the coldest month is –3°C or below.	
E			Average temperature of the warmest month is below 10°C.	**E** Polar Climates
	T		Average temperature of the warmest month is greater than 0°C and less than 10°C.	
	F		Average temperature of the warmest month is 0°C or below.	

▲ **Figure 17.4 Köppen system of climate classification.** (Photo A by Michael Collier; photo B by Witold Skrypczak/Alamy; photo C by; photo D by Martin Shields/Alamy; photo E by J. G. Paren/Science Source)

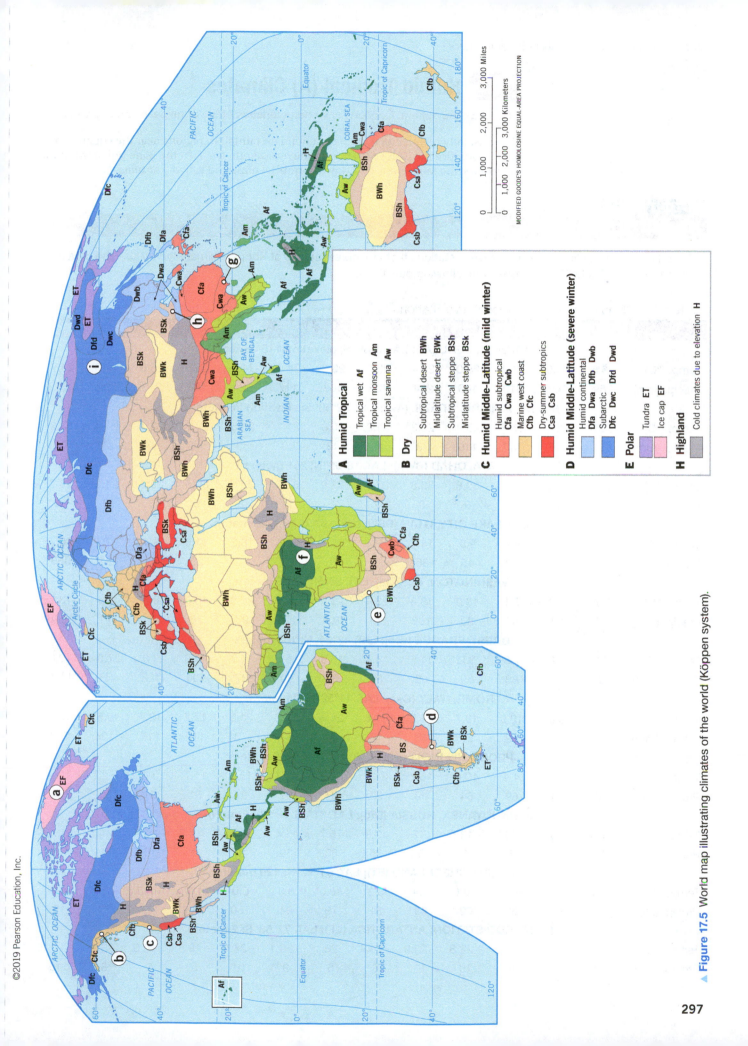

▲ **Figure 17.5** World map illustrating climates of the world (Köppen system).

A Humid Tropical

Tropical wet **Af**
Tropical monsoon **Am**
Tropical savanna **Aw**

B Dry

Subtropical desert **BWh**
Midlatitude desert **BWk**
Subtropical steppe **BSh**
Midlatitude steppe **BSk**

C Humid Middle-Latitude (mild winter)

Humid subtropical
Cfa Cwa Cwb
Marine west coast
Cfb Cfc
Dry-summer subtropics
Csa Csb

D Humid Middle-Latitude (severe winter)

Humid continental
Dfa Dwa Dfb Dwb
Subarctic
Dfc Dwc Dfd Dwd

E Polar

Tundra **ET**
Ice cap **EF**

H Highland

Cold climates due to elevation **H**

MODIFIED GOODE'S HOMOLOSINE EQUAL-AREA PROJECTION

17.5 Humid Tropical (A) Climates

■ Describe the general locations and characteristics of the principal climate groups.

The constantly high temperatures and abundant rainfall in regions near the equator combine to produce the lush vegetation of the *tropical rain forest*. In the zone poleward of the wet tropics, the rain forests give way to a tropical grassland called a *savanna*.

ACTIVITY 17.5

Humid Tropical (A) Climates

Table 17.3 contains climate data for several stations that are representative of Köppen climate types. Use the data in Figure 17.4 and Table 17.3 to answer the following questions.

Table 17.3 Climate Data for Representative Stations

	J	F	M	A	M	J	J	A	S	O	N	D	YEAR
IQUITOS, PERU (Af); LAT. 3°39′ S; elevation 115 meters
Temp. (°C)	25.6	25.6	24.4	25.0	24.4	23.3	23.3	24.4	24.4	25.0	25.6	25.6	24.7
Precip. (mm)	259	249	310	165	254	188	168	117	221	183	213	292	2619
RIO DE JANEIRO, BRAZIL (Aw); LAT. 22°50′ S; elevation 26 meters
Temp. (°C)	25.9	26.1	25.2	23.9	22.3	21.3	20.8	21.1	21.5	22.3	23.1	24.4	23.2
Precip. (mm)	137	137	143	116	73	43	43	43	53	74	97	127	1086
FAYA, CHAD (BWh); LAT. 18°00′ N; 251 meters
Temp. (°C)	20.4	22.7	27.0	30.6	33.8	34.2	33.6	32.7	32.6	30.5	25.5	21.3	28.7
Precip. (mm)	0	0	0	0	0	2	1	11	2	0	0	0	16
SALT LAKE CITY, UTAH (BSk); LAT. 40°46′ N; 1288 meters
Temp. (°C)	−2.1	0.9	4.7	9.9	14.7	19.4	24.7	23.6	18.3	11.5	3.4	−0.2	10.7
Precip. (mm)	34	30	40	45	36	25	15	22	13	29	33	31	353
WASHINGTON, DC (Cfa); LAT. 38°50′ N; 20 meters
Temp. (°C)	2.7	3.2	7.1	13.2	18.8	23.4	25.7	24.7	20.9	15.0	8.7	3.4	13.9
Precip. (mm)	77	63	82	80	105	82	105	124	97	78	72	71	1036
BREST, FRANCE (Cfb); LAT. 48°24′ N; 103 meters
Temp. (°C)	6.1	5.8	7.8	9.2	11.6	14.4	15.6	16.0	14.7	12.0	9.0	7.0	10.8
Precip. (mm)	133	96	83	69	68	56	62	80	87	104	138	150	1126
ROME, ITALY (Csa); LAT. 41°52′ N; 3 meters
Temp. (°C)	8.0	9.0	10.9	13.7	17.5	21.6	24.4	24.2	21.5	17.2	12.7	9.5	15.9
Precip. (mm)	83	73	52	50	48	18	9	18	70	110	113	105	749
PEORIA, ILLINOIS (Dfa); LAT. 40°52′ N; 180 meters
Temp. (°C)	−4.4	−2.2	4.4	10.6	16.7	21.7	23.9	22.7	18.3	11.7	3.8	−2.2	10.4
Precip. (mm)	46	51	69	84	99	97	97	81	97	61	61	51	894
VERKHOYANSK, RUSSIA (Dfd); LAT. 67°33′ N; 137 meters
Temp. (°C)	−46.8	−43.1	−30.2	−13.5	2.7	12.9	15.7	11.4	2.7	−14.3	−35.7	−44.5	−15.2
Precip. (mm)	7	5	5	4	5	25	33	30	13	11	10	7	155
IVITTUUT GREENLAND (ET); LAT. 61°12′ N; 129 meters
Temp. (°C)	−7.2	−7.2	−4.4	−0.6	4.4	8.3	10.0	8.3	5.0	1.1	−3.3	−6.1	0.7
Precip. (mm)	84	66	86	62	89	81	79	94	150	145	117	79	1132
MCMURDO STATION, ANTARCTICA (EF); LAT. 77°53′ S; 2 meters
Temp. (°C)	−4.4	−8.9	−15.5	−22.8	−23.9	−24.4	−26.1	−26.1	−24.4	−18.8	−10.0	−3.9	−17.4
Precip. (mm)	13	18	10	10	10	8	5	8	10	5	5	8	110

1. What temperature criterion is used for defining an A climate?

_____

2. Use **Figure 17.6** to plot the monthly temperature (using a red line) and precipitation data (using light blue bars) for Iquitos, Peru, an A climate, given in Table 17.3. (*Note:* Plot temperatures first.)

Use the Iquitos, Peru, climograph to answer Questions 3–6.

3. What is the *annual temperature range* (the difference between the highest and lowest average monthly temperatures) for Iquitos?

_____°C

4. Why are month-to-month temperature changes so small for A climates?

_____

5. Notice that Iquitos receives an average of 2619 mm of precipitation per year. How many inches of precipitation per year would this equal? How many feet? (*Hint:* Refer to the conversion tables located on the inside back cover of this manual.)

2619 mm = _____ in = _____ ft

6. Is the precipitation at Iquitos concentrated in one season or distributed fairly evenly throughout the year?

_____

Use the world climate map shown in Figure 17.5 to answer Questions 7 and 8.

7. In what latitude belt are A climates located?

Latitude belt: _____

8. Considering the locations of A climates, would weather fronts produced by the interaction of continental polar (cP) and maritime tropical (mT) air be responsible for the precipitation? Explain your answer.

_____

_____

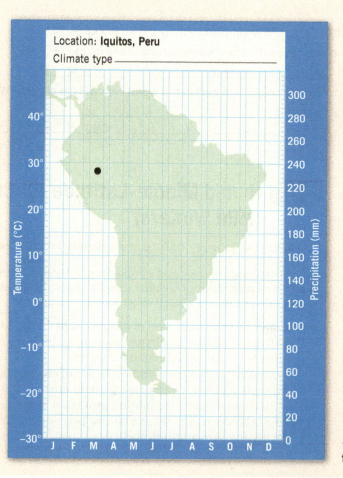

◀ **Figure 17.6** Climograph for Iquitos, Peru.

17.6 Dry (B) Climates

■ Describe the general locations and characteristics of the principal climate groups.

Dry climates cover more land area than any other climate group. The characteristic features of dry climates are their meager yearly rainfall and the fact that their precipitation is very unreliable. Climate scientists define a dry climate as one in which the *yearly precipitation is less than the potential water loss by evaporation.* Dryness is not only related to annual rainfall, it is also a function of evaporation, which in turn is closely related to temperature. As temperatures climb, potential evaporation also increases.

ACTIVITY 17.6
Dry (B) Climates

1. Why does the Köppen classification use three formulas to determine the boundary between dry and humid climates?

_____

_____

2. What names are applied to the following climate types?

BW: _____

BS: _____

3. Are deserts located closer to or farther from the wet tropics (Af) than steppes (see Figure 17.5)?

_____

4. Using Figure 17.5, find at least two desert locations that are classified BWk. What does the letter *k* in BWk tell you about this type of desert climate?

_____

To answer Questions 5 and 6, refer to the world climate map shown in Figure 17.5.

5. At what latitudes (north and south) are the most extensive arid areas located?

_____

6. The Sahara Desert in North Africa is the largest area in the world with a BWh climate. What are some other regions that have this climate?

_____

17.7 Humid Middle-Latitude (C) Climates (Mild Winters)

■ Describe the general locations and characteristics of the principal climate groups.

A large percentage of the world's population is located in areas with C climates. These climates are characterized by weather contrasts brought about by changing seasons. Within the C group of climates, several subgroups are recognized:

- Located on the eastern sides of continents, in the 25°–40° latitude range, the **humid subtropical climate** dominates the southeastern United States, as well as other similarly situated areas around the world.
- Situated on the western (windward) side of continents, from about 40°–65° north and south latitude, is a climate region dominated by the onshore flow of oceanic air. In North America, the **marine west coast climate** extends from near the U.S.–Canada border northward as a narrow belt into southern Alaska.
- The **dry-summer subtropical climate** is typically located along the west sides of continents between latitudes 30° and 45°. This includes much of the coast

of California and the coastal areas around the Mediterranean. Because this dry-summer subtropical climate is particularly extensive in the latter region, the name *Mediterranean climate* is often used as a synonym.

ACTIVITY 17.7

Humid Middle-Latitude (C) Climates (Mild Winters)

1. What temperature criteria define C climates?

_____

2. Prepare climographs for Washington, DC, and Rome, Italy, by plotting the data in Table 17.3 on the graphs in **Figure 17.7**. Use your climographs to answer Questions 3–5.

3. In what manner are the temperature curves for the two cities similar?

_____

4. How does the annual distribution of precipitation vary between the two cities?

_____

5. What is the difference between a Cf (Washington, DC) climate and a Cs (Rome, Italy) climate?

_____

_____

Use the world climate map shown in Figure 17.5 to answer Questions 6–8.

6. Name a country in Asia that has a large area of climate similar to that of the southeastern United States.

_____

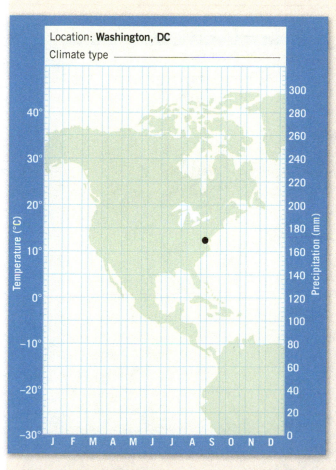

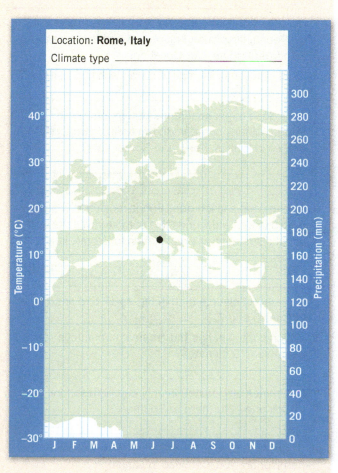

▲ **Figure 17.7** Climographs for Washington, DC, and Rome, Italy.

continued

Activity 17.7 continued

7. What Southern Hemisphere countries have climates similar to that of Washington, DC?

_____

8. Which U.S. state has a climate similar to that of Rome, Italy?

_____

17.8 Humid Middle-Latitude (D) Climates (Severe Winters)

■ **Describe the general locations and characteristics of the principal climate groups.**

The D climates are land-controlled climates, the result of broad continents in the middle latitudes. Harsh winters and a relatively short growing season restrict agricultural activity in much of the area of D climates. The northern portions of D climate regions are covered by coniferous forests, and logging is a significant economic activity.

ACTIVITY 17.8
Humid Middle-Latitude (D) Climates (Severe Winters)

1. What criteria are used to define D climates?

_____

_____

2. Use the data from Table 17.3 to plot a climograph for Peoria, Illinois, on **Figure 17.8** and then use this climograph to answer Questions 3–5.

3. What is the annual temperature range in Peoria, Illinois?

_____ °C

4. How does the annual range of temperature in Peoria, Illinois, compare to the annual temperature ranges in Iquitos, Peru, and Rome, Italy?

_____

_____

5. During what season does Peoria receive its greatest precipitation? How does this compare with the seasonal distribution of precipitation for Rome, Italy?

_____

_____

Use the world climate map in Figure 17.5 to answer Questions 6 and 7.

6. Which continent has the greatest continuous expanse of D climates?

_____

7. D climates are located in only one hemisphere. Which one—Northern or Southern?

_____

8. Suggest a reason D climates are located in only the hemisphere you identified in Question 7.

_____

_____

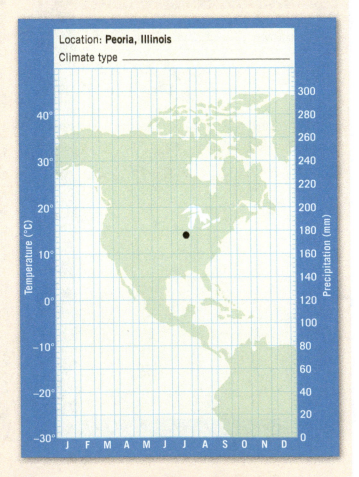

▲ **Figure 17.8** Climograph for Peoria, Illinois.

17.9 Polar (E) Climates

■ **Describe the general locations and characteristics of the principal climate groups.**

The polar climates are regions of cold temperatures and sparse population found at high latitudes and in scattered high mountains at all latitudes. Low evaporation rates allow these areas to be classified as humid, even though the annual precipitation is modest.

ACTIVITY 17.9
Polar (E) Climates

1. What criterion is used to define E climates?

2. Describe the characteristics and locations of the two polar climates.

ET climates:

EF climates:

17.10 Highland Climates

■ **Describe the general locations and characteristics of the principal climate groups.**

Although often not included with the principal climate groups, *high-altitude climates*, or *highland climates*, exist in all climatic regions. They result from the changes in temperature, humidity, and precipitation that occur with elevation and orientation of mountain slopes.

ACTIVITY 17.10
Highland Climates

Compare the climate data for Cruz Loma, Ecuador, and Guayaquil, Ecuador, found in **Table 17.4**. These towns are both located near the equator, less than 200 kilometers (120 miles) apart.

1. Use the data in Table 17.4 and the Köppen classification chart in Figure 17.4 to determine whether the climate for Guayaquil, Ecuador, is an A, B, C, D, or E climate.

Table 17.4 Climate Data for Cruz Loma and Guayaquil, Ecuador

	J	F	M	A	M	J	J	A	S	O	N	D	YEAR
CRUZ LOMA, ECUADOR; LAT. 0°, LONG. 79°W													
Temp. (°C)	6.1	6.6	6.6	6.6	6.6	6.1	6.1	6.1	6.1	6.1	6.6	6.6	6.4
Precip. (mm)	198	185	241	236	221	122	36	23	86	147	124	160	1779
GUAYAQUIL, ECUADOR; LAT. 2°, LONG. 80°W													
Temp. (°C)	27.6	27.1	27.6	27.5	26.6	25.4	24.5	25.1	25.1	25.3	26.4	26.5	25.8
Precip. (mm)	195	261	237	157	31	5	6	0	0	2	1	22	917

continued

Activity 17.10 continued

2. What could explain the temperature differences between Guayaquil and Cruz Loma?

3. Why would you expect the vegetation in the area around Cruz Loma to be different from that found at Guayaquil, Ecuador, a city located at the same latitude but on the coast?

4. Examine Figure 17.5. Where is the greatest continuous expanse of high-altitude (highland) climate located?

17.11 Climates of North America

■ Use climate data to classify climates according to the Köppen system.

Now that you have examined the characteristics of the climate groups in the Köppen system, you have the background you need to classify climates. A flowchart is provided to help you though the classification process. This activity is intended to reinforce your understanding of climate controls and to demonstrate how scientists classify climates around the globe.

ACTIVITY 17.11
Climates of North America

1. Table 17.5 includes data for eight stations in North America that represent different climate types. Temperatures are given in degrees Celsius, and precipitation is presented in millimeters. Use Figure 17.5 and the flowchart in Figure 17.9 to determine the Köppen classification for each station.

Table 17.5 Selected Climate Data for North America

STATION NUMBER		J	F	M	A	M	J	J	A	S	O	N	D	YR.
1	Temp. (°C)	1.7	4.4	7.9	13.2	18.4	23.8	25.8	24.8	21.4	14.7	6.7	2.8	13.8
	Precip. (mm)	10	10	13	13	20	15	30	32	23	18	10	13	207
2	Temp. (°C)	−10.4	−8.3	−4.6	3.4	9.4	12.8	16.6	14.9	10.8	5.5	−2.3	−6.4	3.5
	Precip. (mm)	18	25	25	30	51	89	64	71	33	20	18	15	459
3	Temp. (°C)	10.2	10.8	13.7	17.9	22.2	25.7	26.7	26.5	24.2	19.0	13.3	10.0	18.4
	Precip. (mm)	66	84	99	74	91	127	196	168	147	71	53	71	1247
4	Temp. (°C)	−23.9	−17.5	−12.5	−2.7	8.4	14.8	15.6	12.8	6.4	−3.1	−15.8	−21.9	−3.3
	Precip. (mm)	23	13	18	8	15	33	48	53	33	20	18	15	297
5	Temp. (°C)	18.9	20.0	21.1	22.8	25.0	26.7	27.2	27.8	27.2	25.0	21.1	20.0	23.6
	Precip. (mm)	51	48	58	99	163	188	172	178	241	208	71	43	1520
6	Temp. (°C)	−4.4	−2.2	4.4	10.6	16.7	21.7	23.9	22.7	18.3	11.7	3.8	−2.2	10.4
	Precip. (mm)	46	51	69	84	99	97	97	81	97	61	61	51	894
7	Temp. (°C)	12.8	13.9	15.0	16.1	17.2	18.8	19.4	22.2	21.1	18.8	16.1	13.9	15.9
	Precip. (mm)	53	56	41	20	5	0	0	2	5	13	23	51	269
8	Temp. (°C)	12.8	15.0	18.9	21.1	26.1	31.1	32.7	33.9	31.1	22.2	17.7	13.9	23.0
	Precip. (mm)	10	9	6	2	0	0	6	13	10	10	3	8	77

Station 1: _____   Station 2: _____   Station 3: _____

Station 4: _____   Station 5: _____   Station 6: _____

Station 7: _____   Station 8: _____

2. Once you have classified these stations, determine a likely location for each, based on factors such as mean annual temperature, temperature range, and total precipitation. Place the number of each station in the appropriate circle on the map in **Figure 17.10**.

3. What climate group covers the greatest area in North America (A, B, C, D, or E)?

_____

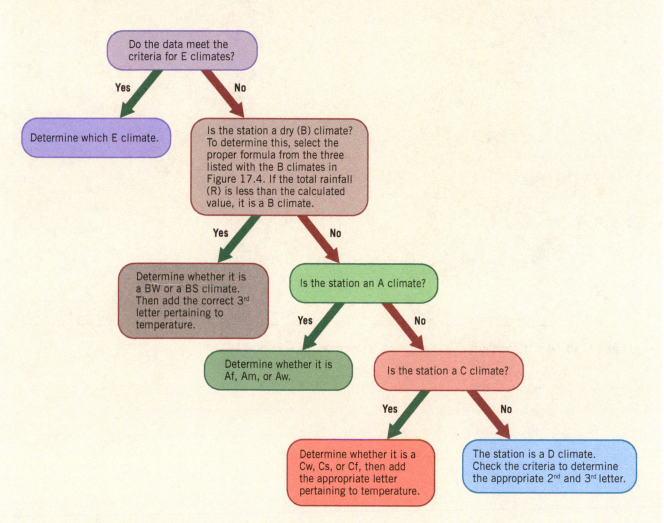

▲ **Figure 17.9** Flowchart used to aid in climate classification.

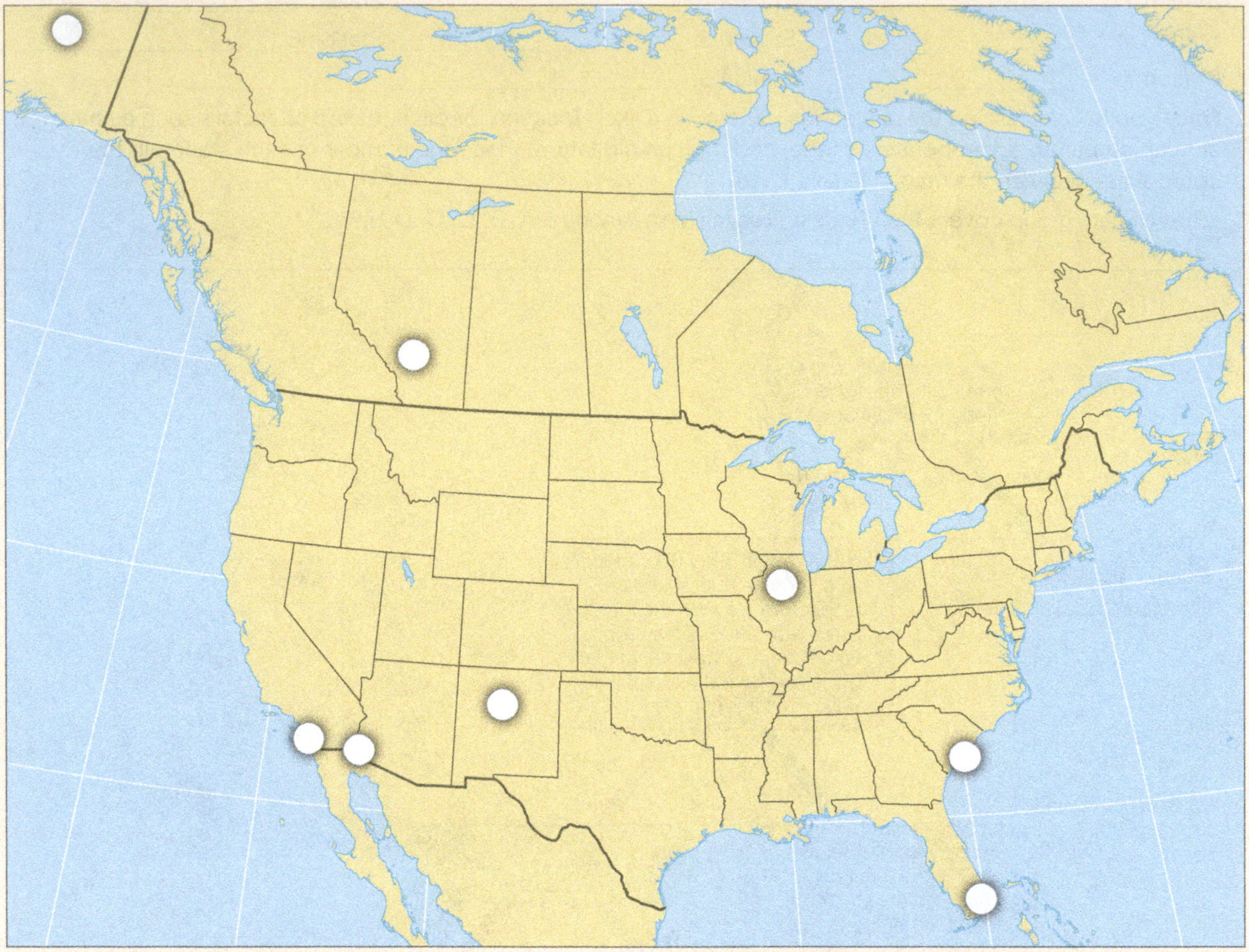

▲ **Figure 17.10** Map of North America to accompany Question 2 in Activity 17.11.

MasteringGeology™

Looking for additional review and lab prep materials? Go to **www.masteringgeology.com** for Pre-Lab Videos, Geoscience Animations, RSS Feeds, Key Term Study Tools, The Math You Need, an optional Pearson eText, and more.

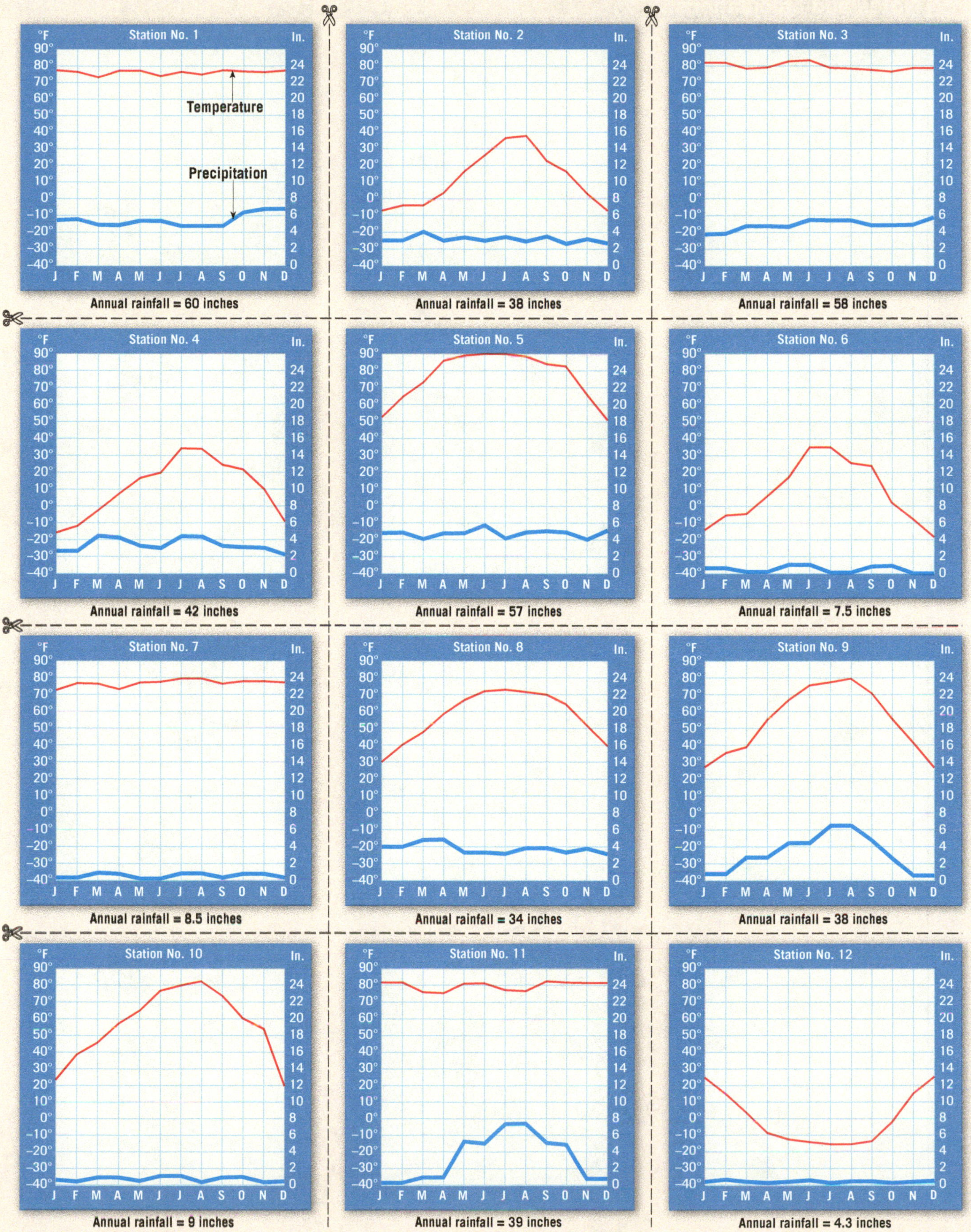

▲ **Figure 17.3** Generalized climographs that illustrate a variety of climate types.

Notes and calculations:

Global Climates

Name _____   Course/Section _____

Date _____   Due Date _____

1. List the six major climate controls.

   _____

   _____

   _____

   _____

   _____

   _____

2. Which control has the greatest influence on global temperatures?

   _____

3. Would a place with a marine climate experience a lower or higher annual temperature range than a place at the same latitude that has a continental climate?

   _____

4. What is a climograph?

   _____

5. List some key words to describe the characteristics of each of the following Köppen climate groups.

   A climates: _____

   _____

   B climates: _____

   _____

   C climates: _____

   _____

   D climates: _____

   _____

   E climates: _____

   _____

   Highland climates: _____

   _____

6. Indicate the name of the Köppen climate group best described by each of the following statements.

   Vast areas of northern coniferous forests:

   _____

   Smallest annual range of temperature:

   _____

   Highest annual precipitation:

   _____

   Mean temperature of the warmest month is below 10°C:

   _____

   The result of high elevation and mountain slope orientation:

   _____

   Potential evaporation exceeds precipitation:

   _____

   Very little change in monthly precipitation and temperature throughout the year:

   _____

Notes and calculations:

Astronomical Observations

LEARNING OBJECTIVES

Each statement represents an important learning objective that relates to one or more sections of this lab. After you complete this exercise you should be able to:

- **Describe the changing position of the Sun as it rises and sets and how the angle of the Sun at noon changes daily.**
- **Explain how the position of the Moon changes over a period of several weeks.**
- **Explain how the positions of stars in the night sky relate to Earth's rotational motion.**

MATERIALS

meterstick (or yardstick)	small weight (such as a steel nut)
protractor	ruler
string	star chart

PRE-LAB VIDEO https://goo.gl/LQqUB5

Prepare for lab! Prior to attending your laboratory session, view the pre-lab video. Each video provides valuable background that will contribute to your understanding and success in lab.

INTRODUCTION

Scientific inquiry often starts with the systematic collection of data from which hypotheses—and occasionally theories—are developed. The study of astronomy begins with carefully observing and recording the changing positions of the Sun, Moon, planets, and stars, using both the unaided eye and a telescope.

PART 4 Astronomy

18.1 Measuring the Position of the Sun

■ Describe the changing position of the Sun as it rises and sets and how the angle of the Sun at noon changes daily.

Many people are unaware that the Sun rises and sets at different locations on the horizon each day. Because our planet **revolves** around the Sun, the orientation of Earth's axis to the Sun continually changes. As a result, the location of the rising and setting Sun and the **altitude** (angle above the horizon) of the Sun at noon change throughout the year.

ACTIVITY 18.1A
Sunset Observations

Use the following procedure to observe and record the Sun's location on the horizon at sunset:

Step 1: Several minutes prior to sunset, *estimate* where the Sun will set on the horizon. Draw the prominent features (buildings, trees, etc.) to the north and south of the Sun's approximate setting position on the first sunset data sheet in Figure 18.1.

CAUTION Never look directly at the Sun; eye damage may result.

Step 2: As the Sun sets, draw its position on the data sheet relative to the fixed features on the horizon. From this same location repeat Steps 1 and 2 for three additional sunsets.

SUNSET DATA SHEETS

South   North
HORIZON

Date of observation _____   Time of observation _____

South   North
HORIZON

Date of observation _____   Time of observation _____

South   North
HORIZON

Date of observation _____   Time of observation _____

South   North
HORIZON

Date of observation _____   Time of observation _____

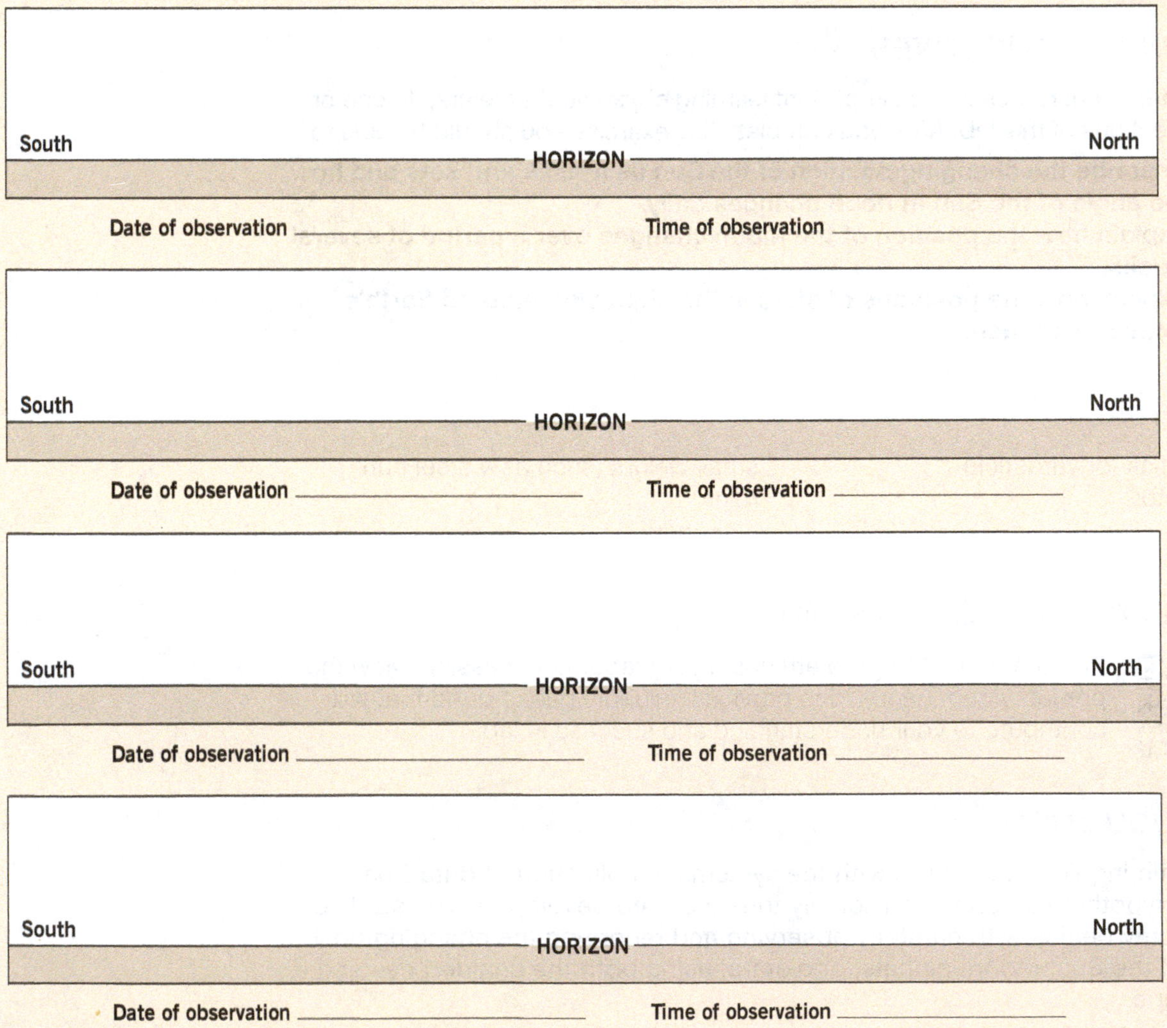

▲ **Figure 18.1** Sunset data sheets.

Step 3: Note the date and time of your observation on the data sheet.

Step 4: Return to the same location several days later. Repeat your observation and record the results on a new data sheet.

1. Following Steps 1–4, record at least four separate observations of the setting Sun on Figure 18.1. Gather the data over a period of several weeks; *wait 4 or 5 days between observations*.

2. After completing your observations, describe the changing location of the Sun at sunset.

_____

_____

_____

_____

ACTIVITY 18.1B
Measuring the Noon Sun Angle

A method for measuring the altitude (angle) of the noon Sun above the horizon is illustrated in **Figure 18.2**. Follow the steps below to determine the altitude of the Sun at noon on several days:

Step 1: Place a yardstick (or meterstick or ruler) perpendicular to the ground or a tabletop.

Step 2: When the Sun is at its highest position in the sky (noon standard time or 1 P.M. daylight saving time), measure the length of the shadow.

Step 3: Divide the height of the measuring stick by the length of the shadow.

Step 4: Determine the Sun angle by referring to **Table 18.1**. Locate the number on the table that comes closest to your answer from Step 3 and then record the angle.

Step 5: Repeat the measurement at the same time on 4 days, spaced over 4 or 5 weeks.

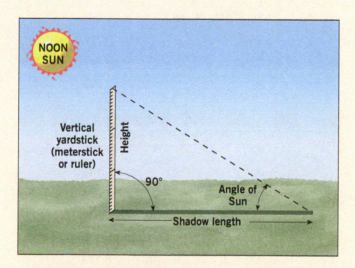

▲ **Figure 18.2** Illustration depicting how to measure the angle of the Sun above the horizon at noon. With each observation, be certain that the yardstick, meterstick, or ruler is perpendicular (at a 90° angle) to the ground or tabletop.

continued

Activity 18.1B continued

Table 18.1 Data Table for Determining the Noon Sun Angle[*]

IF $\dfrac{\text{HEIGHT OF STICK}}{\text{LENGTH OF SHADOW}}$ IS	THEN SUN ANGLE IS	IF $\dfrac{\text{HEIGHT OF STICK}}{\text{LENGTH OF SHADOW}}$ IS	THEN SUN ANGLE IS
0.2679	15°	1.235	51°
0.2867	16°	1.280	52°
0.3057	17°	1.327	53°
0.3249	18°	1.376	54°
0.3443	19°	1.428	55°
0.3640	20°	1.483	56°
0.3839	21°	1.540	57°
0.4040	22°	1.600	58°
0.4245	23°	1.664	59°
0.4452	24°	1.732	60°
0.4663	25°	1.804	61°
0.4877	26°	1.881	62°
0.5095	27°	1.963	63°
0.5317	28°	2.050	64°
0.5543	29°	2.145	65°
0.5774	30°	2.246	66°
0.6009	31°	2.356	67°
0.6249	32°	2.475	68°
0.6494	33°	2.605	69°
0.6745	34°	2.748	70°
0.7002	35°	2.904	71°
0.7265	36°	3.078	72°
0.7536	37°	3.271	73°
0.7813	38°	3.487	74°
0.8098	39°	3.732	75°
0.8391	40°	4.011	76°
0.8693	41°	4.332	77°
0.9004	42°	4.705	78°
0.9325	43°	5.145	79°
0.9657	44°	5.671	80°
1.0000	45°	6.314	81°
1.0360	46°	7.115	82°
1.0720	47°	8.144	83°
1.1110	48°	9.514	84°
1.1500	49°	11.430	85°
1.1920	50°		

[*] Select the number nearest to the quotient determined by dividing the height of the stick by the length of the shadow. Read the corresponding Sun angle.

1. Record the dates and results of your measurements in the following spaces.

   Date: _____   Noon Sun angle: _____ °

   Date: _____   Noon Sun angle: _____ °

   Date: _____   Noon Sun angle: _____ °

   Date: _____   Noon Sun angle: _____ °

2. Did the altitude of the noon Sun increase or decrease over the period of your measurements?

   The altitude (angle) _____.

3. How many degrees did the noon Sun angle change over the period of your observations?

   Change: _____ °

4. What was the approximate average change of the noon Sun angle per day?

   Change: _____ ° per day

5. Based on your answer to Question 4, how many degrees will the noon Sun angle change over a 6-month period?

   Change: _____ °

18.2   Measuring the Position of the Moon

■ **Explain how the position of the Moon changes over a period of several weeks.**

Most people have noticed that the shape of the illuminated portion of the Moon changes regularly. However, very few people take the time to systematically record and explain why these changes occur.

ACTIVITY 18.2
Measuring the Position of the Moon

1. Record at least four observations of the Moon by completing the following steps:

   Step 1: On a Moon observation data sheet provided in Figure 18.3, indicate the approximate east–west position of the Moon in the sky by drawing a circle at the appropriate location.

   Step 2: Shade the circle to indicate the shape of the illuminated portion of the Moon that you observe.

   Step 3: Record the date and time of your observation.

   Step 4: Keep in mind that the approximate time between moonrise on the eastern horizon and moonset on the western horizon is 12 hours. Estimate when the Moon may have risen and when it may set. Write your estimates on the data sheet.

   Step 5: Repeat your observations/recordings at least four times, at intervals of 4 to 5 days, using a new data sheet each time.

   Answer Questions 2–5 after you have completed all your observations of the Moon.

2. Describe what happened to the size and shape of the illuminated portion of the Moon over the period of your observations.

   _____

   _____

3. Did the Moon move farther eastward or westward in the sky with each successive observation?

   _____

continued

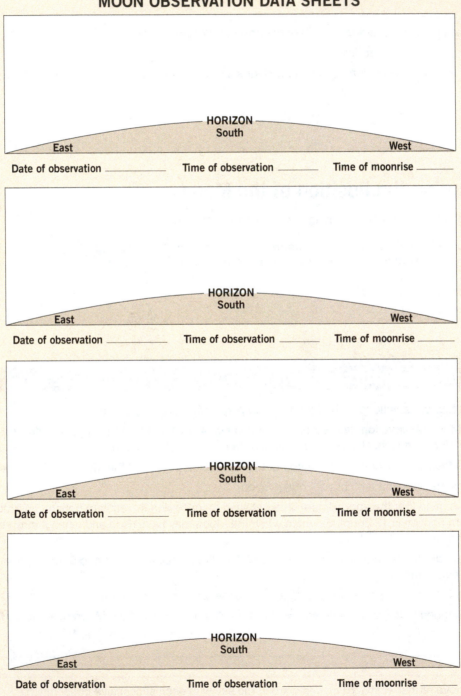

Activity 18.2 continued

4. Did moonrise and moonset occur earlier or later with each successive observation?

_____

5. Based on your observations and your answers to Questions 3 and 4, does the Moon revolve around Earth from east to west or west to east?

Moon revolves around Earth from _____.

MOON OBSERVATION DATA SHEETS

HORIZON
South
East   West
Date of observation _____ Time of observation _____ Time of moonrise _____

HORIZON
South
East   West
Date of observation _____ Time of observation _____ Time of moonrise _____

HORIZON
South
East   West
Date of observation _____ Time of observation _____ Time of moonrise _____

HORIZON
South
East   West
Date of observation _____ Time of observation _____ Time of moonrise _____

▲ **Figure 18.3** Moon observation data sheets.

18.3 Observing the Stars

■ **Explain how the positions of stars in the night sky relate to Earth's rotational motion.**

Throughout history, people have been recording the positions and nightly movements of stars that result from Earth's **rotation** on its axis, as well as the seasonal changes in the constellations as Earth **revolves** around the Sun. Ancient observers offered many explanations for the changes before the nature of planetary motions was understood in the seventeenth century.

ACTIVITY 18.3
Observing the Stars

Complete this activity on a clear, moonless night, preferably in an area away from bright city lights.

1. List the colors of the stars you observe.

_____

To answer Questions 2–4 select a bright star that is about 45° above the southern horizon and observe its movement over a period of 1 hour.

2. With your arm extended, measure how much the position of the star has changed by using your fist.

Approximately _____ fist widths

3. Does the star appear to have moved eastward or westward over a period of 1 hour?

_____

4. How is the movement of the star you observed in Question 3 related to the direction of Earth's rotation?

_____

_____

Use a star chart to locate several constellations and the North Star (Polaris).

5. Use **Figure 18.4** to sketch the pattern of stars for each of two constellations you were able to locate. Write the name of each constellation under your sketch.

6. Use **Figure 18.5** to construct a simple **astrolabe** and use it to measure the angle of the North Star (Polaris) above the horizon as accurately as possible.

_____° above the horizon

Over a period of several hours, observe the motion of the stars in the vicinity of Polaris.

7. Write a brief summary of the motion of the stars in the vicinity of Polaris.

_____

_____

If possible, return to the same location you chose for Question 2 several weeks later, at the same time, to answer Questions 8–10.

Constellation Star Pattern

Constellation name: _____

Constellation Star Pattern

Constellation name: _____

▲ **Figure 18.4** Constellation sketches.

continued

Activity 18.3 continued

8. Did the same star you observed overhead in Question 2 remain overhead, move eastward, or move westward?

_____

9. How is the change in position of the star you observed overhead several weeks earlier related to the revolution of Earth?

_____

_____

_____

10. Using the astrolabe you constructed in Question 6, repeat your measurement of the angle of the North Star (Polaris) above the horizon. List your new measurement and compare it to the measurement you obtained several weeks earlier. Explain your result(s).

_____

_____

_____

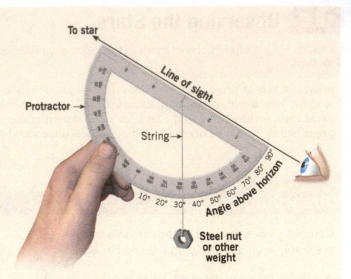

▲ **Figure 18.5** Simple astrolabe, an instrument used to measure the angle of an object above the horizon. The angle is read where the string crosses the outer edge of the protractor—32° in the example illustrated here. Notice that the angle above the horizon is the difference between 90° and the angle printed on the protractor.

MasteringGeology™

Looking for additional review and lab prep materials? Go to www.masteringgeology.com for Pre-Lab Videos, Geoscience Animations, RSS Feeds, Key Term Study Tools, The Math You Need, an optional Pearson eText, and more.

Notes and calculations:

Astronomical Observations

Name _____

Date _____

Course/Section _____

Due Date _____

1. On **Figure 18.6**, prepare a sketch illustrating your observed positions of the setting Sun during a span of several weeks. Show the reference features on the horizon that you used. Label each position of the Sun with the date of the observation. Write a brief summary of your observations below the diagram.

4. Did the Moon rise earlier or later each night you observed it?

_____

5. List the different colors of stars you observed.

_____

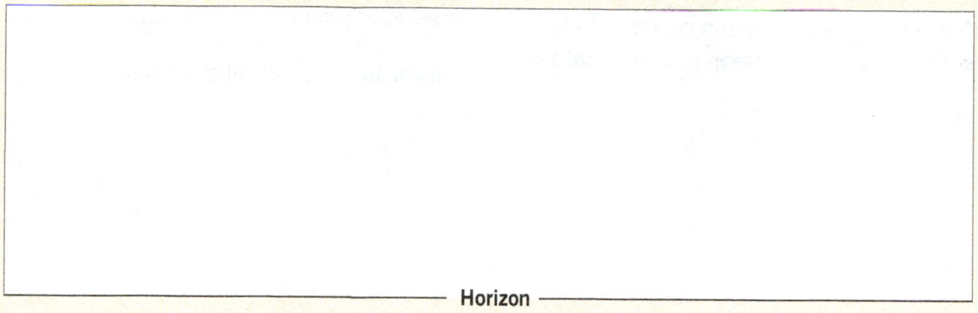

――――――――― Horizon ―――――――――

Summary: _____

_____

▲ **Figure 18.6** Sunset observations that accompany Question 1.

2. From Question 1 in Activity 18.1B, list the noon Sun angle that you calculated for the first day and last day of your measurements.

Date of observation: _____

Noon Sun angle on the first day: _____ °

Date of observation: _____

Noon Sun angle on the last day: _____ °

3. On **Figure 18.7**, draw two sketches of the Moon—the first illustrating the Moon as you saw it on your first lunar observation and the second as you saw it on your last observation. Label the date and time of each.

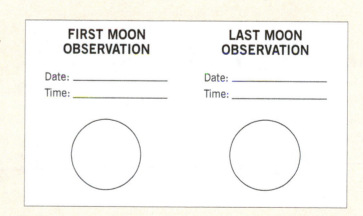

FIRST MOON OBSERVATION

Date: _____

Time: _____

LAST MOON OBSERVATION

Date: _____

Time: _____

▲ **Figure 18.7** Sketches of the Moon accompany Question 3.

continued

6. Approximately how many widths of your fist, with your arm extended, will a star appear to move in 1 hour? Toward which direction do the stars appear to move throughout the night? What is the reason for the apparent motion?

_____

_____

_____

7. What was your measured angle of the North Star (Polaris) above the horizon at your location? Did the angle change over a several-week period? Explain.

_____

_____

_____

8. On **Figure 18.8** sketch and name the pattern of stars for any constellation you have been able to locate in the sky.

Constellation Star Pattern

Constellation name: _____

▲ **Figure 18.8** Constellation sketch.

Patterns in the Solar System

LEARNING OBJECTIVES

Each statement represents an important learning objective that relates to one or more sections of this lab. After you complete this exercise you should be able to:

- Explain how the nebular theory accounts for many of the differences observed between the terrestrial and Jovian planets.
- Compare and contrast the sizes of the terrestrial and Jovian planets.
- Construct a scale model of the solar system to analyze the spacing of the planets.
- Describe the inclinations of planetary orbits.
- Compare the planets in terms of their masses and describe the relationship between a planet's mass and its number of moons.
- Describe the basic motions of the planets.
- List and describe the factors that cause the planets to have different surface temperatures.

MATERIALS

ruler
calculator
black containers with Styrofoam
   lids and thermometers

colored pencils
two 4-meter lengths of cash register paper
meterstick
light source (150-watt incandescent bulb)

PRE-LAB VIDEO https://goo.gl/LQtrWA

Prepare for lab! Prior to attending your laboratory session, view the pre-lab video. Each video provides valuable background that will contribute to your understanding and success in lab.

INTRODUCTION

The Sun is at the center of a revolving system, trillions of miles across, that consists of eight planets, their satellites, and numerous smaller asteroids, comets, and meteoroids. An estimated 99.85 percent of the mass of our solar system is

PART 4 Astronomy

contained in the Sun. Collectively, the planets account for most of the remaining 0.15 percent. Starting from the Sun, the planets are Mercury, Venus, Earth, Mars, Jupiter, Saturn, Uranus, and Neptune.

19.1 The Nebular Theory

■ **Explain how the nebular theory accounts for many of the differences observed between the terrestrial and Jovian planets.**

The **nebular theory**, which explains the formation of the solar system, states that the Sun and planets formed from a rotating cloud of interstellar gases (mainly hydrogen and helium) and dust called the **solar nebula**. As the solar nebula contracted due to gravity, most of the material collected in the center to form the hot *protosun*. The remaining material formed a flattened rotating disk, within which matter gradually cooled and condensed into grains and clumps of icy, rocky material. Repeated collisions resulted in most of the material clumping together into larger and larger chunks that eventually became asteroid-sized objects called **planetesimals**.

The composition of planetesimals was largely determined by their proximity to the protosun. Temperatures were highest in the inner solar system and decreased toward the outer edge of the rotating disk. Therefore, between the present orbits of Mercury and Mars, the planetesimals were composed of materials with high melting temperatures—metals and rocky substances. Through repeated collisions and accretion, these asteroid-sized rocky bodies combined to form the four inner planets—Mercury, Venus, Earth, and Mars.

The planetesimals that formed beyond the orbit of Mars, where temperatures were low, contained high percentages of ices—water, carbon dioxide, ammonia, and methane—as well as some rocky and metallic debris, mainly iron and nickel. It was mainly from these planetesimals that the four outer planets eventually formed. The accumulation of ices accounts, in part, for the large sizes and low densities of the outer planets. The two most massive planets, Jupiter and Saturn, had surface gravities sufficient to attract and retain large quantities of hydrogen and helium, the lightest elements.

Figure 19.1 is an artistic view of the solar system, in which the planets are not drawn to scale. Because all of the planets formed from the same flattened rotating disk, their orbits all occur in nearly the same plane, called the *plane of the ecliptic*.

▼ **SmartFigure 19.1** Tutorial
The solar system, showing the orbits of the planets.

TUTORIAL
https://goo.gl/QbTY4R

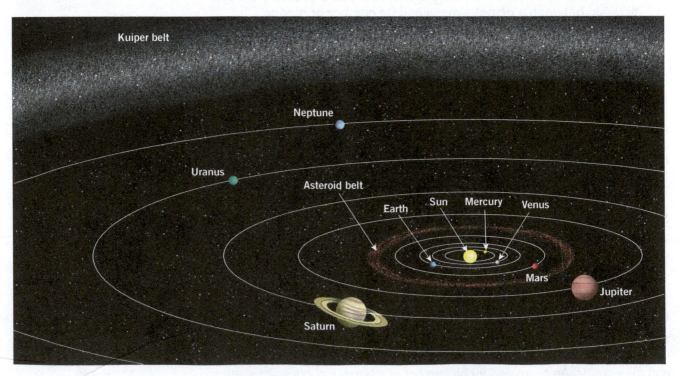

ACTIVITY 19.1 ///
The Nebular Theory

1. Briefly describe the nebular theory.

_____

_____

_____

2. Describe the overall composition of the inner and outer planets.

Inner planets: _____

Outer planets: _____

3. Explain how the nebular theory accounts for the differences in composition of the inner and outer planets.

_____

_____

4. Name two planets that were massive enough to attract and retain large quantities of hydrogen and helium, the lightest elements.

_____ and _____

19.2   Comparing the Sizes of the Planets

■ Compare and contrast the sizes of the terrestrial and Jovian planets.

Because the four inner planets—Mercury, Venus, Earth, and Mars—are rocky objects with solid surfaces, they are collectively called the **terrestrial (Earth-like) planets**. By comparison, the outer planets—Jupiter, Saturn, Uranus, and Neptune—are similar in size and composition and form a group called the **Jovian (Jupiter-like) planets**.

ACTIVITY 19.2 ///
Comparing the Sizes of the Planets

1. Table 19.1 lists many of the characteristics of the planets in our solar system. Draw a line on both the upper and lower parts of Table 19.1 to separate the terrestrial planets from the Jovian planets. Label both lines *Asteroid belt*.

2. On both parts of the table write the word *terrestrial* next to Mercury, Venus, Earth, and Mars and the word *Jovian* next to Jupiter, Saturn, Uranus, and Neptune.

3. To visually compare the relative sizes of the planets and the Sun, complete the following steps, using a 4-meter-long piece of cash register paper:

   Step 1: Determine the radius of each planet, in kilometers, by dividing its diameter (in kilometers) by 2. List your answers in the "Radius" column of Table 19.2.

   Step 2: Use a scale of 1 cm = 2000 km to determine the scale model radius of each planet and record your answer in the "Scale Model Radius" column of Table 19.2.

   Step 3: Place an X about 10 cm from one end of the cash register paper and label it *Starting point*.

   Step 4: Using the scale model radius in Table 19.2 and a meterstick, begin at the starting point and mark the radius of each planet with a line on the paper. Use a different colored pencil for the terrestrial planets than for the Jovian planets. Label each line with the corresponding planet's name.

   Step 5: The diameter of the Sun is approximately 1,350,000 kilometers. Using the scale 1 cm = 2000 km, determine the scale model radius of the Sun. Mark the Sun's radius on the cash register paper, using a different colored pencil than you used for the two planet groups. Label the line *Sun*.

continued

Activity 19.2 continued

Table 19.1 Planetary Data

PLANET	SYMBOL	AU*	MEAN DISTANCE FROM SUN MILLIONS OF MILES	MILLIONS OF KILOMETERS	ORBITAL PERIOD	INCLINATION OF ORBIT	ORBITAL VELOCITY mi/s	km/s
Mercury	☿	0.39	36	58	88d	7°00′	29.5	47.5
Venus	♀	0.72	67	108	225d	3°24′	21.8	35.0
Earth	⊕	1.00	93	150	365.25d	0°00′	19.5	29.8
Mars	♂	1.52	142	228	687d	1°51′	14.9	24.1
Jupiter	♃	5.20	483	778	12yr	1°18′	8.1	13.1
Saturn	♄	9.54	886	1427	30yr	2°29′	6.0	9.6
Uranus	♅	19.18	1783	2870	84yr	0°46′	4.2	6.8
Neptune	♆	30.06	2794	4497	165yr	1°46′	3.3	5.3

PLANET	PERIOD OF ROTA-TION	DIAMETER MILES	KILOME-TERS	RELATIVE MASS (EARTH = 1)	AVERAGE DENSITY (g/cm³)	POLAR FLATTEN-ING (%)	MEAN TEMPERA-TURE (°C)	NUMBER OF KNOWN SATELLITES††
Mercury	59d	3015	4878	0.06	5.4	0.0	167	0
Venus	243d	7526	12,104	0.82	5.2	0.0	464	0
Earth	23h56m04s	7920	12,756	1.00	5.5	0.3	15	1
Mars	24h37m23s	4216	6794	0.11	3.9	0.5	−65	2
Jupiter	9h56m	88,700	143,884	317.87	1.3	6.7	−110	67
Saturn	10h30m	75,000	120,536	95.14	0.7	10.4	−140	62
Uranus	17h14m	29,000	51,118	14.56	1.2	2.3	−195	27
Neptune	16h07m	28,900	50,530	17.21	1.7	1.8	−200	14

* AU = astronomical unit, Earth's mean distance from the Sun.
†† Includes all satellites discovered as of December 2017.

Use Table 19.1 to answer Questions 4–9.

4. Which is the largest terrestrial planet? What is its diameter?

Planet: _____

Diameter: _____ km

5. Which is the smallest Jovian planet? What is its diameter?

Planet: _____

Diameter: _____ km

Table 19.2 Planetary Radii with Scale Model Equivalents

PLANET	RADIUS	SCALE MODEL RADIUS
Mercury	_____ km	_____ cm
Venus	_____ km	_____ cm
Earth	_____ km	_____ cm
Mars	_____ km	_____ cm
Jupiter	_____ km	_____ cm
Saturn	_____ km	_____ cm
Uranus	_____ km	_____ cm
Neptune	_____ km	_____ cm

6. How many times larger is the diameter of the smallest Jovian planet than the diameter of the largest terrestrial planet?

_____ times larger

7. Write a general statement that compares the sizes of the terrestrial planets to the sizes of the Jovian planets.

_____

_____

8. How much larger is the diameter of the Sun compared to the diameter of Earth? How much larger is the diameter of the Sun compared to the diameter of Jupiter?

The Sun's diameter is _____ times larger than Earth's diameter.

The Sun's diameter is _____ times larger than Jupiter's diameter.

9. The diameter of the dwarf planet Pluto is approximately 2300 km. Is Pluto's diameter about one-third, one-half, or twice the diameter of Mercury? _____

19.3 Spacing of the Planets

■ **Construct a scale model of the solar system to analyze the spacing of the planets.**

Although many ancient astronomers tried to establish a pattern for the distribution of the planets, it was not until the mid-1700s that astronomers found a simple mathematical relationship that described the arrangement of the known planets.

ACTIVITY 19.3
Spacing of the Planets

1. Prepare a scale model of the solar system, using the following steps:

   Step 1: Obtain a 4-meter length of cash register paper and a meterstick from your instructor.

   Step 2: Place an *X* about 10 centimeters from one end of the cash register paper and label it *Sun*.

   Step 3: Using the mean distances of the planets from the Sun, in miles, found in Table 19.1 and the following scale, draw a small circle for each planet at its appropriate distance from the Sun. Use a different colored pencil for the terrestrial planets than for the Jovian planets and write the name of each planet next to its position.

                      SCALE
            1 millimeter = 1 million miles
            1 centimeter = 10 million miles
            1 meter = 1000 million miles

   Step 4: Label the area 258 million miles from the Sun *Asteroid belt*.

2. What feature of the solar system separates the terrestrial planets from the Jovian planets?

_____

3. Refer to your diagram and summarize the spacing for each of the two groups of planets.

   Terrestrial planets spacing: _____

   Jovian planets spacing: _____

4. Briefly describe the spacing of the planets in the solar system.

_____

_____

5. Which Jovian planet varies the most from the general pattern of spacing?

_____

19.4 Inclinations of Planetary Orbits

■ Describe the inclinations of planetary orbits.

The orbits of the planets all lie in nearly the same plane, called the **plane of the ecliptic**—an imaginary surface that passes though the Sun and Earth's orbit. The column labeled "Inclination of Orbit" in Table 19.1 lists how many degrees the orbit of each planet is inclined from the plane of the ecliptic. Notice in Table 19.1 that because the plane of the ecliptic is defined by Earth's orbit, the two align perfectly—meaning that Earth's orbit has *zero* inclination.

ACTIVITY 19.4
Inclinations of Planetary Orbits

1. Is the inclination of the orbit of Mercury 4°, 7°, or 10°? Of the remaining planets, are they inclined less than 4°, 7°, or 10°?

   Mercury is inclined _____°.

   The other planets are inclined less than _____°.

2. Considering the nebular origin of the solar system, suggest a reason the orbits of the planets are nearly all in the same plane.

   _____

3. The orbit of the dwarf planet Pluto is inclined 17°. When compared to the eight planets, describe the inclination of Pluto's orbit.

   _____

19.5 Mass and Density of Planets

■ Compare the planets in terms of their masses and describe the relationship between a planet's mass and its number of moons.

Mass is a measure of the quantity of matter an object contains. In Table 19.1, the masses of the planets are given in relationship to the mass of Earth. For example, the mass of Mars is 0.108, which means that it is only about 11 percent as massive as Earth. On the other hand, the Jovian planets are all many times more massive than Earth.

Because the gravitational attraction of a body is directly related to its mass, planets with high masses exhibit high surface gravities. Furthermore, while the mass of an object is measured by the amount of material it contains, the **weight** of an object depends on the gravitational attraction between the object and the body (planet) it is on. For example, astronauts *weigh* much less on the Moon than on Earth even though their *mass* remains the same.

Density is the mass per unit volume of a substance. In Table 19.1, the average densities of the planets are expressed in grams per cubic centimeter. As a reference, the density of water is approximately 1 g/cm^3.

ACTIVITY 19.5A
Comparing Planetary Masses

Using the masses of the planets provided in Table 19.1, complete the following.

1. What is the most massive planet in the solar system? How many times more massive is this planet than Earth?

   Most massive planet: _____

   This planet is _____ times more massive than Earth.

2. Which is the least massive planet? What is its mass as a percentage of Earth's mass?

Least massive planet: _____

This planet is only _____ percent as massive as Earth.

3. Given the fact that the gravitational attraction of a body is directly related to its mass, which planet exerts the greatest gravitational attraction? Which planet exerts the least attraction?

Greatest gravitational attraction: _____

Least gravitational attraction: _____

4. The surface gravities of Mars and Jupiter are, respectively, about 0.4 and 2.5 times that of Earth. What would be the approximate weight of a 200-pound person on these planets?

This person would weigh _____ pounds on Mars.

This person would weigh _____ pounds on Jupiter.

5. Which of the two groups of planets is more likely to attract and hold low-density gaseous material, such as hydrogen and helium?

_____

ACTIVITY 19.5B

Comparing the Number of Moons to Planetary Masses

The column labeled "Number of Known Satellites" in Table 19.1 indicates the number of known moons orbiting each planet.

1. Write a brief statement comparing the number of known moons orbiting the terrestrial planets to the number orbiting the Jovian planets.

_____

_____

2. What is the relationship between a planet's number of moons and its mass? Suggest a reason for your answer.

_____

_____

_____

ACTIVITY 19.5C

Comparing a Planet's Diameter to Its Density

1. Use the data in Table 19.1 to complete the graph in **Figure 19.2** by following these instructions:

a. Using two different colored pencils (one for the terrestrial planets and a different one for the Jovian planets), plot a point on the graph for each planet at the place where its diameter and density intersect.

b. Label each point with the planet's name.

2. What relationship exists between a planet's size (diameter) and its density?

_____

3. The average density of Earth's crust is about 2.8 g/cm³. Is the average density of the terrestrial planets greater or less than the density of Earth's crust?

_____

continued

Activity 19.5C continued

4. The average density of Earth is about 5.5 g/cm³. Since the density of Earth's crustal rocks is only about 2.8 g/cm³, what does this indicate about the density of Earth's interior?

_____

_____

5. What could explain the fact that Mars has a somewhat lower average density than the other terrestrial planets?

_____

_____

6. Compare the densities of the Jovian planets to the density of water. (*Note:* Water has a density of about 1 g/cm³.)

_____

_____

_____

7. Which of the planets has a density less than that of water and, therefore, would "float"?

_____

_____

8. Explain why Jupiter is such a massive object and yet has such a low density. (*Hint:* Remember that density is the relationship between mass and volume.)

_____

_____

9. Write a general statement comparing the densities of the terrestrial planets to the densities of the Jovian planets.

_____

_____

10. Why are the densities of the terrestrial and Jovian planets so different? (*Hint:* See Section 19.1.)

_____

_____

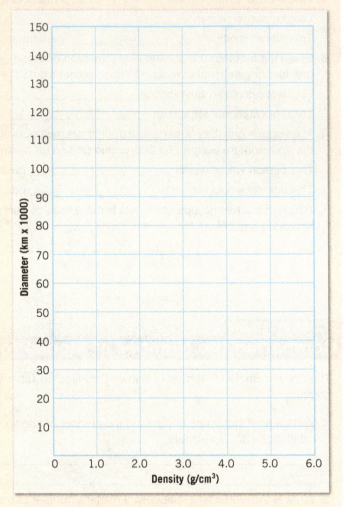

▲ **Figure 19.2** Diameter versus density graph.

19.6 Rotation and Orbital Period

■ **Describe the basic motions of the planets.**

All the planets, with the exception of Venus, rotate about their axis in a counterclockwise direction. Venus exhibits a very slow clockwise rotation. The time that it takes for a planet to complete one 360° rotation on its axis is called its **period of rotation**. The units used to express a planet's period of rotation shown in Table 19.1 are Earth hours and days.

The **orbital period**, or **revolution**, is the time it takes for a planet to complete one orbit around the Sun. The units used to express a planet's orbital period are Earth days and years. Without exception, the direction that the planets orbit the Sun is counterclockwise (from a Northern Hemisphere point of view).

ACTIVITY 19.6A
Rotation and Orbital Period

Use the planetary data in Table 19.1 to answer the following questions.

1. How long is the period of rotation of Mercury? (*Note:* The small *d* by the number stands for "Earth days.")
_____ days

2. How long is the period of rotation of Jupiter?
_____ hr

3. Compare the periods of rotation of the terrestrial planets to those of the Jovian planets.

_____

_____

4. Jupiter rotates once on its axis approximately every 10 hours. If an object were on the equator of Jupiter and rotating with it, it would travel approximately 280,000 miles (the distance around Jupiter's equator) in about 10 hours. Calculate the equatorial rotational velocity of Jupiter, using the following formula:

$$\text{Velocity} = \frac{\text{Distance}}{\text{Time}} = \underline{\hspace{3cm}} \frac{\text{miles}}{\text{hour}}$$

5. The distance around Earth's equator is about 24,000 miles. Use the formula in Question 4 to calculate the approximate rotational velocity at Earth's equator.
_____ mph

6. How many times faster is Jupiter's equatorial rotational velocity than Earth's?
_____ times faster

7. In 1 Earth year, how many orbits (trips around the Sun) will Mercury complete? What fraction of an orbit will Neptune accomplish?

Mercury: _____ orbits in 1 Earth yr

Neptune: _____ of an orbit in 1 Earth yr

8. How many rotations on its axis will Mercury complete in 1 of its years?

Mercury: _____ rotations in 1 Mercury yr

9. On average, do the terrestrial planets have longer or shorter periods of rotation than the Jovian planets? Are the terrestrial planet years longer or shorter than those of the Jovian planets?

Terrestrial planet days are _____.

Terrestrial planet years are _____.

ACTIVITY 19.6B
Kepler's Third Law of Planetary Motion

In the early 1600s, Johannes Kepler established three laws of planetary motion. According to Kepler's third law, the orbital period of a planet, measured in Earth years, is related to its distance from the Sun in astronomical units. One **astronomical unit (AU)** is defined as the average distance from the Sun to Earth—93 million miles, or 150 million kilometers. The law states that a planet's orbital period squared is equal to its mean solar distance cubed: $p^2 = d^3$. Applying Kepler's third law, what would be the period of revolution of a hypothetical planet that is 4 AUs from the Sun? Show your calculation in the following space.

19.7 Temperatures on Terrestrial Planets

■ **List and describe the factors that cause the planets to have different surface temperatures.**

The surface temperature of a planet is mainly influenced by three factors: the nature of its surface, the composition and thickness of its atmosphere, and its distance from the Sun. In the following activity, you will use a laboratory setup similar to what is shown in **Figure 19.3**. This experiment will help you determine how distance from the Sun affects a planet's surface temperature.

▲ **Figure 19.3** Temperatures on terrestrial planets lab equipment setup.

ACTIVITY 19.7
Distance Versus Temperature

Complete the following steps using the laboratory equipment supplied by your instructor.

Step 1: Work in groups of four students. Have each group member obtain *identical* light (heat) sources and *identical* black containers with covers and thermometers.

Step 2: Have each member of the group conduct a different experiment simultaneously. One student should do the experiment with the covered can and thermometer 15 cm from the light source, another with the can 30 cm from the light source, the third 45 cm, and the fourth 60 cm.

Step 3: Record the starting temperature for each container on **Table 19.3**; the temperatures should all be the same.

Step 4: For each of the four setups, turn on the light. After 10 minutes, record the temperature of the container in Table 19.3.

Step 5: Plot the temperatures from Table 19.3 on the graph in **Figure 19.4**, using the temperatures on the left side of the graph. Connect the points and label the line you constructed *Temperature change with distance*.

1. Based on the experiment you performed, briefly describe the relationship between the temperature of an object and its distance from a heat source.

_____

_____

Table 19.3 Temperature Data

DISTANCE FROM LIGHT SOURCE (cm)	STARTING TEMPERATURE (°C)	10-MINUTE TEMPERATURE
15	_____	_____
30	_____	_____
45	_____	_____
60	_____	_____

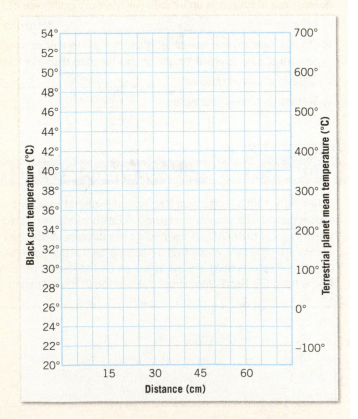

▲ **Figure 19.4** Temperatures on terrestrial planets graph.

2. Refer to the "Mean Temperature" column in Table 19.1. Plot the mean temperatures of the *terrestrial* planets at their appropriate distances from the Sun on the graph in Figure 19.4. Use a scale of 40 cm = 1 AU and the temperature scale on the right axis of the graph. Label each point with the planet's name. Connect the points and label the line *Mean terrestrial planet temperatures*.

3. Compare the line labeled *Mean terrestrial planet temperatures* on the graph in Figure 19.4 to the line labeled *Temperature change with distances*. The first line represents the actual mean temperatures of the planets, and the second line shows how temperature varies with distance based on experimental data. How are the lines similar? How are they different?

_____

_____

_____

_____

4. Briefly suggest the reason(s) for the difference(s) between the two lines on the graph in Figure 19.4. (*Hint:* Examine Figure 14.4, page 237, which explains the role that atmosphere has on the temperature of a planet.)

_____

_____

_____

_____

_____

MasteringGeology™

Looking for additional review and lab prep materials? Go to www.masteringgeology.com for Pre-Lab Videos, Geoscience Animations, RSS Feeds, Key Term Study Tools, The Math You Need, an optional Pearson eText, and more.

Notes and calculations:

Patterns in the Solar System

Name _____   Course/Section _____

Date _____   Due Date _____

1. Define *planetesimal*.

_____
_____
_____

_____

2. Outline the steps in the formation of our solar system, according to the nebular theory (**Figure 19.5**).

_____
_____
_____
_____
_____
_____

3. Use the nebular theory to explain why the planets revolve around the Sun in the same direction.

_____
_____
_____

4. By what criteria are planets considered either terrestrial or Jovian?

_____
_____
_____
_____
_____
_____
_____
_____

5. Briefly describe the spacing of the planets in the solar system.

_____
_____
_____
_____
_____
_____

▲ **Figure 19.5** This artist's rendering is of an asteroid belt filled with rocks and dusty debris that orbits a star similar to the Sun. It is from material like this that the terrestrial planets are thought to have formed. (NASA/JPL/T. Pyle)

6. Compare the terrestrial planets and the Jovian planets, based on the following characteristics:

Diameter: _____

_____

_____

Density: _____

_____

_____

Period of rotation: _____

_____

_____

Number of moons: _____

_____

_____

Mass: _____

_____

_____

7. What accounts for the large density differences between the terrestrial and Jovian planets?

_____

_____

_____

8. Using Kepler's third law, calculate the period of revolution of a hypothetical planet that is 10 AUs from the Sun.

_____

_____

_____

9. How does a planet's distance from the Sun affect the amount of solar radiation the planet receives?

_____

_____

_____

10. Explain how the nebular theory accounts for the differences in the composition of the inner planets compared to the outer planets.

_____

_____

11. On Figure 19.6, prepare a sketch, using circles, to illustrate the locations of the planets Mercury, Venus, Earth, and Mars at their approximate distances from the Sun. Assuming that you are looking down on the Northern Hemisphere of each planet, draw a curved arrow on each planet to illustrate its direction of rotation and a straight arrow that shows its direction of revolution around the Sun.

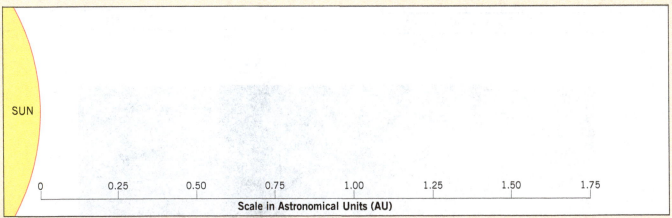

▲ Figure 19.6 Spacing and motion of the terrestrial planets.

Locating the Planets

LEARNING OBJECTIVES

Each statement represents an important learning objective that relates to one or more sections of this lab. After you complete this exercise you should be able to:

- **Describe the motions of the planets as viewed from Earth.**
- **Explain how the orbital period of a planet is related to its distance from the Sun.**
- **Describe the cause of the apparent retrograde motion of Mars, as observed from Earth.**
- **Using a model of the solar system, determine whether a planet can be seen from Earth on a specified date.**

MATERIALS

ruler   colored pencils
calculator

PRE-LAB VIDEO https://goo.gl/1W5Gme

 Prepare for lab! Prior to attending your laboratory session, view the pre-lab video. Each video provides valuable background that will contribute to your understanding and success in lab.

INTRODUCTION

The ability to recognize patterns in the night sky is one of the first skills that astronomers develop. It relies on an understanding of the motions of celestial objects and the proficient use of astronomical charts. This exercise examines a "working" model of the inner solar system. Using this model, you will investigate the movements of the planets that are visible to the unaided eye. The information in this exercise can also be used to locate the positions of the outer planets.

PART 4 Astronomy

20.1 A Model of the Inner Solar System

■ Describe the motions of the planets as viewed from Earth.

Figure 20.1 is a model of the inner solar system that shows the orbits of Mercury, Venus, Earth, and Mars, as viewed from above Earth's Northern Hemisphere. Earth's orbit has been subdivided into months and select days. (Red marks indicate the first day of each month) The outer circle in Figure 20.1, called the **reference circle**, can be used to determine the position (measured in degrees) of a planet in its orbit around the Sun for any day of the year.

The location of a planet can also be described by its position in a region of the sky called the **zodiac** ("zone of animals"). The zodiac consists of 12 **constellations**, which are shown

▼ **Figure 20.1** Model of the inner solar system, showing the orbits of Mercury, Venus, Earth, and Mars to scale, as viewed from above Earth's Northern Hemisphere.

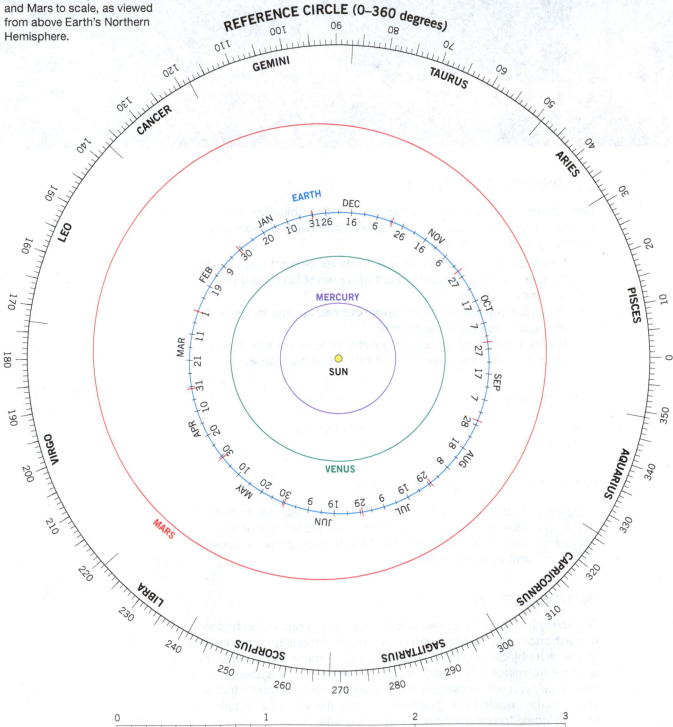

on the reference circle. These background stars (constellations) are relatively fixed and serve as a reference map for describing the positions of the planets, which change continually as they orbit the Sun.

ACTIVITY 20.1A

Locating a Planet Using the Reference Circle

Table 20.1 shows the locations (bearings) of Mercury, Venus, and Mars for the years 2017 through 2020 on the first day of each month. The locations are expressed in degrees and correspond to the equivalent degree designations found in the reference circle in Figure 20.1. For example, the position of Mercury on the first day of January 2017 was 117 degrees.

Use the following steps to determine a planet's position:

Step 1: Use Table 20.1 to determine the location of a planet, expressed in degrees, on a specified date. (*Note:* The data can be used to estimate a planet's position on dates other than the first of each month.)

Step 2: Mark the location you determine in Step 1 on the reference circle on Figure 20.1.

Step 3: Use a ruler, or any other straight edge, to connect the Sun with the position you marked in Step 2.

Table 20.1 Planetary Heliocentric Longitudes (in degrees)

2017	MERCURY	VENUS	MARS	2019	MERCURY	VENUS	MARS
January	117°	50°	16°	January	226°	140°	42°
February	239	100	35	February	316	191	59
March	323	145	51	March	100	235	74
April	129	196	68	April	231	285	89
May	242	244	83	May	320	332	104
June	338	293	98	June	124	21	118
July	145	340	112	July	240	69	132
August	253	30	126	August	335	120	146
September	356	79	140	September	146	170	159
October	164	128	153	October	251	218	172
November	264	178	167	November	352	268	186
December	10	227	180	December	161	315	200
2018	**MERCURY**	**VENUS**	**MARS**	**2020**	**MERCURY**	**VENUS**	**MARS**
January	181°	276°	194°	January	262°	4°	214°
February	276	325	208	February	12	54	229
March	21	9	222	March	174	101	244
April	189	59	237	April	271	151	261
May	278	107	253	May	22	200	278
June	44	158	271	June	190	249	297
July	200	206	288	July	279	297	315
August	290	256	307	August	45	346	335
September	68	305	327	September	204	35	354
October	213	352	346	October	291	83	13
November	303	42	5	November	70	134	32
December	88	90	23	December	214	182	49

Source: Data from West Virginia University/Johns Hopkins/Tomchin Planetarium.

continued

Activity 20.1A continued

Step 4: Place a dot representing the planet's position where the ruler crosses the *planet's orbit*. Label the dot with the date.

1. Using Table 20.1, determine the location, in degrees, of the following planets for the dates indicated. (One location is provided for reference.)

PLANET	MAY 1, 2019	JULY 1, 2019
Mercury	320°	_____°
Venus	_____°	_____°
Mars	_____°	_____°

2. Using the data, you compiled in Question 1, plot the May 1, 2019, locations of Mercury, Venus, Earth, and Mars on Figure 20.1. (*Note:* Earth's position can be read directly from its orbit.)

3. Using the data in Table 20.1, *estimate* how many degrees the following planets move in their orbit over a period of 1 month:

PLANET	MOVEMENT IN 1 MONTH
Mercury	_____°
Venus	_____°
Earth	_____°
Mars	_____°

Locating a Planet Among the Constellations

A planet is often referred to as being "in" a constellation. When viewing a planet from Earth, this refers to the particular constellation that is positioned *directly behind* the planet—but is, in fact, millions of miles away. The following steps are used to determine in which constellation a planet is located:

Step 1: Using Figure 20.1, mark the position of Earth and the planet to be observed for the same date.

Step 2: Beginning at the position of Earth, draw a line that extends through the position of the planet until it intersects the outer reference circle.

Step 3: Note the constellation that lies in the planet's background.

1. Determine the constellation in which each of the following planets can be observed (from Earth) on May 1, 2019:

CONSTELLATION ON MAY 1, 2019

Mercury _____

Venus _____

Mars _____

2. On Figure 20.1, plot the locations of Mercury, Venus, Earth, and Mars for July 1, 2019, and label each *7/1/19*. (These are the positions of the planets 2 months after those you plotted in Question 2 in Activity 20.1A.)

3. If you had observed Mercury, Venus, and Mars each night over the 2-month period May 1 to July 1, how would the position of each have changed in the sky relative to the background of stars? *Example:* Mercury traveled about 280° and moved from the constellation Pisces to the constellation Cancer.

Venus traveled _____° and moved from the constellation _____ to the constellation _____.

Mars traveled _____° and moved from the constellation _____ to the constellation _____.

20.2 Comparing Orbital Periods of the Planets

■ **Explain how the orbital period of a planet is related to its distance from the Sun.**

Orbital motion, or **revolution,** is a planet's motion around the Sun. Without exception, the orbital direction of the planets is counterclockwise around the Sun when the point of view is from above the Northern Hemisphere.

The time it takes a planet to complete one orbit around the Sun, called the **orbital period**, is also the length of its year. The units used to express a planet's orbital period are Earth days and years. The length of a planets' orbital period is directly related to its distance from the Sun

ACTIVITY 20.2
Comparing Orbital Periods of the Planets

1. On Figure 20.1, draw an arrow on the orbit of each planet to show its direction of its orbit around the Sun.

2. Refer to Table 19.1 in Exercise 19 (page 324), and record the distance from the Sun in astronomical units (average distance from Earth to Sun) and the orbital period for the planets listed below:

PLANET	DISTANCE (AU)	ORBITAL PERIOD
Mercury	_____	_____
Venus	_____	_____
Earth	_____	_____
Mars	_____	_____
Jupiter	_____	_____
Neptune	_____	_____

3. Using the information you compiled in Question 2, write a brief statement comparing the orbital period of a planet and its distance from the Sun.

_____

_____

_____

4. How many orbits will Mercury complete in 1 Earth year (365 days)?

_____

5. What fraction of its orbit will Mars complete in 1 Earth year?

_____

20.3   Retrograde Motion

■ **Describe the cause of the apparent retrograde motion of Mars, as observed from Earth.**

If you were able to look in the southern sky at the same time each night and note the position of Mars as compared to the constellations of stars, you would expect to find Mars a little farther to the east each night. However, every 2 years or so, there are a couple of months when Mars's motion in the night sky seems to change direction and move from east to west compared to the constellations of stars. The fact that Earth periodically over-takes Mars, as well as the other planets more distant from the Sun, makes the motion of the outer planets *appear* to move backward during that period. The apparent backward, or westward, movement of a planet when viewed from Earth is called **retrograde motion**. A similar observation is made when one vehicle passes another going the same direction on a highway. For a very brief time, relative to a fixed background, the vehicle being passed *appears* to move backward.

ACTIVITY 20.3
Retrograde Motion

Figure 20.2 illustrates six positions of Earth and Mars over a period of approximately 6 months. Complete the following steps, using Figure 20.2:

continued

Activity 20.3 continued

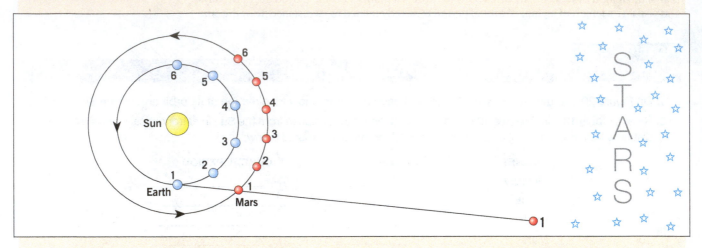

▲ **Figure 20.2** Diagram of the Sun, Earth, and Mars for illustrating retrograde motion. (Diagram *not* drawn to scale).

Step 1: On Figure 20.2, draw a line connecting Earth to Mars at each of the six positions in the orbits. Extend each line from Mars to the background of stars shown on the right side of the figure. Position 1 has been completed.

Step 2: Mark the place where each line intersects the background of stars and label it with the number of the position.

Step 3: Using a continuous line, connect the six numbered positions on the background of stars in order, 1 through 6.

1. Describe the apparent motion of Mars, as viewed from Earth, relative to the background of stars.

   _____

   _____

2. At any time, did Mars appear to move backward in its orbit?

   _____

3. Why does Mars exhibit retrograde motion?

   _____

   _____

4. Assume, as Ptolemy erroneously did in C.E..141, that our planet is stationary, and the planets and Sun orbit Earth. In this model of the solar system, how could the apparent backward motion of Mars be explained?

   _____

   _____

Figure 20.3 shows the apparent position of Mars in 2016, as it appeared from Earth, compared to the constellations of stars. The numbers next to Mars are dates—1/23 represents January 13, 2016; 2/12 represents Februarys 12, 2016; and so forth. Use Figure 20.3 to complete Questions 5–8.

5. As viewed from Earth, on what date was Mars roughly centered on the constellation Libra?

   _____, 2016

6. On about what date did Mars appear to change direction?

   _____, 2016

7. On what date did Mars resume its original eastward motion?

   _____, 2016

8. Approximately how long was the time span when Mars appeared to exhibit retrograde motion?

   _____

   _____

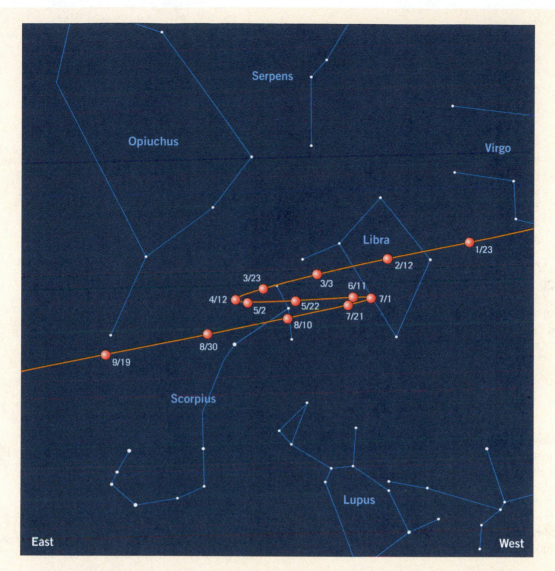

▲ **Figure 20.3** Illustration showing the position of Mars as seen in Earth's night sky from Pasadena, California, every 20 days from January 23 to October 19, 2016. Sky charts can be confusing because they are drawn as if you were sitting or lying down with your legs pointed south and you are looking up at the sky. In that position, east would be on your left and west on your right.

20.4 Viewing a Planet from Earth

■ **Using a model of the solar system, determine whether a planet can be seen from Earth on a specified date.**

Observing a planet requires a dark sky. In addition, the observer must be on the half of Earth facing the planet. Noon is experienced at the place on Earth facing directly toward the Sun, while midnight occurs at the location on the opposite side of Earth. Places experiencing sunrise and sunset are located between the noon and midnight positions.

ACTIVITY 20.4

Viewing a Planet from Earth

Use Figure 20.1 and the following steps to determine the best times for viewing a planet from Earth:

Step 1: On Figure 20.1, indicate the position of the planet to be observed and Earth's position on the desired date.

continued

Activity 20.4 continued

Step 2: At Earth's position, draw a circle the size of a dime and mark the locations of noon, sunset, midnight, and sunrise, as shown in **Figure 20.4**.

Step 3: Draw a line through the circle so that it separates the half of Earth facing the planet to be observed from the half facing away. In Figure 20.4 a red line has been drawn as an example.

Step 4: Keeping in mind that an Earth-bound observer must be on the half of Earth facing the planet to see it, note the two places where the line from Step 3 intersects Earth's surface. These will be the times the planet is first visible (*rises*) and when it is no longer visible (*sets*).

Use the May 1, 2019, positions of the planets plotted on Figure 20.1 to answer the following questions.

1. Using the steps listed above, estimate the approximate rising and setting times for each of the following planets on May 1, 2019:

PLANET	RISE	SET
Venus	_____	_____
Mars	_____	_____

2. On May 1, 2019, what are the best times to observe the following planets with the unaided eye? (*Hint:* Times must be between sunset, 6 Y.Y., and sunrise, 6 MY.)

   Venus: _____

   Mars: _____

3. On May 1, 2019, is Mercury visible in the evening, in the morning, or neither?

   _____

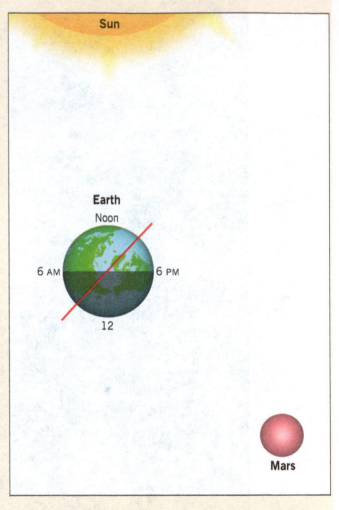

▲ **Figure 20.4** Illustration used to determine the best times for viewing Mars from Earth on a particular date.

4. Use Table 20.1 and what you have learned in this exercise to construct a model of the inner solar system for today's date on Figure 20.1. Show the positions of the four inner planets and label them with today's date.

5. Based on your model, will tonight—between 6 Y.Y. and midnight—be a good or poor time to view Mars?

   _____

MasteringGeology™

Looking for additional review and lab prep materials? Go to **www.masteringgeology.com** for Pre-Lab Videos, Geoscience Animations, RSS Feeds, Key Term Study Tools, The Math You Need, an optional Pearson eText, and more.

Locating the Planets

Name _____

Date _____

Course/Section _____

Due Date _____

1. Label the orbits of Mercury, Venus, Earth, and Mars on **Figure 20.5**.

2. Using data from Table 20.1, mark the positions of Mercury, Venus, and Mars for December 1, 2019, on Figure 20.5.

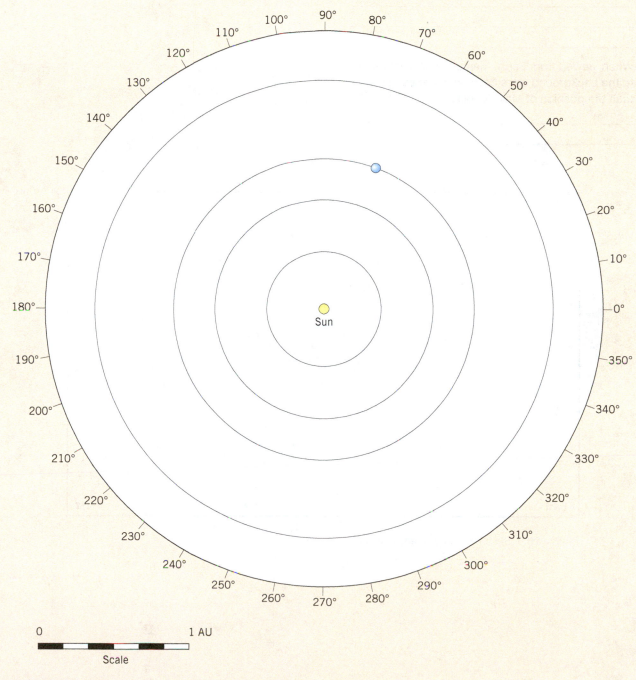

▲ **Figure 20.5** Sketch to accompany Questions 1 and 2.

3. Why does Mars periodically exhibit retrograde motion?

_____

_____

_____

4. Does Venus exhibit retrograde motion? Explain.

_____

_____

_____

5. Describe what is meant when a planet is said to be "in" a particular constellation.

_____

_____

_____

6. When viewed from Earth, Mercury's position relative to the background stars changes more each year than the position of Mars. Explain.

_____

_____

_____

_____

7. Describe the relative positions of Earth, Mars, and the Sun at the time when Mars is easiest to see from Earth.

_____

_____

_____

_____

8. **Figure 20.6** is a simple sketch that shows the path of Mars over several months, as viewed from Earth. Place several arrows to show the direction that Mars is moving along this path.

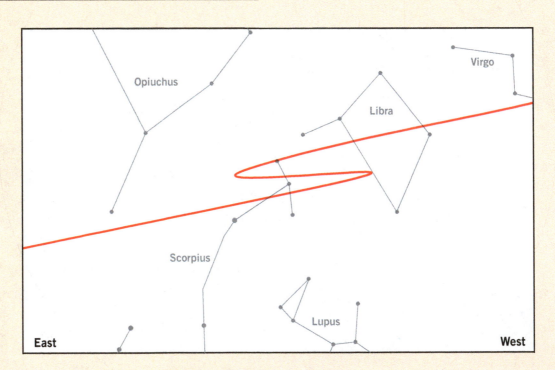

▲ **Figure 20.6** Sketch to accompany Question 8.

Examining the Terrestrial Planets

LEARNING OBJECTIVES

Each statement represents an important learning objective that relates to one or more sections of this lab. After you complete this exercise you should be able to:

- Describe the major geologic processes that have shaped the landscapes of the terrestrial planets.
- Identify landforms on each of the terrestrial planets and the Moon.
- Describe the procedure for determining the relative ages of a planet's surface features.
- List the factors that influence the size of impact craters.

MATERIALS

ruler	calculator	meterstick
colored pencils	sand and sandbox	small projectiles

PRE-LAB VIDEO ▶ https://goo.gl/hBfdrC

 Prepare for lab! Prior to attending your laboratory session, view the pre-lab video. Each video provides valuable background that will contribute to your understanding and success in lab.

INTRODUCTION

This exercise examines the four main geologic processes—*volcanism*, *tectonic activity*, *gradation*, and *impact cratering*—that have shaped the topography of the terrestrial planets: Mercury, Venus, Earth, and Mars. Volcanism and tectonic activity are driven by a planet's internal forces and cause portions of a planet's surface to gradually become elevated. By contrast, gradation and impact cratering are external processes. *Gradation*, which involves weathering, mass wasting, and erosion, tends to lower elevated surface features and reduce the overall relief of a region.

PART 4 Astronomy

21.1 Geologic Processes

■ Describe the major geologic processes that have shaped the landscapes of the terrestrial planets.

The forces that drive Earth's evolution and shape its topography also operate elsewhere in the solar system. Because the four processes listed in the introduction produce distinctive landforms, identifying and analyzing the surface features they generate makes it possible to unravel the geologic history of Earth and other planetary surfaces.

ACTIVITY 21.1A

Volcanism

During **volcanism**, molten rock (**magma**) and its entrapped gases erupt onto a planet's surface as **lava**. Volcanic cones with summit *craters*, or large *calderas*, and vast plains covered by lava flows are examples of volcanic landforms. We will use Mount Shasta, California, which consists of overlapping cones, including the prominent secondary cone Shastina, as our model for looking at volcanism on Earth. **Figure 21.1** provides two views of Mount Shasta, the most massive volcano in the Cascade Range.

A. B.

▲ **Figure 21.1** Mount Shasta, California. (**A.** Courtesy of U.S. Geological Survey. **B.** Photo by University of Washington Libraries, Special Collections, John Shelton Collection, KC8663)

1. On the aerial views in Figure 21.1A, B, label the summit (background) and Shastina (foreground).

2. Measure the width of Mount Shasta in Figure 21.1A along line A–B.

_____ km

3. Lava flows and volcanic debris flows extend far beyond the bases of most volcanoes. Outline the margin of the large, dark-colored lava flow in Figure 21.1B.

4. Measure the width of the outer margin of the lava flow along line C–D in Figure 21.1B.

_____ km

ACTIVITY 21.1B
Tectonic Activity

Tectonic activity involves the deformation of crustal rock by fracturing, faulting, and/or folding. Tectonic activity is responsible for the major structural features of a planet, including ocean basins and continents. Landforms produced by tectonic activity include *mid-ocean ridges*, *rift valleys*, *fault-generated scarps* (cliffs), and *mountainous terrains*. On Earth most tectonic activity occurs along plate boundaries.

Figure 21.2 is an aerial view of a portion of the Valley and Ridge Province of the Appalachian Mountains, which is composed of folded sedimentary strata.

1. Label two of the ridges and the intervening valley between them on the image.

2. Measure the width of the area shown on this image.

_____ km

3. What geologic process *deformed* the rock exposed at the surface: volcanism, tectonic activity, gradation, or impact cratering?

_____ deformed the rock exposed at the surface.

▲ **Figure 21.2** Valley and Ridge Province, central Pennsylvania. (Courtesy of NASA)

ACTIVITY 21.1C
Gradation

Gradation is a general term that refers to the external processes that lower landforms. Rock is broken down (weathered), moved to lower elevations by gravity (mass movement), and/or carried away by erosional agents such as running water, wind, and glacial ice to locations where it is deposited. Earth's atmosphere plays a key role in these processes.

1. **Figure 21.3** is an image of a landscape on Earth that has been dissected by running water. In your own words, describe the drainage pattern in this area.

_____

_____

_____

▶ **Figure 21.3** Satellite image of a landscape eroded by running water. (Courtesy of NASA)

continued

Activity 21.1C continued

2. Is the erosional agent that created the sand dunes in **Figure 21.4** running water, wind, or glacial ice?

_____ created the sand dunes.

3. Did the erosional agent in Figure 21.4 move from left to right or right to left?

The erosional agent moved from _____ to _____.

▶ **Figure 21.4** Sand dunes.
(Photo by Michael Collier)

©2019 Pearson Education, Inc.

ACTIVITY 21.1D
Impact Cratering

The fourth geologic process acting on a planet's surface, **impact cratering**, is the result of rapidly moving *meteoroids* striking a planetary surface. Large craters often have *central peaks* and slump blocks around the rim. In addition, they are usually surrounded by a thick blanket of material called *ejecta*. By contrast, small craters are simple structures with well-defined rims and bowl-shaped depressions. Although impact cratering was important in Earth's early evolution, it has had a minimal role in modifying modern landscapes compared to the other processes.

Figure 21.5 is an image of Meteor Crater, an impact crater in Arizona that is about 50,000 years old. Meteor Crater has a diameter of approximately 1200 meters. Scientists estimate that it was produced by an object about 50 meters across.

1. How many times larger is the diameter of Meteor Crater than the object that produced it?

_____

times larger

2. Suggest a reason why such a comparatively small object can form such a large hole.

_____

_____

3. Despite being one of the best-preserved and youngest impact craters on Earth, Meteor Crater shows signs of erosion. Describe this evidence.

_____

_____

4. It is possible that Meteor Crater once had a central peak. If it existed, what might have happened to this feature?

_____

5. Label at least three large blocks of ejecta on the image.

▲ **Figure 21.5** Meteor Crater. (Courtesy of NASA)

21.2 Landforms on the Moon

■ **Identify landforms on each of the terrestrial planets and the Moon.**

The Moon has two major types of terrains that can be easily seen with the unaided eye—bright highlands and dark smooth regions that resemble seas. The **lunar highlands** are elevated several kilometers above the dark areas called **maria basins**. The maria basins are large impact basins that were subsequently filled with numerous layers of fluid basaltic lava

ACTIVITY 21.2
Landforms on the Moon

1. On the central image of **Figure 21.6**, page 350, label the lunar highlands **A**, label the maria basins **B**, label the large craters called *rayed craters* that have bright rays of ejecta extending far beyond their crater rims **C**, and label the small basins (craters) that have been flooded with basaltic lava **D**. Letters may be used more than once.

2. Which of the areas of the Moon—the bright highlands or the dark maria—are more heavily cratered?

   The _____ are more heavily cratered.

3. Figure 21.6 shows the side of the Moon that faces Earth. Does approximately 20, 40, or 70 percent of this side of the Moon consist of lunar maria?

   _____%

4. Figure 21.6A is an image of Euler Crater, located in the western part of Mare Imbrium. Label its central peak, the ejecta surrounding the crater, and the crater rim.

5. Measure the diameter of Euler Crater from rim to rim.

   _____ km

6. How many times larger or smaller is Euler Crater compared to Meteor Crater?

   Euler Crater is about _____ times _____ than Meteor Crater.

21.3 Landforms on Mercury

■ **Identify landforms on each of the terrestrial planets and the Moon.**

Mercury, the planet closest to the Sun, is only slightly larger than Earth's Moon. Because Mercury lacks an atmosphere, it experiences noontime temperatures approaching 800°F (427°C)—one of the most extreme environments in the solar system.

ACTIVITY 21.3
Landforms on Mercury

Use the images in **Figure 21.7**, page 351, to answer the following questions.

1. How does the surface of Mercury compare to that of Earth's Moon? List at least one similarity and one difference.

   _____
   _____

2. Figure 21.7A shows a close-up view of a portion of Mercury's surface. Figure 21.7A resembles what landform on the Moon?

   _____

3. Figure 21.7B resembles what feature on the Moon? What is the most likely rock type that covers the central part of this structure?

   _____
   _____

continued

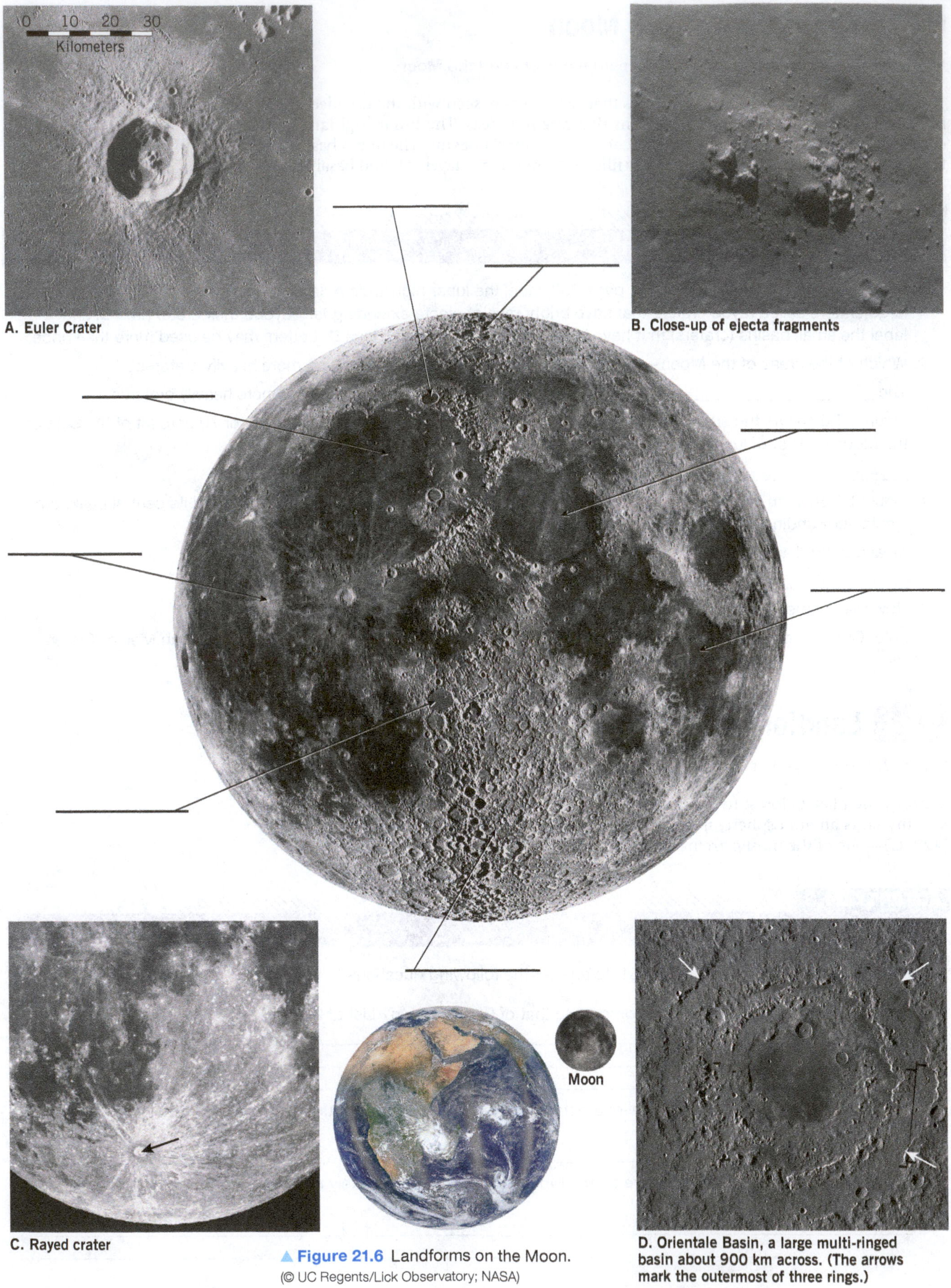

A. Euler Crater

B. Close-up of ejecta fragments

Moon

C. Rayed crater

▲ Figure 21.6 Landforms on the Moon.
(© UC Regents/Lick Observatory; NASA)

D. Orientale Basin, a large multi-ringed basin about 900 km across. (The arrows mark the outermost of three rings.)

A.

B.

Mercury

C.

D.

▲ **Figure 21.7** Landforms on Mercury. (Courtesy of NASA)

Activity 21.3 continued

4. The arrows on Figure 21.7C point to the trace of a cliff-like structure that has an offset of more than 1.6 kilometers (1 mile). Which of the four major geologic processes created this structure?

_____

5. Label the central peak on each of two of the craters in Figure 21.7D.

6. Of the four major processes that alter a planet's surface, which appears to be the least noticeable on Mercury?

_____

7. Why does Mercury show little evidence of erosion due to running water, wind, or ice?

_____

21.4   Landforms on Venus

■ **Identify landforms on each of the terrestrial planets and the Moon.**

Venus, second only to the Moon in brilliance in the night sky, is similar to Earth in size, density, mass, and location in the solar system. However, the similarities end there. The Venusian atmosphere consists of 97 percent carbon dioxide, a greenhouse gas, which is responsible for surface temperatures that reach 475°C (900°F). Although strong winds occur aloft, surface winds average only a few meters per second. Surface pressures are 92 times greater than those on Earth.

Venus is shrouded in thick, opaque clouds that completely hide the surface from the view of traditional cameras. Nevertheless, using radar, which can penetrate clouds, the *Magellan* spacecraft has provided thousands of radar images of the Venusian surface that reveal its varied topography (**Figure 21.8**). In general, radar images show rough topography (fractured rock, rough lava flows, and ejecta) as bright regions, while areas covered by smooth lava flows appear dark. About 80 percent of the surface of Venus consists of low-lying plains covered by lava flows.

ACTIVITY 21.4
Landforms on Venus

Use the Venusian images in Figure 21.8 to answer the following questions.

1. Figure 21.8A shows extensive lava flows on Venus that originated from the volcano Ammavaru, which lies about 300 kilometers west (left) of the image. Outline the margin of the lava flow, which appears bright in this radar image. (*Hint:* Remember that rough lava flows appear bright in radar images, and smooth flows appear dark.)

2. Figure 21.8B shows prominent circular volcanic structures, which are thought to be similar to lava domes on Earth. On Venus, these structures are referred to as *pancake domes*. Measure the diameter of one of the two largest pancake domes.

_____ km

3. How does the diameter of the volcanic dome compare to the diameter of Mount Shasta, the volcanic structure that you measured in Question 2 in Activity 21.1A?

_____

4. Do the volcanic domes in Figure 21.8B have steep conical tops or broad flat tops?

_____

5. Are the outer margins of these volcanic structures steeply sloping or gently sloping?

_____

6. Based on your answer to Question 5, were these structures produced by fluid or viscous lavas?

_____

continued

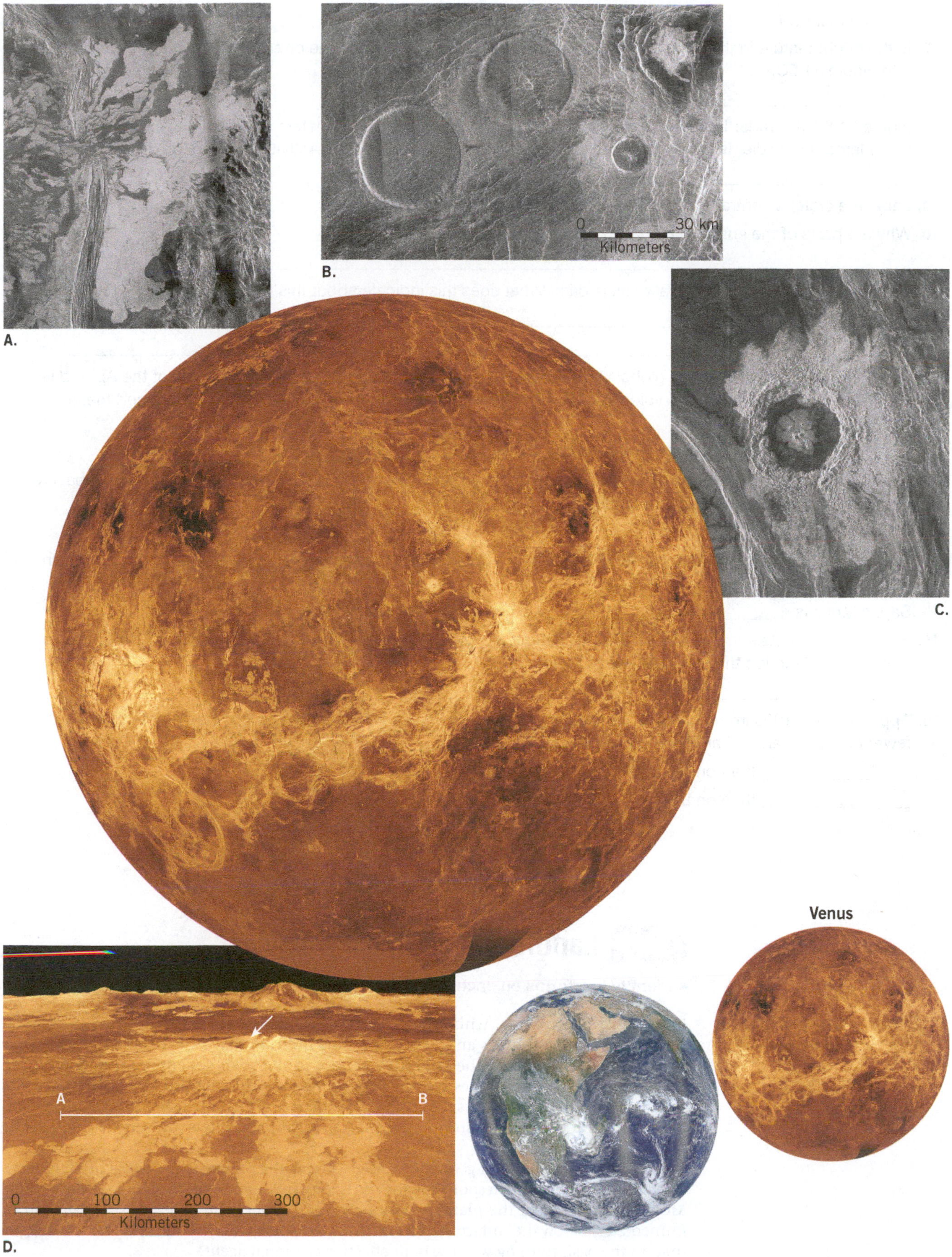

A.

0   30 km
Kilometers

B.

C.

A   B

0   100   200   300
Kilometers

D.

Venus

▲ **Figure 21.8** Landforms on Venus. (Courtesy of NASA)

continued

Activity 21.4 continued

7. Is the bright feature in the upper-right corner of Figure 21.8B a volcanic cone or an impact crater? (*Hint:* Compare it to Figure 21.8C.)

   _____

8. Figure 21.8C is a radar image of Dickinson Crater, an impact crater 69 kilometers (42 miles) wide. How many times larger or smaller is Dickinson Crater than Meteor Crater (described in Activity 21.1D, page 348)?

   _____

9. Label the crater's central peak, crater floor, and ejecta on Figure 21.8C.

10. Why are parts of the image in Figure 21.8C bright white in color?

    _____

11. The area surrounding the central peak is dark. What does this indicate about this feature?

    _____

    _____

12. The bright areas on the central global view of Venus show highly fractured ridges and canyons of the Aphrodite Terra. What geologic process—volcanism, tectonic activity, gradation, or impact cratering—produced these features?

    _____

13. Figure 21.8D shows a false-color satellite image of the volcano Sapas Mons. Using the scale provided, measure the width of the base of Sapas Mons along line A–B.

    _____ km

14. Approximately how many times wider is Sapas Mons than Mount Shasta, which you examined in Question 2 in Activity 21.1A?

    Sapas Mons is _____ times wider than Mount Shasta.

15. From its base to its summit, Mount Shasta is about 3.5 kilometers high, whereas Sapas Mons is 1.5 kilometers high. Briefly describe the shape of Sapas Mons.

    _____

16. Approximately 1000 impact craters have been identified on Venus. Is this fewer or more than on Mercury? Is this fewer or more than on Earth?

    _____ than on Mercury

    _____ than on Earth

21.5 Landforms on Mars

■ **Identify landforms on each of the terrestrial planets and the Moon.**

Mars is similar to Earth, with a thin atmosphere, polar ice caps, and seasons. Numerous large shield volcanoes and vast lava plains indicate that Mars has had an extensive volcanic history. Some of the largest volcanoes in the solar system are found on Mars. However, all of the volcanoes on Mars are apparently extinct. Unlike Earth, Mars does not experience plate tectonics, but it does exhibit many tectonic features caused by compressional and extensional stresses. Furthermore, impact craters are very evident on the Martian surface.

Erosion is the dominant geologic process altering the present-day surface of Mars. Mass movement and wind are responsible for landslides and numerous sand dunes, respectively. Moreover, because of the planet's extremely low surface temperatures, the water on or immediately below the surface of Mars is frozen. However, many Martian landforms suggest that, in the past, running water was an effective erosional agent.

ACTIVITY 21.5
Landforms on Mars

After you examine the images of Mars in **Figure 21.9**, page 356, complete the following.

1. The long canyon system near the center of the full-disk image of Mars is Valles Marineris. Measure the length of this feature.

   _____ km

2. How does the length of Valles Marineris compare to the width of the 48 contiguous states?

   _____

3. Based on your answer to Question 2, do you think Valles Marineris is larger or smaller than Arizona's Grand Canyon?

   _____

4. The image shown in Figure 21.9A is a close-up of a valley wall of Valles Marineris. Name the erosional agent thought to be responsible for producing the gullies in this image.

   _____

Referring to the impact crater in Figure 21.9B, complete Questions 5–7.

5. Label the crater rim and the central peak on the image.

6. Because the subsurface of some regions on Mars contains abundant ice, impacts often generate ejecta with a mud-like consistency. Outline the ejecta surrounding this crater.

7. Compare the appearance of the ejecta blanket that surrounds this crater with the ejecta of Dickinson Crater shown in Figure 21.8C.

   _____

   _____

Use Figure 21.9C, which shows Olympus Mons, one of four huge volcanoes in a region called Tharsis, to complete Questions 8 and 9.

8. What is the approximate width of Olympus Mons?

   _____ km

9. How does the width of Olympus Mons compare to the width of Mount Shasta, shown in Figure 21.1A, page 346?

   _____

Use Figure 21.9D, which shows channels on the surface of Mars, to complete Questions 10 and 11.

10. Describe the pattern of channels marked with the letter A in Figure 21.9D.

    _____

    _____

11. Which of these processes—wind, water, glacial ice, or mass movement—most likely produced the pattern of channels marked with the letter A in Figure 21.9D?

    _____

21.6 Determining the Relative Ages of Landforms

■ **Describe the procedure for determining the relative ages of a planet's surface features.**

Although the Moon and planets formed at the same time, approximately 4.6 billion years ago, due to variations in the rates of geologic activity, their current surfaces differ significantly in age. When comparing planetary surfaces, impact craters provide a means for determining relative ages. In general, older surfaces show more impact craters per unit area In addition, old craters lack ejecta blankets and rays because these features are obliterated by subsequent bombardment, and some old craters become partially buried by younger lava flows.

A.

B.

C.

0   100   200
Kilometers

D.

0   1000   2000
Kilometers

Mars

▲ **Figure 21.9** Landforms on Mars. (Courtesy of NASA)

ACTIVITY 21.6 ///
Determining the Relative Ages of Landforms

Use Figure 21.6 (page 350) to answer the following questions.

1. Based on crater density, which region—the lunar highlands or the maria basins—is older?

   The _____ are older.

2. Rocks brought back from the lunar maria by the *Apollo* astronauts are between about 3.2 and 3.8 billion years old. Are the lunar highlands older or younger than 3.2–3.8 billion years? Briefly explain.

   _____

   _____

3. Are rayed craters that have bright ejecta deposits around them older or younger than the regions that surround them? How did you arrive at your answer?

   _____

   _____

4. List these lunar features in order, from oldest to youngest: maria, highlands, rayed craters.

   Oldest: _____

   Middle: _____

   Youngest: _____

5. Based on crater density, is the surface of Mercury (Figure 21.7D, page 351) older than, younger than, or about the same age as the maria (Figure 21.6D, page 350) on the Moon?

   The surface of Mercury is _____ the maria on the Moon.

6. Does the surface of Mars (Figure 21.9) appear to be older than, younger than, or about the same age as the Moon's highlands? Does the surface of Mars appear to be older than, younger than, or about the same age as the surface of Mercury?

   The surface of Mars appears to be _____ the lunar highlands.

   The surface of Mars appears to be _____ the surface of Mercury.

7. Which of the four planetary surfaces that you have investigated appears to have the most active geologic history? Which appears to have the least active geologic history?

   Most active: _____

   Least active: _____

8. Based on crater density, list the surfaces of the four terrestrial planets, from oldest to youngest.

   Oldest: _____

   Second oldest: _____

   Third oldest: _____

   Youngest: _____

9. Describe how the surface of Mercury is likely to change during the next billion years.

   _____

   _____

21.7 Impact Cratering Experiment

■ **List the factors that influence the size of impact craters.**

As preceding sections have demonstrated, impact cratering is one of the most recognizable processes responsible for altering the surfaces of Earth's Moon and the terrestrial planets.

ACTIVITY 21.7

Impact Cratering Experiment

Observe the equipment shown in **Figure 21.10** and conduct the following experiment:

Step 1: Gather the equipment necessary to conduct the impact cratering experiment.

Step 2: Add sand to the sandbox. Use a ruler to flatten the surface of the sand.

Step 3: Write a hypothesis that describes what you think the relationship will be between an impact crater's diameter and the mass and velocity of the object that produces it.

_____

_____

_____

Step 4: One at a time, drop each of the projectiles from heights of 0.5 meter, 1.0 meter, and 1.5 meters on the sand in the box and measure the diameter, in millimeters, of the crater produced each time. Make sure to flatten the surface of the sand between trials. Repeat the experiment three times and average your results. Record the average for each of the trials on the data sheet in **Table 21.1**.

1. Examine your data closely and state your conclusions concerning the general relationships between crater size and both the mass and the velocity of the object that produced the crater.

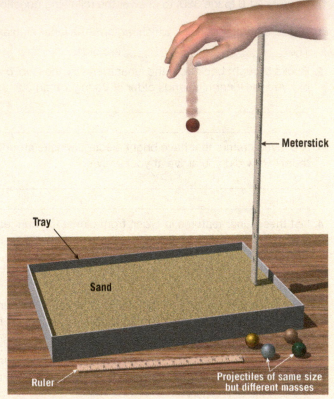

▲ **Figure 21.10** Impact cratering lab equipment.

_____

_____

_____

_____

2. Write a general statement that evaluates your impact-cratering hypothesis with reference to your conclusions.

_____

_____

_____

Table 21.1 Impact Crater Data Sheet

PROJECTILE TYPE		CRATER DIAMETER		
		0.5 M HEIGHT	1.0 M HEIGHT	1.5 M HEIGHT
DESCRIPTION	MASS	CRATER DIAMETER (mm)	CRATER DIAMETER (mm)	CRATER DIAMETER (mm)

MasteringGeology™

Looking for additional review and lab prep materials? Go to **www.masteringgeology.com** for Pre-Lab Videos, Geoscience Animations, RSS Feeds, Key Term Study Tools, The Math You Need, an optional Pearson eText, and more.

Examining the Terrestrial Planets

Name _____

Date _____

Course/Section _____

Due Date _____

1. List the four major geologic processes that have shaped the surfaces of the terrestrial planets and the Moon.

   _____

   _____

   _____

   _____

2. Does the surface of the Moon most resemble the surface of Mercury, Venus, Earth, or Mars?

   _____

3. Which planet's surface has been mapped exclusively by radar? Why is radar mapping used?

   _____

   _____

   _____

4. Which planet—Mercury, Venus, Earth, or Mars—has the largest known volcano in the solar system?

   _____

5. On which planet other than Earth is it most likely that water-deposited sediments occur? Explain your choice.

   _____

   _____

6. Notice the teardrop-shaped elevated landform near the center of **Figure 21.11** that extends behind an impact crater. This landform was generated because the raised, rocky rim of the crater diverted floodwater and protected the ground behind it from erosion.

   a. In what general direction was the floodwater flowing—upper left to lower right, or lower right to upper left?

      _____

   b. On what planet—Mercury, Venus, Earth, or Mars—are these landforms most likely found? Explain.

      _____

      _____

7. **Figure 21.12** is an image of Tycho Crater, a relatively young crater about 86 kilometers (53 miles)

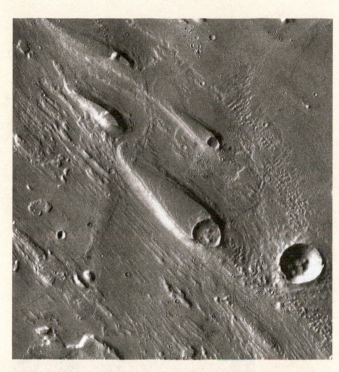

▲ **Figure 21.11** Image to accompany Question 6. (Courtesy of NASA)

▲ **Figure 21.12** Image to accompany Question 7. (Courtesy of NASA)

wide and 4.8 kilometers (3 miles) deep, located on the Moon.

a. Label the central peak on the image.

b. Label the crater rim on the image.

c. Describe the size of Tycho Crater compared to Meteor Crater, shown in Figure 21.5.

_____

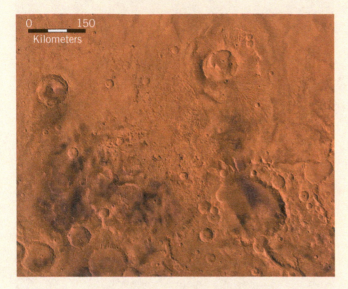

▲ **Figure 21.13** Image to accompany Question 9.
(Courtesy of NASA)

8. Briefly describe how the relative ages of various landforms can be determined using crater density.

_____

_____

_____

_____

9. Compare the bottom half of **Figure 21.13** to the top half. Which surface is older? Explain how you arrived at your conclusion.

The _____ half is older.

_____

_____

10. **Figure 21.14** is a three-dimensional image of Venus's largest volcano, Maat Mons. The vertical scale has been exaggerated 10 times, and Maat Mons is much wider than it is tall—almost 400 kilometers wide and less than 10 kilometers high.

a. Based on this information, what type of volcano is Maat Mons?

_____

b. The bright areas surrounding the volcano are lava flows. What is the likely composition of these lava flows?

_____

_____

▲ **Figure 21.14** Image to accompany Question 10 (Courtesy of NASA)

Motions of the Earth–Moon System

LEARNING OBJECTIVES

Each statement represents an important learning objective that relates to one or more sections of this lab. After you complete this exercise you should be able to:

- **Recognize and name each of the phases of the Moon.**
- **Explain the difference between the synodic and sidereal cycles of the Moon.**
- **Sketch the positions of Earth, the Moon, and the Sun during solar and lunar eclipses.**

MATERIALS

ruler
pencil

PRE-LAB VIDEO ▶ https://goo.gl/vQUupM

 Prepare for lab! Prior to attending your laboratory session, view the pre-lab video. Each video provides valuable background that will contribute to your understanding and success in lab.

INTRODUCTION

Earth has one natural satellite—the Moon. In addition to accompanying Earth in its annual trek around the Sun, the Moon orbits Earth about once each month. When viewed from a Northern Hemisphere perspective, the Moon moves counterclockwise (eastward) around Earth. The motions of the Earth–Moon system constantly change the relative positions of the Sun, Earth, and Moon. The results are some of the most noticeable astronomical phenomena—the *phases of the Moon* and the occasional *eclipses of the Sun and Moon.*

PART 4 **Astronomy**

22.1 Phases of the Moon

■ **Recognize and name each of the phases of the Moon.**

The changing phases of the Moon were among the earliest astronomical phenomena to be understood. The lunar phases result from the motion of the Moon and sunlight that is reflected from the Moon's surface. Half of the Moon is always illuminated by the Sun, but the portion that is visible to an observer on Earth changes daily. When the Moon is located between the Sun and Earth, the bright side of the Moon *cannot* be seen from Earth—a phase called the *new-Moon* ("no moon") phase. Conversely, when the Moon lies on the side of Earth opposite the Sun, all of its bright side is visible, producing the *full-Moon* phase. At any position between these extremes, only a fraction of the Moon's illuminated half is visible.

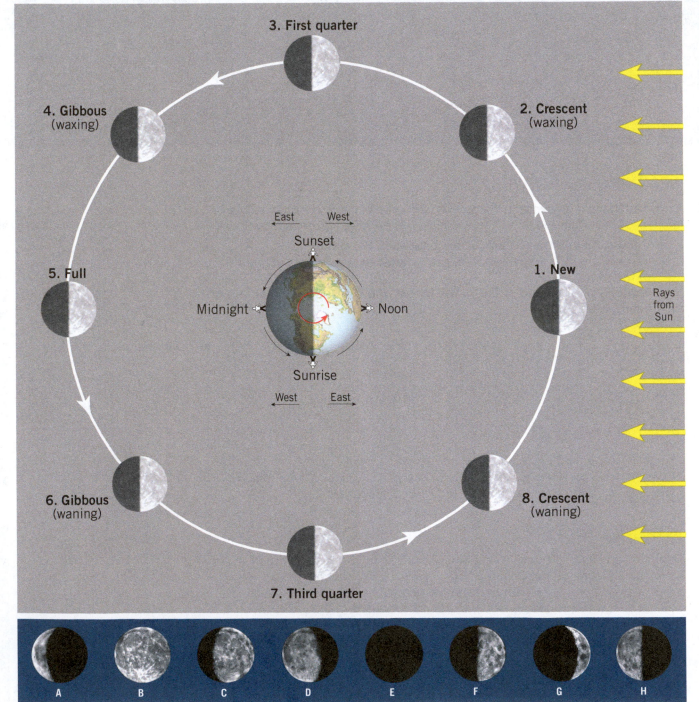

PHASES FROM EARTH

▲ **Figure 22.1** The lunar cycle, as viewed from above the Northern Hemisphere of Earth.

ACTIVITY 22.1A ///
Phases of the Moon

Figure 22.1 is a view of the Earth–Moon system from above the Northern Hemisphere of Earth. Eight positions of the Moon, during its monthly journey around Earth, are illustrated. Complete the following questions using Figure 22.1.

1. For each of the eight numbered positions of the Moon, draw a line through the Moon that separates the half of the Moon that is facing Earth from the half that cannot be seen.

2. With a pencil, darken the portion (half) of the Moon that is *not* visible from Earth in each of the eight numbered positions.

3. As the Moon journeys from position 1 to position 5, does the portion of its *illuminated side visible from Earth* increase or decrease?

_____

4. As the Moon journeys from position 5 to position 8, does the portion of its *illuminated side visible from Earth* increase or decrease?

_____

5. Match each of the eight images (A–H) of the lunar phases as seen from Earth with the Moon's corresponding position (1–8) in the space provided below.

POSITION NUMBER	LETTER OF PHASE SEEN FROM EARTH	NAME OF PHASE
1.	_____	_____
2.	_____	_____
3.	_____	_____
4.	_____	_____
5.	_____	_____
6.	_____	_____
7.	_____	_____
8.	_____	_____

ACTIVITY 22.1B ///
Observing the Phases of the Moon

Observe the time of day (noon, sunset, and so on) represented by the four positions of an observer on Earth in Figure 22.1. Find the observer in the "noon" position. Because of Earth's counterclockwise rotation, about 6 hours later, that observer will be in the "sunset" position and will see the Sun setting in the west.

Use the eight positions of the Moon in its orbit and the times represented on Earth to answer the following questions. (*Note:* Except during the new-Moon phase, you can assume that the Moon is visible from Earth during some daylight hours.)

1. Is the full Moon highest in the sky to an Earth-bound observer at noon, sunset, midnight, or sunrise?

_____

2. Can a full Moon be observed from Earth at noon? Explain the reason for your answer.

_____

_____

3. Does the third-quarter Moon appear highest in the sky at noon, sunset, midnight, or sunrise?

The third-quarter Moon is highest at _____.

continued

Activity 22.1B continued

4. Does a full Moon set—that is, become no longer visible—at noon, sunset, midnight, or sunrise? Should an observer look eastward or westward to see the setting full Moon? (*Hint:* Earth's rotation causes the Moon to rise and set.)

   A full Moon sets at _____.

   An observer must look _____ to see the setting full Moon.

5. Can the first- and third-quarter lunar phases be observed during daylight hours? Explain the reason for your answer.

   _____

   _____

6. At approximately what times does the first-quarter Moon rise and set?

   Rise: _____

   Set: _____

7. At approximately what times does the third-quarter Moon rise and set?

   Rise: _____

   Set: _____

8. Throughout the lunar cycle, the Moon moves further eastward in the sky. Therefore, to an observer on Earth, does the time of day when the Moon is highest in the sky become progressively earlier or later?

   The time becomes progressively _____.

9. Assume that a crescent-phase Moon is observed in the early evening in the western sky. During the next few days, will the Moon be rising earlier or later? Will the visible portion of the Moon become progressively larger or smaller?

   The Moon will be rising _____.

   The visible portion of the Moon will become progressively _____.

22.2 Synodic and Sidereal Months

■ **Explain the difference between the synodic and sidereal cycles of the Moon.**

The time interval required for the Moon to complete a full cycle of phases is 29.5 days, a period of time called the **synodic month**. This complete cycle of the phases of the Moon (that is, one new Moon to the next new Moon) is the basis of the word "month" (or "moonth"). Although the cycle of phases requires 29.5 days, the true period of the Moon's 360° revolution around Earth takes only 27.3 days and is known as the **sidereal month**. The difference of approximately 2 days results from the fact that as the Moon revolves around Earth, the Earth–Moon system is also orbiting the Sun.

ACTIVITY 22.2

Synodic and Sidereal Months

Figure 22.2 illustrates the monthlong motions of the Earth–Moon system around the Sun.

1. On Month 1 in Figure 22.2, indicate the dark half of the Moon on each of the eight lunar positions by shading the area with a pencil.

2. Select from the eight lunar positions (labeled 1–8) in Month 1 and indicate which of them represents each of the following lunar phases:

PHASE	LUNAR POSITION (MONTH 1)
New moon	_____
Third-quarter moon	_____
Full moon	_____
First-quarter moon	_____

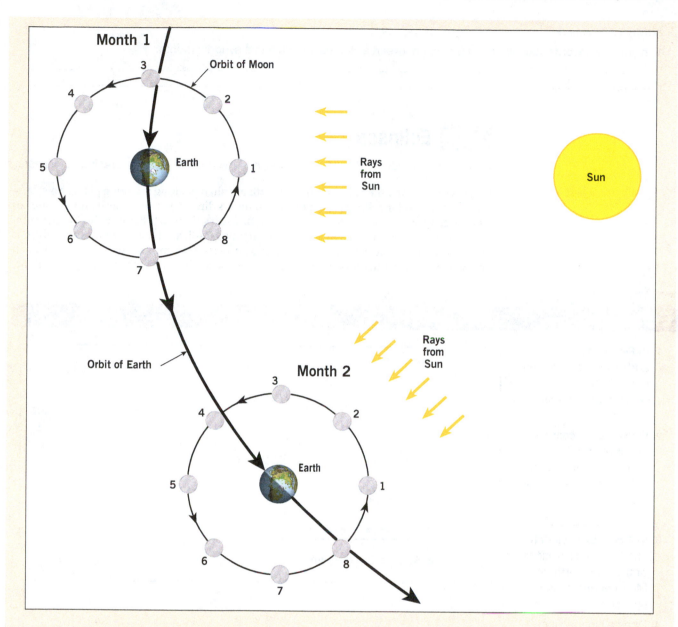

▲ **Figure 22.2** Monthly motion of the Earth–Moon system around the Sun, viewed from above the Northern Hemisphere.

3. On Month 1, label the position of the new-Moon phase with the words *new Moon*.

4. Begin with position 1 (new-Moon phase) in Month 1 and imagine that the Moon is revolving 360° around Earth. At the same time, the Moon is following Earth's orbit toward Month 2. After the 360° revolution is complete, the Moon returns to position 1, and it is 1 month later (Month 2). Circle position 1 on the Month 2 diagram.

5. What is the name for one complete 360° revolution of the Moon around Earth? How many days does this take?

The moon will have completed a _____ month.

This takes _____ days.

6. Is position 1 in Month 2 the new-Moon phase?

_____

7. In what position does the new-Moon phase occur in Month 2?

Position: _____

8. When the Moon moves in its orbit from position 1 to position 2 and is once again in the new-Moon phase, what type of month will have been completed? How many days does this take?

The Moon will have completed a _____ month.

This takes _____ days.

continued

Activity 22.2 continued

9. In your own words, explain the difference between a sidereal month and a synodic month.

_____

_____

22.3 Eclipses

■ **Sketch the positions of Earth, the Moon, and the Sun during solar and lunar eclipses.**

Eclipses occur when the Sun, Moon, and Earth are aligned along the same plane. An eclipse can be either a **solar eclipse** (when the Moon moves directly between Earth and the Sun) or a **lunar eclipse** (when the Moon moves within Earth's shadow). Because the Moon's orbit is slightly tilted from the imaginary plane containing Earth's orbit around the Sun, which is called the **plane of the ecliptic**, an eclipse *does not* occur each month. An eclipse can happen only when a new or full Moon occurs while the Moon's orbit crosses the plane of the ecliptic.

ACTIVITY 22.3
Eclipses

1. In Figure 22.3, sketch and label the positions of the Sun, Earth, and Moon during a solar eclipse.

2. Does a solar eclipse occur during the new-Moon, first-quarter, or full-Moon phase of the Moon?

_____

3. In Figure 22.4, sketch and label the positions of the Sun, Earth, and Moon during a lunar eclipse.

4. Does a lunar eclipse occur during the new-Moon, third-quarter, or full-Moon phase of the Moon?

_____

5. Suggest a reason Earth does not experience a solar and lunar eclipse every month.

_____

▲ **Figure 22.3** Solar eclipse diagram.

▲ **Figure 22.4** Lunar eclipse diagram.

_____

_____

MasteringGeology™

Looking for additional review and lab prep materials? Go to www.masteringgeology.com for Pre-Lab Videos, Geoscience Animations, RSS Feeds, Key Term Study Tools, The Math You Need, an optional Pearson eText, and more.

Motions of the Earth–Moon System

Name _____

Date _____

Course/Section _____

Due Date _____

1. Write a description of how the phases of the Moon change during a full lunar cycle. Begin with the new-Moon phase.

   _____

   _____

   _____

   _____

   _____

2. Each of the four photographs in **Figure 22.5** was taken when the Moon was at its highest position in the sky. In the space provided below each photo, write the name of the phase represented and the time of day when the picture was taken.

3. Explain the difference between a sidereal month and a synodic month.

   _____

   _____

   _____

4. How does the crescent phase that *precedes* the new-Moon phase differ from the crescent phase that *follows* the new-Moon phase? (*Hint:* See the bottom of Figure 22.1.)

   _____

   _____

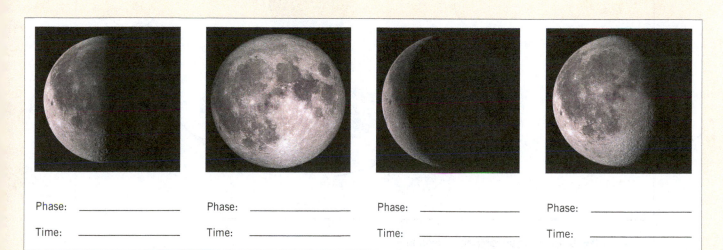

Phase: _____   Phase: _____   Phase: _____   Phase: _____

Time: _____   Time: _____   Time: _____   Time: _____

▲ **Figure 22.5** Photos to accompany Question 2. (NASA)

continued

5. What phase of the Moon occurs approximately 1 week after the new Moon? 2 weeks?

   1 week: _____

   2 weeks: _____

6. Does a solar eclipse occur during the new-Moon, first-quarter, or full-Moon phase of the Moon?

   _____

7. Does a lunar eclipse occur during the new-Moon, third-quarter, or full-Moon phase of the Moon?

   _____

8. **Figure 22.6** illustrates the Earth–Moon system viewed from above the Northern Hemisphere. First, label the positions of the Moon that represent the new-Moon, first-quarter, third-quarter, full-Moon, gibbous, and crescent phase of the Moon. Next, using Figure 22.1, page 362, as a guide, shade the circles representing the Moon to depict the portion of the Moon that would be visible from Earth during each of these phases.

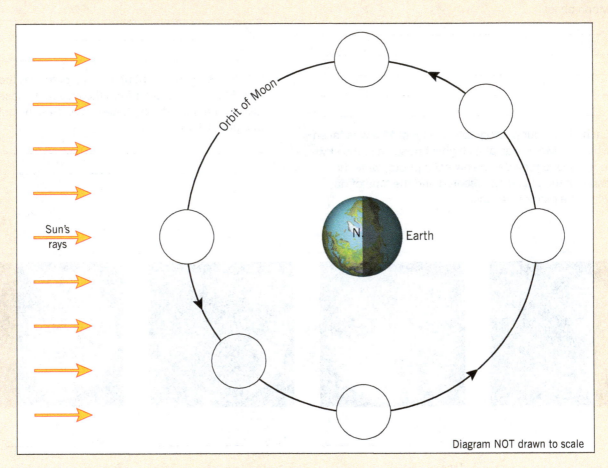

▲ **Figure 22.6** Diagram to accompany Question 8.

Location and Distance on Earth

LEARNING OBJECTIVES

Each statement represents an important learning objective that relates to one or more sections of this lab. After you complete this exercise you should be able to:

- **Distinguish between latitude and longitude and explain how latitude and longitude are used to locate places on Earth.**
- **Apply knowledge of Earth's grid system to accurately locate a place or feature.**
- **Contrast great circles and small circles.**
- **Determine the shortest distance between any two places on Earth's surface.**
- **Describe how distance can be measured along a parallel of latitude.**
- **Explain the relationship between longitude and solar time.**
- **Explain the relationship between latitude and the angle of the North Star (Polaris) above the horizon.**

MATERIALS

ruler
protractor
50- to 80-cm length of string

calculator
globe
world map or atlas

PRE-LAB VIDEO ▶ https://goo.gl/NqpUxa

Prepare for lab! Prior to attending your laboratory session, view the pre-lab video. Each video provides valuable background that will contribute to your understanding and success in lab.

INTRODUCTION

A *geographic coordinate system* is a system that enables every location on Earth to be specified by a set of numbers or letters. In this exercise you will learn about one of the most common coordinate systems, called *Earth's grid*, which uses lines of longitude and latitude. In addition, you will learn how this coordinate system is used to measure distances between places on Earth.

PART 5 Earth Science Skills

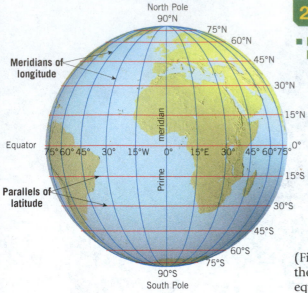

North Pole
90°N

Meridians of longitude

Equator

Parallels of latitude

Prime meridian

South Pole
90°S

▲ Figure 23.1 Earth's grid system. Parallels of latitude are shown in red; meridians of longitude are shown in blue.

23.1 Earth's Grid System

■ Distinguish between latitude and longitude and explain how latitude and longitude are used to locate places on Earth.

Globes and maps display a system of north–south and east–west lines, called **Earth's grid**, that forms the basis for locating points on Earth's surface (**Figure 23.1**). The grid is much like a large sheet of graph paper that has been laid over the surface of Earth Using the system is very similar to using a graph: The position of a point is determined by the intersection of two lines.

The *latitude* of a location on Earth is the distance of that location north or south of the **equator** (Figure 23.1). The lines (circles) of the grid that extend around Earth in an east–west direction are called **parallels of latitude**. As the name implies, these circles are parallel to one another. Two places on Earth, the **North Pole** and the **South Pole**, are exceptions; they are points of latitude rather than circles.

Longitude is the distance of a place east or west of the **prime meridian** (Figure 23.1). **Meridians of longitude** are lines (half circles) that extend from the North Pole to the South Pole. Adjacent meridians are farthest apart at the equator and converge (come together) toward the poles.

Earth's shape is nearly spherical. Since parallels and meridians mark distances on a sphere, their designation, like distance around a circle, is given in *degrees* (°). When more precision is necessary, a degree can be divided into 60 equal parts, called *minutes* ('), and a minute of an angle can be divided into 60 parts, called *seconds* ("). Thus, 31°10'20" means 31 degrees, 10 minutes, 20 seconds.

ACTIVITY 23.1
Earth's Grid System

1. Are lines of latitude oriented parallel to the equator or the prime meridian?

   _____

2. Are lines (circles) of latitude called parallels or meridians?

   _____

3. Are adjacent meridians farthest apart at the equator or at the poles?

   _____

4. Distances on meridians and parallels are given in *degrees* (°). Degrees of longitude or latitude can be subdivided into how many—10, 30, 60, or 100—equal parts called *minutes* (')? Minutes can be further divided into how many—10, 30, 60, or 100—equal parts, called *seconds* (")?

   Degrees can be subdivided into _____ equal parts, called *minutes* (').

   Minutes can be divided into _____ equal parts, called *seconds* (").

23.2 Determining Latitude

■ Distinguish between latitude and longitude and explain how latitude and longitude are used to locate places on Earth.

The equator is a circle drawn on a globe that is equally distant from the North Pole and the South Pole. It divides the globe into two equal halves, called **hemispheres**. The equator serves as the beginning point for determining latitude and is assigned the value 0° latitude.

Latitude is distance north and south of the equator, measured as an angle in degrees from the center of Earth (**Figure 23.2**). Latitude begins at the equator, extends north to the North Pole, designated 90°N latitude (a 90° angle measured north from the equator), and also extends south to the South Pole, designated 90°S latitude. The poles and all parallels of latitude, with the exception of the equator, must be designated either *N* (if they are *north* of the equator) or *S* (if they are *south* of the equator).

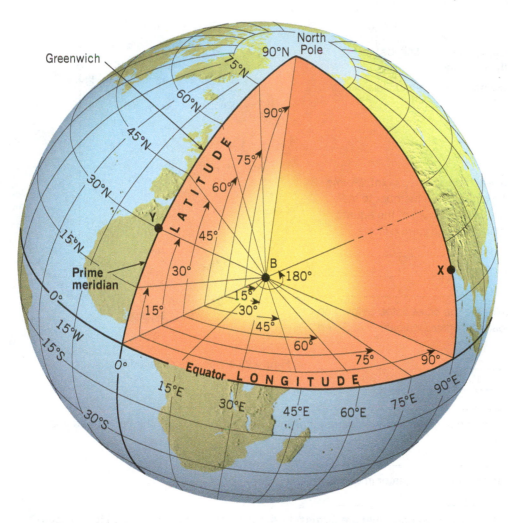

◀ **Figure 23.2** Measuring latitude and longitude. The angle measured from the equator to the center of Earth (B) and then northward to the parallel where Point Y is located is 30°. Therefore, the latitude of Point Y is 30°N. All points on the same parallel as Y are designated 30°N latitude. The angle measured from the prime meridian where it crosses the equator to the center of Earth (B) and then eastward to the meridian where Point X is located is 90°. Therefore, the longitude of Point X is 90°E. All points on the same meridian as X are designated 90°E longitude.

ACTIVITY 23.2
Determining Latitude

Figure 23.3 represents an idealized Earth, with Point B its center. Use Figure 23.3 and refer to Figure 23.2 to complete Questions 1–5.

1. Draw a line on the diagram in Figure 23.3 that represents the equator and label it.

2. Draw a line from Point B (the center of Earth) to Point C in the Northern Hemisphere. Using a protractor, measure the angle (∠ABC) from the equator to Point C.

   a. ∠ABC = _____ °

   b. The latitude of Point C is _____. (Remember to designate north with *N* or south with *S*.)

3. Draw a line on Figure 23.3 parallel to the equator that intersects Point C. All points on this line are the same latitude. Label this parallel of latitude in degrees N or S.

4. Using a protractor, measure ∠ABD on Figure 23.3. Then draw a line parallel to the equator that also goes through Point D. Label the line with its proper latitude.

5. Repeat the steps in Question 4 for ∠ABE.

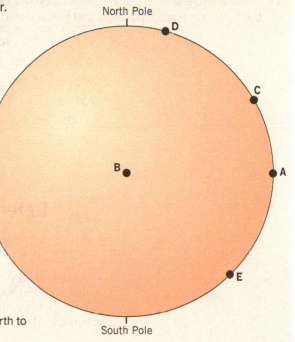

▶ **Figure 23.3** Idealized Earth to illustrate latitude.

continued

Activity 23.2 continued

6. Use the diagram that illustrates parallels of latitude in **Figure 23.4** to complete the following.

   a. Draw and label the following additional parallels of latitude on the figure:

   5°N latitude

   10°S latitude

   25°N latitude

   b. Write out the latitude for each designated point. (Points A and B are completed examples.) Remember to indicate whether the point is north or south of the equator by writing an *N* or *S* and include the word *latitude*.

   Point A: <u>30°N latitude</u>   Point D: _____

   Point B: <u>5°S latitude</u>   Point E: _____

   Point C: _____   Point F: _____

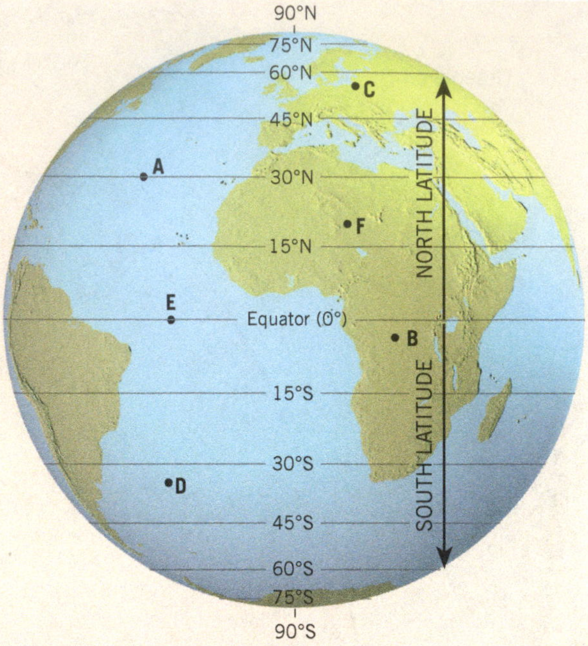

7. Use a globe or an atlas to locate the cities listed below and give their latitude to the nearest degree. Indicate *N* or *S* and include the word *latitude*.

   Moscow, Russia: _____

   Durban, South Africa: _____

   Your college campus city: _____

8. Using a globe or an atlas, give the name of a city or feature that is equally as far south of the equator as your college campus city is north.

   _____

9. What is the farthest one can be from the equator in degrees of latitude?

   _____ °

10. What are the names of the two places on Earth that are farthest from the equator to the north and to the south?

    _____

    _____

11. There are five special parallels of latitude marked and named on most globes. Use a globe or an atlas to locate the following special parallels and indicate the name given to each:

NAME OF PARALLEL

66°30′N latitude: _____

23°30′N latitude: _____

0°00′ latitude: _____

23°30′S latitude: _____

66°30′S latitude: _____

▲ **Figure 23.4** Parallels of latitude.

23.3 Determining Longitude

- Distinguish between latitude and longitude and explain how latitude and longitude are used to locate places on Earth.

Meridians are the north–south lines (half circles) on the globe that converge at the poles and are farthest apart along the equator. Notice on the globe that all meridians are alike. The choice of a zero, or beginning, meridian was arbitrary. The meridian that was chosen by international agreement in 1884 to be 0° longitude passes through the Royal Astronomical

Observatory at Greenwich, England, located near London. This internationally accepted reference for longitude is named the **prime meridian**.

Longitude is distance east and west of the prime meridian, measured as an angle in degrees (see Figure 23.5 below). Longitude begins at the prime meridian (0° longitude) and extends to the east and to the west, halfway around Earth, to the 180° meridian, which is directly opposite the prime meridian. All meridians, with the exception of the prime meridian and the 180° meridian, must be designated either E (if they are *east* of the prime meridian) or W (if they are *west* of the prime meridian).

ACTIVITY 23.3
Determining Longitude

1. Locate the prime meridian on a globe or map.

2. Meridians can be drawn at *any* interval. How many degrees of longitude separate each of the meridians on the globe or map you are using?

   _____ °

3. Refer to **Figure 23.5**. Write out the longitude for each designated point as was done for Point A below. Remember to indicate east (*E*) or west (*W*) of the prime meridian and include the word *longitude*.

   Point A: _30°E longitude_   Point D: _____

   Point B: _____   Point E: _____

   Point C: _____   Point F: _____

4. What is the approximate longitude of the city in which you were born?

   _____

5. Use a globe or an atlas to locate the places listed below and identify their longitude to the nearest degree. Indicate *E* or *W* and include the word *longitude*.

   Wellington, New Zealand: _____

   Honolulu, Hawaii: _____

   Your college campus: _____

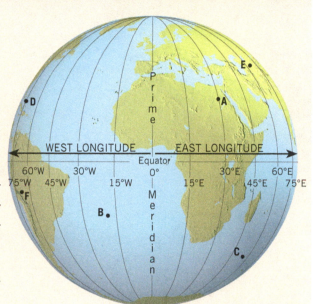

▲ **Figure 23.5** Meridians of longitude.

6. Using a globe or an atlas, identify a location that is at the same longitude as your college campus but in the opposite hemisphere.

   _____

7. The farthest a place can be east or west of the prime meridian is how many degrees of longitude—45, 90, 180, or 360?

   _____ °

23.4 Using Earth's Grid System

■ **Apply knowledge of Earth's grid system to accurately locate a place or feature.**

Earth's grid system consists of north–south lines, called *parallels of latitude*, and east–west lines, called *meridians of longitude*, that form the basis for locating points on Earth. When using this grid system, the position of a point is determined by the intersection of these two lines.

ACTIVITY 23.4

Using Earth's Grid System

1. Using **Figure 23.6**, record the latitude and longitude of each of the following points. Point A has been completed for reference. Remember to indicate the direction (*N, S, E,* or *W*) and *latitude* or *longitude*. Standard practice dictates that *latitude* is listed first, followed by *longitude*.

   Point A: ___30°N___ latitude, ___60°E___ longitude

   Point B: _____ latitude, _____ longitude

   Point C: _____ latitude, _____ longitude

   Point D: _____ latitude, _____ longitude

   Point E: _____ latitude, _____ longitude

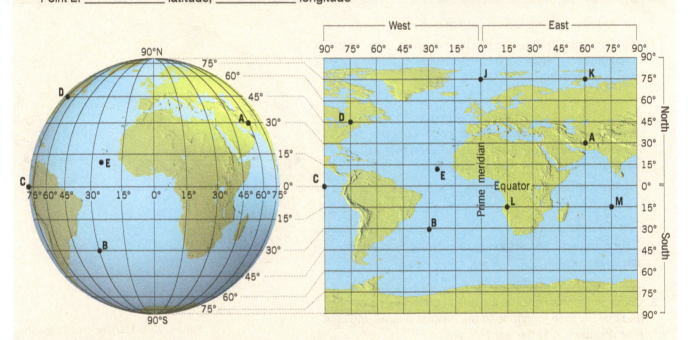

▲ **Figure 23.6** Locating places using Earth's grid system.

2. Locate the following points on Figure 23.6. Place a dot on the figure at the proper location and label each point with the designated letter.

   Point F: 15°S latitude, 75°W longitude

   Point G: 45°N latitude, 0° longitude

   Point H: 30°S latitude, 60°E longitude

   Point I: 0° latitude, 30°E longitude

3. Use a globe, a map, or an atlas to determine the latitude and longitude of the following cities.

   Kansas City, Missouri: _____

   Miami, Florida: _____

   Oslo, Norway: _____

   Auckland, New Zealand: _____

   Quito, Ecuador: _____

   Baghdad, Iraq: _____

4. Beginning with a world map and then proceeding to a regional map, determine the major city or significant feature at each of the following locations:

   19°28′N latitude, 99°09′W longitude: _____

   41°52′N latitude, 12°37′E longitude: _____

   1°30′S latitude, 33°00′E longitude: _____

23.5 Great Circles and Small Circles

■ Contrast great circles and small circles.

Any plane that is passed through the center of a sphere bisects that sphere (divides it into equal halves) and creates a **great circle** where it intersects the surface of the sphere (**Figure 23.7A**). Some characteristics of a great circle are:

- A great circle is the largest circle that can be drawn on a sphere; it divides the sphere into two equal parts, called *hemispheres*.
- A great circle can pass through any two places on Earth's surface. Because great circles do not necessarily follow parallels or meridians, an infinite number of them can be drawn on a globe.
- The shortest distance between two places on Earth's surface is along the great circle that passes through those two places.
- If Earth were a perfect sphere, 1° of angle along a great circle would cover identical distances everywhere. But Earth is slightly *flattened* at the poles and *bulges* at the equator, which results in slight differences in lengths of degrees. Generally, however, *1° of angle along a great circle equals approximately 111 kilometers, or 69 miles.*

Any circle on the globe that does not meet the characteristics of a great circle is considered a **small circle** (**Figure 23.7B**). A small circle *does not* divide the globe into two equal parts and *is not* the shortest distance between two places on Earth's surface.

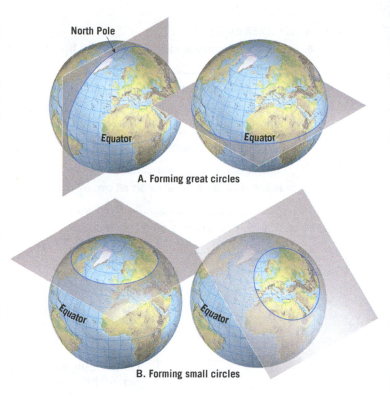

A. Forming great circles

B. Forming small circles

▲ **Figure 23.7 A.** A great circle results from the intersection of Earth's surface with any plane that passes through the center of Earth. **B.** A small circle results from the intersection of Earth's surface with any plane that does not pass through the center of Earth.

ACTIVITY 23.5A
Great Circles

Keeping in mind the characteristics of great circles, examine a globe and answer the following questions.

1. Estimate several great circles on the globe by wrapping a piece of string anywhere around the globe such that it divides the globe into *two equal halves*. Notice that there are an infinite number of great circles that can be marked on the globe.

2. Which parallel of latitude is a great circle?

_____

3. Each meridian of longitude is a half circle. If each meridian is paired with the meridian on the opposite side of the globe, a circle is formed. Which meridians, paired with their opposite meridians, are great circles?

_____

ACTIVITY 23.5B
Small Circles

Keeping in mind the characteristics of small circles, examine a globe and complete the following.

1. In general, which parallels of latitude are small circles?

_____

2. Which two latitudes are actually points rather than circles?

_____

continued

Activity 23.5B continued

3. In general, which meridians, paired with their opposite meridians, are small circles?

_____

4. Refer to the characteristics of great and small circles to complete the following.

a. Are all meridians halves of great or small circles?

_____ circles

b. Are all parallels, with the exception of the equator, great or small circles?

_____ circles

c. Is the equator a great or small circle?

_____ circle

d. Are the poles points or circles of latitude?

_____ of latitude

23.6 Determining Distance Along a Great Circle

■ Determine the shortest distance between any two places on Earth's surface.

Determining the distance between two places on Earth when both are on the same meridian, or the equator, requires two steps:

Step 1: Determine the number of degrees between the two places (degrees of longitude on the equator or degrees of latitude on a meridian).

Step 2: Multiply the number of degrees by 69 miles or 111 kilometers, the approximate number of miles or kilometers per degree for any great circle.

ACTIVITY 23.6A
Determining Distance Along a Meridian or the Equator

Use a globe and the steps outlined above to answer the following questions.

1. Approximately how many miles would you travel if you went from 10°W longitude to 40°E longitude along the equator?

_____ mi

2. Approximately how many kilometers is London, England, directly north of the equator?

_____ km

ACTIVITY 23.6B
Globe and String Method for Determining Distance

Determining the shortest distance between two places on Earth that are *not* both on the same meridian or the equator requires the following four steps:

Step 1: On a globe, determine the great circle that intersects both places.

Step 2: Stretch a piece of string along the great circle between the two places on the globe and mark the distance between them on the string with your fingers (Figure 23.8A).

Step 3: While still marking the distance with your fingers, place the string on the equator, with one end on the prime meridian. Determine the number of degrees along the great circle between the two places by measuring the marked string's length in degrees of longitude along the equator, which is also a great circle (Figure 23.8B).

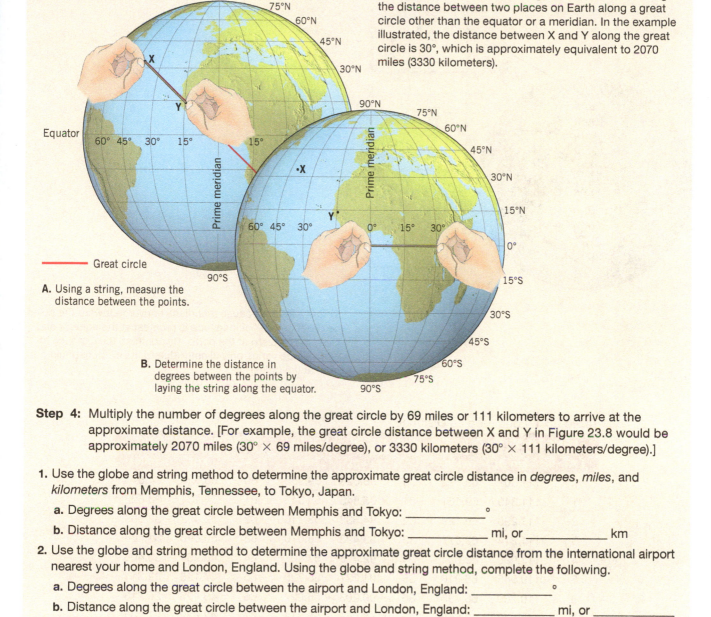

▼ **Figure 23.8** Globe and string method for determining the distance between two places on Earth along a great circle other than the equator or a meridian. In the example illustrated, the distance between X and Y along the great circle is 30°, which is approximately equivalent to 2070 miles (3330 kilometers).

—————— Great circle

A. Using a string, measure the distance between the points.

B. Determine the distance in degrees between the points by laying the string along the equator.

Step 4: Multiply the number of degrees along the great circle by 69 miles or 111 kilometers to arrive at the approximate distance. [For example, the great circle distance between X and Y in Figure 23.8 would be approximately 2070 miles (30° × 69 miles/degree), or 3330 kilometers (30° × 111 kilometers/degree).]

1. Use the globe and string method to determine the approximate great circle distance in *degrees*, *miles*, and *kilometers* from Memphis, Tennessee, to Tokyo, Japan.

   a. Degrees along the great circle between Memphis and Tokyo: _____ °

   b. Distance along the great circle between Memphis and Tokyo: _____ mi, or _____ km

2. Use the globe and string method to determine the approximate great circle distance from the international airport nearest your home and London, England. Using the globe and string method, complete the following.

   a. Degrees along the great circle between the airport and London, England: _____ °

   b. Distance along the great circle between the airport and London, England: _____ mi, or _____ km

   c. What would be the approximate flight time (from liftoff to landing) to London, England, assuming an average speed of 550 miles per hour?

   _____ hr

23.7 Determining Distance Along a Parallel

■ Describe how distance can be measured along a parallel of latitude.

Since all parallels except the equator are small circles, the length of 1° of longitude along a parallel other than the equator will always be *less* than 111 kilometers, or 69 miles (**Figure 23.9**). **Table 23.1** shows the length of 1° of longitude at various latitudes on Earth.

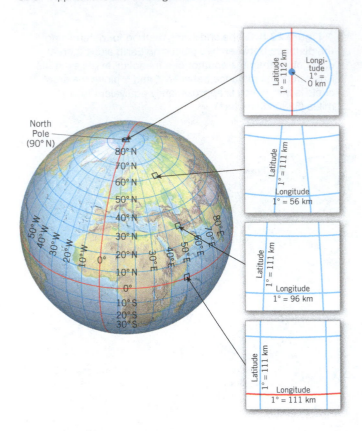

◀ **Figure 23.9** Because meridians converge toward the poles, the distance of 1° of longitude is greatest at the equator and diminishes to zero at the poles. By contrast, the distance of 1° of latitude varies only slightly (due to the slight flattening of Earth at the poles).

Table 23.1 Longitude as Distance

	LENGTH OF 1° LONG.			LENGTH OF 1° LONG.			LENGTH OF 1° LONG.	
°LAT.	km	miles	°LAT.	km	miles	°LAT.	km	miles
0	111.367	69.172	30	96.528	59.955	60	55.825	34.674
1	111.349	69.161	31	95.545	59.345	61	54.131	33.622
2	111.298	69.129	32	94.533	58.716	62	52.422	32.560
3	111.214	69.077	33	93.493	58.070	63	50.696	31.488
4	111.096	69.004	34	92.425	57.407	64	48.954	30.406
5	110.945	68.910	35	91.327	56.725	65	47.196	29.314
6	110.760	68.795	36	90.203	56.027	66	45.426	28.215
7	110.543	68.660	37	89.051	55.311	67	43.639	27.105
8	110.290	68.503	38	87.871	54.578	68	41.841	25.988
9	110.003	68.325	39	86.665	53.829	69	40.028	24.862
10	109.686	68.128	40	85.431	53.063	70	38.204	23.729
11	109.333	67.909	41	84.171	52.280	71	36.368	22.589
12	108.949	67.670	42	82.886	51.482	72	34.520	21.441
13	108.530	67.410	43	81.575	50.668	73	32.662	20.287
14	108.079	67.130	44	80.241	49.839	74	30.793	19.126
15	107.596	66.830	45	78.880	48.994	75	28.914	17.959
16	107.079	66.509	46	77.497	48.135	76	27.029	16.788
17	106.530	66.168	47	76.089	47.260	77	25.134	15.611
18	105.949	65.807	48	74.659	46.372	78	23.229	14.428
19	105.337	65.427	49	73.203	45.468	79	21.320	13.242
20	104.692	65.026	50	71.727	44.551	80	19.402	12.051
21	104.014	64.605	51	70.228	43.620	81	17.480	10.857

Table 23.1 Longitude as Distance   (*continued*)

°LAT.	km	miles	°LAT.	km	miles	°LAT.	km	miles
	LENGTH OF 1° LONG.			LENGTH OF 1° LONG.			LENGTH OF 1° LONG.	
22	103.306	64.165	52	68.708	42.676	82	15.551	9.659
23	102.565	63.705	53	67.168	41.719	83	13.617	8.458
24	101.795	63.227	54	65.604	40.748	84	11.681	7.255
25	100.994	62.729	55	64.022	39.765	85	9.739	6.049
26	100.160	62.211	56	62.420	38.770	86	7.796	4.842
27	99.297	61.675	57	60.798	37.763	87	5.849	3.633
28	98.405	61.121	58	59.159	36.745	88	3.899	2.422
29	97.481	60.547	59	57.501	35.715	89	1.950	1.211
30	96.528	59.955	60	55.825	34.674	90	0.000	0.000

ACTIVITY 23.7
Determining Distance Along a Parallel

1. Examine a globe. What do you observe about the distance around Earth along each parallel as you get farther away from the equator?

_____

2. Use Table 23.1 to determine the length of 1° of longitude at each of the following parallels:

LENGTH OF 1° OF LONGITUDE

15° latitude: _____ km, _____ mi

30° latitude: _____ km, _____ mi

45° latitude: _____ km, _____ mi

80° latitude: _____ km, _____ mi

3. Use the Earth's grid illustrated in Figure 23.6 to determine the distances between the following points:

Distance between Points D and G:

_____ degrees × mi/degree = _____ mi

Distance between Points B and H:

_____ degrees × km/degree = _____ km

4. Memphis, Tennessee, and Tokyo, Japan, are both located at about 35°N latitude. Use a globe or world map to determine how many degrees of longitude separate Memphis, Tennessee, and Tokyo, Japan.

_____° of longitude

5. From Table 23.1, what is the length of 1° of longitude at latitude 35°N?

_____ mi

6. How many miles is Tokyo, Japan, *directly* west of Memphis, Tennessee?

_____ mi

7. In Activity 23.6B, Question 1, you determined the great circle distance between Memphis, Tennessee, and Tokyo, Japan. How many miles shorter is the great circle route between these cities than the east–west distance along a parallel that you determined in Question 6?

The great circle route is _____ mi shorter.

23.8 Longitude and Solar Time

■ Explain the relationship between longitude and solar time.

Time can be kept in two ways. **Standard time**, the system used throughout most of the world, divides the globe into 24 standard time zones. Using standard time, when it is 3:27 P.M. in New York City, it is also 3:27 P.M. in Baltimore (same time zone), but it is 2:27 P.M. in Chicago and 1:27 P.M. in Denver (different time zones).

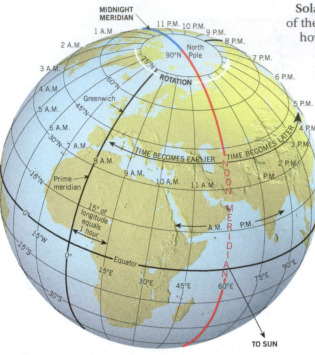

▲ **Figure 23.10** Determining longitude using solar time.

Solar time, or "Sun time," is a measure of time based on the position of the Sun. Solar time, like standard time, is based on a 24-hour day; however, noon at any location occurs when the Sun is at its highest point each day. Midnight in solar time occurs 12 hours later.

The following will assist you in understanding the relationship between solar time and longitude:

- Earth rotates on its axis from west to east (eastward), or counterclockwise when viewed from above the North Pole (**Figure 23.10**).
- It is noon on the meridian that is directly facing the Sun (the Sun has reached its highest position in the sky, called the *zenith*), and it is midnight on the meridian on the opposite side of Earth.
- The time interval from noon on one day to noon on the next day averages 24 hours and is known as the *mean solar day*.
- Earth turns through 360° of longitude in 1 mean solar day, which is equivalent to 15° of longitude per hour, or 1° of longitude every 4 minutes.
- Places that are east or west of each other, regardless of the distance, have different solar times. For example, people located to the east of the noon meridian have already experienced noon; their time is afternoon (P.M.—*post* ["after"] *meridiem*). People living west of the noon meridian have yet to reach noon; their time is before noon (A.M.—*ante* ["before"] *meridiem*. *Time becomes later going eastward and earlier going westward.*

ACTIVITY 23.8A
Longitude and Solar Time

Use the relationship between solar time and longitude described above to answer the following questions. Circle either A.M. or P.M. with each answer.

1. What would be the solar time of a person living 15° of longitude west of the noon meridian?

   _____ (A.M., P.M.)

2. What would be the solar time of a person living 1° of longitude west of the noon meridian?

   _____ (A.M., P.M.)

3. What would be the solar time of a person located 4° of longitude east of the noon meridian?

   _____ (A.M., P.M.)

4. If it is noon, solar time, at 70°W longitude, what is the solar time at each of the following locations?

   72°W longitude: _____

   65°W longitude: _____

   90°W longitude: _____

   110°E longitude: _____

ACTIVITY 23.8B
Using a Chronometer to Determine Longitude

The invention of reliable clocks, called *chronometers*, that could keep time on lengthy voyages over rough seas allowed navigators to *accurately* determine their east–west position, or longitude, for the first time. Using solar time and reliable clocks, sailors navigated the vast oceans with confidence regarding their position and the position of any location along their route that they might want to revisit. (We will examine how sailors determined latitude in the next section.)

A shipboard chronometer is set to keep the time at a known place on Earth, usually the prime meridian. If it is noon by the Sun where the ship is located, and at that same instant the chronometer indicates that it is 8 A.M. on the prime meridian, the ship must be 60° longitude from the prime meridian (4 hours difference × 15° per hour), and it must be located to the east because the ship's time is later (Figure 23.10). The difference in time need not be in whole hours. A 30-minute difference in time between two places would be equivalent to 7.5° of longitude, 20 minutes would equal 5°, and so forth.

1. Imagine that you are on a ship, and based on the position of the Sun, you determine that it is exactly noon on board. What is the ship's longitude if, at that instant, the time at the prime meridian is each of the following? (*Note:* It may be helpful to sketch a diagram showing the prime meridian, the ship's location east or west of the prime meridian, and the difference in hours.)

   a. 6:00 P.M.: _____

   b. 1:00 A.M.: _____

   c. 2:30 P.M.: _____

23.9 Latitude and the North Star

■ **Explain the relationship between latitude and the angle of the North Star (Polaris) above the horizon.**

Today, most ships use Global Positioning System (GPS) navigational satellites to determine their location. However, early explorers used the angle of the North Star (a star named Polaris) above the horizon to determine their north–south position in the Northern Hemisphere. As shown in **Figure 23.11**, someone standing at the North Pole would look overhead (at a 90° angle above the horizon) to see Polaris. The person's latitude would be 90°N. In contrast, someone standing on the equator, 0° latitude, would observe Polaris on the horizon (at a 0° angle above the horizon).

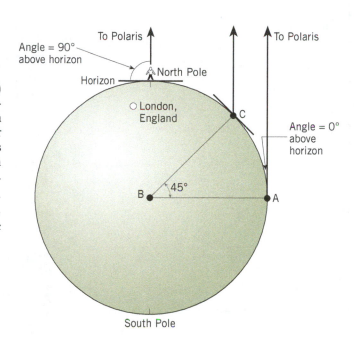

▶ **Figure 23.11** Using the North Star (Polaris) to determine latitude.

ACTIVITY 23.9

Latitude and the North Star

Use Figure 23.11 to answer the following questions.

1. Is the angle of Polaris 45°, 90°, or 180° above the horizon for someone standing at Point C?

   _____ °

2. Describe the relationship between a particular latitude and the angle of Polaris above the horizon at that latitude.

   _____

   _____

3. What is the angle of Polaris above the horizon at the following cities?

   Fairbanks, Alaska: _____ °

   St. Paul, Minnesota: _____ °

   New Orleans, Louisiana: _____ °

MasteringGeology™

Looking for additional review and lab prep materials? Go to www.masteringgeology.com for Pre-Lab Videos, Geoscience Animations, RSS Feeds, Key Term Study Tools, The Math You Need, an optional Pearson eText, and more.

Notes and calculations:

Location and Distance on Earth

Name _____   Course/Section _____

Date _____   Due Date _____

1. On **Figure 23.12**, prepare a diagram illustrating Earth's grid system. Include and label the equator and prime meridian. Explain the system used for locating points on the surface of Earth. Refer to Figure 23.1, page 370, if necessary. _____

_____

_____

▲ **Figure 23.12** Diagram of Earth's grid system to accompany Question 1.

2. Define the following terms:

   Parallel of latitude: _____

   _____

   Meridian of longitude: _____

   Great circle: _____

3. Determine whether each of the following statements is true or false. If a statement is false, reword it to make it a true statement.

   T F **a.** The distance measured north or south of the prime meridian is called latitude.

   T F **b.** Pairing any meridian with its opposite meridian on Earth forms a great circle.

   T F **c.** The equator is the only meridian that is a great circle.

4. Approximately how many miles does 1° equal along a great circle?

   _____ mi

5. What are the latitude and longitude of your home city?

   Latitude: _____

   Longitude: _____

6. Use a globe or map to determine, as accurately as possible, the latitude and longitude of Athens, Greece.

   Latitude: _____

   Longitude: _____

7. Write a brief paragraph describing how to determine the shortest distance between two places on Earth's surface.

   _____

   _____

   _____

8. Approximately how many miles is it from London, England, to the South Pole? (Show your calculation.)

   _____ mi

9. Using **Figure 23.13**, determine and record the latitude and longitude of each of the following points.

   Point A: _____ latitude, _____ longitude

   Point B: _____ latitude, _____ longitude

   Point C: _____ latitude, _____ longitude

   Point D: _____ latitude, _____ longitude

   Point E: _____ latitude, _____ longitude

10. Imagine that you are on a life raft in the Atlantic Ocean, somewhere between London, England, and New York, New York. Fortunately, you managed to save your

globe. You were most recently in London, so your watch is still set for London time. It is noon, solar time, at your location, but your watch indicates that it is 4 P.M. in London. Are you closer to London or to New York City? Explain how you arrived at your answer.

_____

_____

_____

11. What is the relationship between the latitude of a place in the Northern Hemisphere and the angle of Polaris above the horizon at that place?

_____

_____

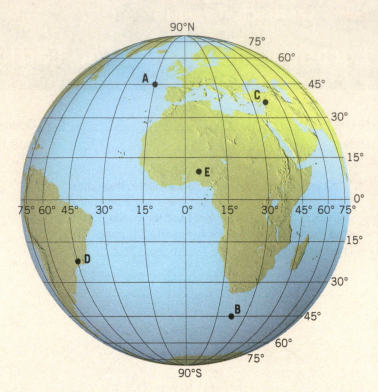

▲ **Figure 23.13** Locating places using Earth's grid to accompany Question 9.

The Metric System, Measurements, and Scientific Inquiry

LEARNING OBJECTIVES

Each statement represents an important learning objective that relates to one or more sections of this lab. After you complete this exercise you should be able to:

- **List the metric system units for length, volume, and mass and measure various items using the metric system.**
- **Convert units within the metric system.**
- **Convert between the metric and English systems of measurement.**
- **Compare Celsius, Fahrenheit, and Kelvin temperature scales.**
- **Understand and use micrometers or nanometers to measure very small objects and astronomical units or light-years to measure great distances.**
- **Determine the approximate density and specific gravity of solid substances.**
- **Conduct a scientific experiment using accepted methods of scientific inquiry.**

MATERIALS

metric ruler	thread	large graduated cylinder
small rock	paper clip	metric balance
calculator	paper cup	small graduated cylinder
measuring tape or meterstick	quarter coin	

PRE-LAB VIDEO https://goo.gl/tXre7A

 Prepare for lab! Prior to attending your laboratory session, view the pre-lab video. Each video provides valuable background that will contribute to your understanding and success in lab.

INTRODUCTION

Earth science involves studying a vast array of phenomena—from the smallest atoms to seemingly immeasurable galaxies. A critical component of scientific investigation is the use of accurate measurements—specifically units of measurement appropriate for the particular feature or phenomenon being studied. For example, scientists use

METER (m)

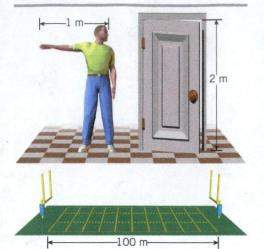

LITER (L)

GRAM (g)

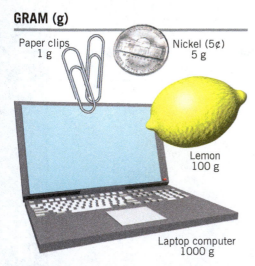

▲ **Figure 24.1** The metric unit of length is the meter (m), the basic unit in the metric system for volume is the liter (L), and the basic unit of mass is the gram (g).

millimeters or centimeters to measure the size of a mineral grain, and they use meters or kilometers to describe the elevation of a mountain peak.

24.1 The Metric System

■ **List the metric system units for length, volume, and mass and measure various items using the metric system.**

The fundamental units of measurement used in science have been established by the *International System of Units* (SI, Système International d'Unités) (**Table 24.1**). You are probably more familiar with the term *metric system*, which is used as a synonym for the International System of Units.

The **metric system** uses only one base unit for each type of measurement. The most commonly used base units are the **meter (m)** as the unit of length, the **liter (L)** as the unit of volume, and the **gram (g)** as the unit of mass (**Figure 24.1**). Some of the other base metric units are also shown in Table 24.1. The metric system is a decimal system (based on fractions or multiples of 10) and, as such, is similar to the U.S. monetary system, where 10 pennies = 1 dime and 10 dimes = 1 dollar. **Table 24.2** lists the prefixes used in the metric system to indicate how many times more (in multiples of 10) or what fraction (in fractions of 10) of the basic unit is present. Therefore, from the information in Table 24.2, you see that 1 *kilo*gram (kg) = 1000 grams, while 1 *milli*gram (mg) = 1/1000 gram.

Table 24.1 Base Units of the SI[1]

QUANTITY MEASURED	UNIT	SYMBOL
Length	Meter	m
Volume	Liter	L
Mass	Gram	g
Thermodynamic temperature	Kelvin	K
Time	Second	s
Quantity of a substance	Mole	mol
Luminous intensity	Candela	cd

[1]From these base units, other units are derived to express quantities such as power (watt [W]), force (newton [N]), energy (joule [J]), and pressure (pascal [Pa]).

Table 24.2 Metric Prefixes and Symbols

PREFIX	SYMBOL[1]	MEANING
giga-	G	1 billion times base unit (1,000,000,000 × base)
mega-	M	1 million times base unit (1,000,000 × base)
kilo-	k	1 thousand times base unit (1000 × base)
hecto-	h	1 hundred times base unit (100 × base)
deka-	da	10 times base unit (10 × base)
BASE UNIT	m (meter)–length L (liter)–volume g (gram)–mass	
deci-	d	one-tenth the base unit (0.1 × base)
centi-	c	one-hundredth the base unit (0.01 × base)
milli-	m	one-thousandth the base unit (0.001 × base)
micro-	μ	one-millionth the base unit (0.000001 × base)
nano-	n	one-billionth the base unit (0.000000001 × base)

[1]When writing in the SI system, periods are not used after the unit symbols, and symbols are not made plural. For example, a length of 50 centimeters would be written as *50 cm*, not *50 cm.* or *50 cms*.

ACTIVITY 24.1

Working with the Metric System

To familiarize yourself with metric units, determine the following measurements with the equipment provided in the lab.

Measuring length:

1. Use a metric measuring tape (or meterstick) to measure your height, in centimeters. _____ cm
2. Use a metric ruler to measure the length of this page to the nearest millimeter. _____ mm
3. Use a metric ruler to measure the length of your shoe to the nearest millimeter. _____ mm

Measuring volume:

4. Use a graduated cylinder to measure the volume of a paper cup to the nearest milliliter. _____ mL

Measuring mass:

5. Two terms that are frequently confused are *mass* and *weight*. Mass is a measure of the amount of matter an object contains. Weight is a measure of the force of gravity on an object. For example, the mass of an object would be the same on both Earth and the Moon. However, because the gravitational force of the Moon is less than that of Earth, the object would weigh less on the Moon. Use a metric balance to measure the mass/weight of each of the following objects. (Follow your instructor's directions for using a metric balance.)

    a. Your lab book: _____ g

    b. A quarter: _____ g

    c. Your shoe: _____ g

    d. Sample of rock: _____ g

    e. Paper clip: _____ g

24.2 Metric Conversions

■ Convert units within the metric system.

One important advantage of the metric system is that it is based on multiples of 10. As shown on the metric conversion diagram in **Figure 24.2**, conversion from one unit to another can be accomplished simply by moving the decimal point to the *left if converting to larger units* or by moving the decimal point to the *right if converting to smaller units.*

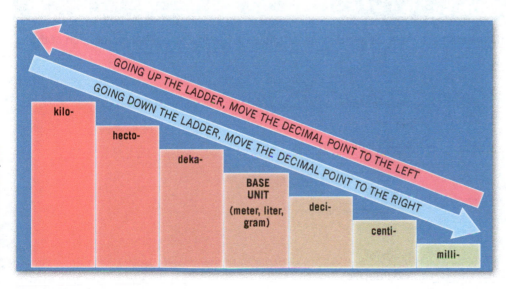

◀ **Figure 24.2** Metric conversion diagram. Beginning at the appropriate step, if going to larger units, move the decimal to the left for each step crossed. When going to smaller units, move the decimal to the right for each step crossed. For example, 1.253 meters (base unit step) would be equivalent to 1253.0 millimeters (decimal moved three steps to the right).

For example, assume that you measure the length of a line and find that it is 12.4 centimeters. In order to convert this length to millimeters, start with 12.4 on the "centi-" step of the diagram. Then move the decimal one place (steps) to the right (the "milli-" step). The length, in millimeters, is 124 millimeters.

ACTIVITY 24.2
Conversions in the Metric System

1. Use the metric conversion diagram in Figure 24.2 to convert the following.

   a. 2.05 meters (m) = _____ centimeters (cm)

   b. 1.50 meters (m) = _____ millimeters (mm)

   c. 9.81 liters (L) = _____ deciliters (dL)

   d. 5.4 grams (g) = _____ milligrams (mg)

   e. 6.8 meters (m) = _____ kilometer (km)

   f. 4214.6 centimeter (cm) = _____ meters (m)

   g. 321.50 grams (g) = _____ kilogram (kg)

   h. 70.73 hectoliter (hL) = _____ dekaliters (daL)

2. Use a metric tape measure (or meterstick) to determine the length of your laboratory table as accurately as possible, to the nearest hundredth of a meter.

   a. Length of table: _____ m

   b. Convert the length of your lab table into millimeters, centimeters, and kilometers:

   _____ mm

   _____ cm

   _____ km

24.3 Metric–English Conversions

■ Convert between the metric and English systems of measurement.

Although the metric system is globally accepted as the standard system of measurement, the United States continues to use the English system as the everyday measure of length, volume, and mass. Therefore, it is useful to be familiar with both systems and to be able to convert from one to the other.

ACTIVITY 24.3
Using Conversion Tables

Use the conversion tables on the inside back cover of this manual to figure out the metric equivalent for each of the following units.

Length conversion:

1. 1 inch (in) = _____ centimeters (cm)

2. 1 meter (m) = _____ feet (ft)

3. 1 mile (mi) = _____ kilometers (km)

Volume conversion:

4. 1 gallon (gal) = _____ liters (L)

5. 1 cubic inch (in) = _____ cubic centimeters (cm)

Mass conversion:

6. 1 ounce (oz) = _____ gram (g)

7. 1 pound (lb) = _____ kilogram (kg)

24.4 Temperature Scales

■ Compare Celsius, Fahrenheit, and Kelvin temperature scales.

We use different systems of temperature measurement to describe the same phenomena. On the Fahrenheit temperature scale, 32°F is the melting point of ice, and 212°F marks the boiling point of water (at standard atmospheric pressure). On the **Celsius scale**, ice melts at 0°C, and water boils at 100°C. On the **Kelvin scale**, ice melts at 273K.

Conversion from one temperature scale to another can be accomplished using either an equation or a graphic comparison scale. A graphic comparison scale can be found on the inside back cover of the lab manual. To convert Celsius degrees to Fahrenheit degrees, the equation is °F = (1.8)°C + 32°. To convert Fahrenheit degrees to Celsius degrees, the equation is °C = (°F − 32)/1.8. To convert Kelvins (K) to Celsius degrees, subtract 273 and add the degree symbol.

ACTIVITY 24.4
Temperature Scales

1. Convert the following temperatures to their equivalents. Do the first four conversions using the appropriate equation, and do the others using the temperature comparison scale on the inside back cover of this manual.

   a. On a cold day, it was 8°F = _____ °C.

   b. Ice melts at 0°C = _____ °F.

   c. Room temperature is 72°F = _____ °C.

   d. A hot summer day was 35°C = _____ °F.

   e. Normal body temperature is 98.6°F = _____ °C.

   f. A warm shower is 27°C = _____ °F.

   g. Hot soup is 72°C = _____ °F.

   h. Water boils at 212°F = _____ K.

2. Use the temperature comparison scale on the inside back cover to answer the following:

   a. The thermometer reads 28°C. Will you need your winter coat? _____

   b. The thermometer reads 10°C. Will the outdoor swimming pool be open today? _____

   c. If your body temperature is 40°C, do you have a fever? _____

   d. The temperature of a cup of cocoa is 90°C. Will it burn your tongue? _____

   e. Your bath water is 15°C. Will you have a scalding, warm, or chilly bath? _____

   f. The thermostat in your home reads 37°C. Are you shivering or perspiring? _____

24.5 Other Units of Measurement

■ Understand and use micrometers or nanometers to measure very small objects and astronomical units or light-years to measure great distances.

Scientists have developed other metric units to measure extremely tiny objects and still others to measure very great distances. Two units commonly used to measure microscopic objects are the *micrometer*, also known as the *micron*, and the *nanometer*. By contrast, astronomers must measure very great distances, such as the distances between planets or the distances to various stars and galaxies. To express such measurements, scientists use *astronomical units* and *light-years*.

ACTIVITY 24.5A
Measuring Very Small Objects

By definition, 1 **micrometer (μm)** equals 0.000001 m (one-millionth of a meter); therefore, there are 1 million micrometers in 1 meter and 10,000 micrometers in 1 centimeter. A **nanometer (nm)** is an even smaller unit that equals 0.000000001 meter (one-billionth of a meter).

continued

Activity 24.5A continued

1. Are there 10, 100, or 1000 nanometers in a micrometer? _____

2. What is the length of a 2.5-centimeter line, expressed in micrometers and nanometers?

_____ μm

_____ nm

3. Some forms of radiation (for example, light) have very small wavelengths, with distances from crest to crest of about 500 nanometers. How many of these waves would it take to equal 1 centimeter?

_____ waves in 1 cm

ACTIVITY 24.5B
Measuring Very Great Distances

The **astronomical unit (AU)** is used for measuring distances within the solar system. One astronomical unit is equal to the average distance of Earth from the Sun, about 150 million kilometers, or about 93 million miles. The **light-year (LY)** is used for measuring distances to the stars and beyond and is defined as the distance that light travels in a vacuum in 1 year—about 6 trillion (6,000,000,000,000) miles.

1. Refer to the scale along the bottom of **Figure 24.3** to determine how many AUs and/or fractions of an AU separate Earth from these planets:

a. Mercury: _____ AUs from Earth

b. Mars: _____ AUs from Earth

c. Jupiter: _____ AUs from Earth

d. Neptune: _____ AUs from Earth

2. The planet Saturn is 1427 million kilometers from the Sun. How many AUs is Saturn from the Sun?
_____ AUs from the Sun

3. Approximately how many kilometers will light travel in 1 year? _____ km per year

4. The nearest star to Earth, excluding our Sun, is named Proxima Centauri. It is about 4.27 light-years away. What is the distance of Proxima Centauri from Earth in both miles and kilometers?

_____ mi, or _____ km

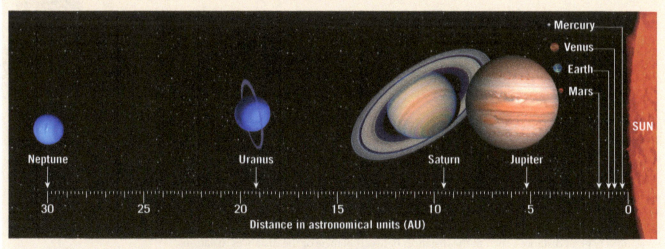

▲ **Figure 24.3** Positions of the planets shown to scale using astronomical units (AU).

24.6 # Density and Specific Gravity

■ **Determine the approximate density and specific gravity of solid substances.**

Two closely related properties of materials, density and specific gravity, are described below:

- **Density** is the mass of a substance per unit volume, usually expressed in grams per cubic centimeter (g/cm^3).

- The **specific gravity** of a solid is the *ratio* of the mass of a given volume of a substance to the mass of an equal volume of some other substance, usually water, at 4°C. For example, granite, a common rock in Earth's crust, has a specific gravity of about 2.7, which means that granite has 2.7 times more mass than an equal volume of water.
- Because specific gravity is a ratio, it is expressed as a number that has no units.
- The specific gravity of a substance is numerically equal to its density because the density of pure water is 1 g/cm^3. Stated another way, granite, which has a specific gravity of 2.7, has a density of 2.7 g/cm^3.

ACTIVITY 24.6
Density and Specific Gravity

The approximate density and specific gravity of a rock or another solid can be determined following the steps below:

Step 1: Determine the mass of the rock, using a metric balance.

Step 2: Fill a graduated cylinder that has its divisions marked in milliliters approximately two-thirds full with water. Note the level of the water in the cylinder in milliliters.

Step 3: Tie a thread to the rock and immerse the rock in the water in the graduated cylinder. Note the new level of the water in the cylinder. (*Note:* You can avoid using the thread by gently sliding the rock into the cylinder without it.)

Step 4: Determine the difference between the beginning level and the after-immersion level of the water in the cylinder.

Step 5: Calculate the density and specific gravity by using the following information and equations.

A milliliter of water has a volume equal to 1 cubic centimeter (cm^3). The difference between the beginning water level and the after-immersion water level in the cylinder equals the volume of the rock, in cubic centimeters. Furthermore, *1 cubic centimeter (1 milliliter) of water has a mass of approximately 1 gram.* Therefore, the difference between the beginning water level and the after-immersion water level in the cylinder is the mass of a volume of water equal to the volume of the rock.

1. Refer to steps 1–5 to determine the density and specific gravity of a small rock sample.

   a. Mass of rock sample: _____ g

   b. Beginning level of water in cylinder: _____ mL

   c. After-immersion level of water: _____ mL

   d. Volume of water displaced by the rock: _____ mL, or _____ cm^3

   e. Volume of rock sample: _____ cm^3

   f. Mass of the water displaced by the rock: _____ g

   g. Density of rock:

   $$\text{Density} = \frac{\text{Mass of rock (g)}}{\text{Volume of rock (cm}^3\text{)}} = \underline{\hspace{3cm}} \text{ g/cm}^3$$

   h. Specific gravity of rock:

   $$\text{Specific gravity} = \frac{\text{Mass of rock (g)}}{\text{Mass of an equal volume of water (g)}} = \underline{\hspace{3cm}}$$

2. As a means of comparison, your instructor may ask you to determine the density and/or specific gravity of other objects. If so, record your results in the following spaces.

   a. Object: _____

      Density: _____ g/cm^3

      Specific gravity: _____

   b. Object: _____

      Density: _____ g/cm^3

      Specific gravity: _____

■ Conduct a scientific experiment using accepted methods of scientific inquiry.

Scientists use a variety of methods to understand natural phenomena, including the following fundamental sequence of steps:

Step 1: Establish a hypothesis—a proposed explanation for a phenomenon that requires further testing.
Step 2: Gather data and conduct experiments to test the hypothesis.
Step 3: Accept, modify, or reject the hypothesis, on the basis of extensive data collection and/or experimentation.

Scientific inquiry requires that experiments be repeated many times by different people before results are widely accepted by the scientific community.

ACTIVITY 24.7
Testing a Hypothesis

The following activity is intended to enhance your understanding of scientific inquiry.

Step 1—Establishing a Hypothesis

Observe all the people in the laboratory and pay particular attention to each individual's height and shoe length.

1. Based on your observations, write a *hypothesis* that relates a person's height to his or her shoe length.
   Hypothesis: _____
   _____
   _____

Step 2—Gathering Data

In Questions 1 and 3 of Activity 24.1, each person in the laboratory measured his or her height and shoe length.

2. *Gather your data* by asking 10 or more people in the lab for their height and shoe length measurements. Enter your data in Table 24.3, recording height to the nearest hundredth of a meter and shoe length to the nearest millimeter.

Step 3—Evaluating the Hypothesis Based Upon the Data

3. *Plot the data* from Table 24.3 on the height versus shoe length graph in Figure 24.4 by locating a person's height on the vertical axis and his or her shoe length on the horizontal axis. Then place a dot on the graph where the two intersect.

Table 24.3 Data Table for Recording Height and Shoe Length Measurements of People in the Lab

PERSON	SHOE LENGTH (NEAREST MILLIMETER)	HEIGHT (NEAREST HUNDREDTH OF A METER)
1	_____ mm	_____ m
2	_____ mm	_____ m
3	_____ mm	_____ m
4	_____ mm	_____ m
5	_____ mm	_____ m
6	_____ mm	_____ m
7	_____ mm	_____ m
8	_____ mm	_____ m
9	_____ mm	_____ m
10	_____ mm	_____ m
11	_____ mm	_____ m
12	_____ mm	_____ m
13	_____ mm	_____ m
14	_____ mm	_____ m
15	_____ mm	_____ m
16	_____ mm	_____ m
17	_____ mm	_____ m
18	_____ mm	_____ m

PERSON	SHOE LENGTH (NEAREST MILLIMETER)	HEIGHT (NEAREST HUNDREDTH OF A METER)
19	_____ mm	_____ m
20	_____ mm	_____ m

4. Describe the pattern of the data points (dots) on the height versus shoe length graph in Figure 24.4. For example, are the points scattered all over the graph, or do they appear to follow a line or trend?

_____

_____

5. Draw a single straight line on the graph that appears to average, or best fit, the pattern of the data points.

6. Describe the relationship of height to shoe length that is illustrated by the line on your graph.

_____

_____

_____

7. Ask several people whose height and shoe length *are not* part of your data set for their height. Then see how accurately your line predicts what their shoe length should be. Do this by marking each person's height on the vertical axis and then follow a line straight across to the right until you intersect the line on the graph. Read the predicted shoe length from the axis directly below the point of intersection.

8. Summarize how accurately your graph predicts a person's shoe length, given *only* his or her height.

_____

_____

_____

9. Using your graph's ability to make predictions as a guide, do you think you should accept, reject, or modify your original hypothesis? Give the reason(s) for your choice.

_____

_____

_____

10. Why would your ability to make a more precise or accurate prediction have been better if you had used the heights and shoe lengths of 10,000 people to construct your graph?

_____

_____

_____

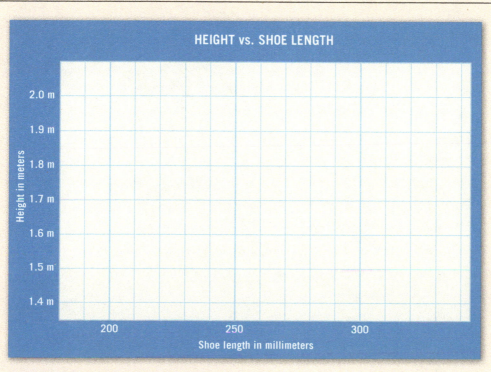

▲ **Figure 24.4** Height versus shoe length graph.

MasteringGeology™

Looking for additional review and lab prep materials? Go to www.masteringgeology.com for Pre-Lab Videos, Geoscience Animations, RSS Feeds, Key Term Study Tools, The Math You Need, an optional Pearson eText, and more.

Notes and calculations:

The Metric System, Measurements, and Scientific Inquiry

Name _____

Course/Section _____

Date _____

Due Date _____

1. List the basic metric unit and symbol used for each of the following measurements:

   Length: _____

   Mass: _____

   Volume: _____

2. Convert the following units:

   a. 2 liters (L) = _____ deciliters (dL)

   b. 600 millimeters (mm) = _____ meter (m)

   c. 72°F = _____ °C

   d. 0.32 kilogram (kg) = _____ grams (g)

   e. 12 grams (g) = _____ milligrams (mg)

3. How many micrometers are there in 3.0 centimeters?

   _____ μm

4. How many waves, each 500 nanometers long, would fit along a 2-centimeter line?

   _____ waves

5. What is the distance, in kilometers, to a star that is 6.5 light-years from Earth?

   _____ km

6. Uranus, one of the most distant planets in our solar system, is 2870 million kilometers from the Sun. What is its distance from the Sun, in astronomical units?

   _____ AU

7. Explain the difference between *density* and *specific gravity*.

   _____

   _____

   _____

8. At the conclusion of the height versus shoe length experiment, did you accept, reject, or modify your original hypothesis? Explain.

   _____

   _____

   _____

9. Use what you have learned about the metric system to determine whether the following statements are *reasonable*. Write *yes* or *no* in each blank. *Do not* convert these units to English equivalents; just *estimate* the value of each.

   a. A man weighs 90 kilograms. _____

   b. A typical classroom door is 1 meter tall. _____

   c. A college student drank 3 kiloliters of coffee last night. _____

   d. The room temperature is 295K. _____

   e. A dime is 1 millimeter thick. _____

   f. Sugar is sold by the milligram. _____

   g. The temperature in Paris today is 80°C. _____

   h. The bathtub has 80 liters of water in it. _____

   i. You will need a coat if the outside temperature is 30°C. _____

   j. A pork roast weighs 18 grams. _____

The Geologic Time Scale

Relative Time Span		
Phanerozoic	Cenozoic	
	Mesozoic	
	Paleozoic	
Precambrian	Proterozoic	
		2500
	Archean	
		~4000
	Hadean*	
		~4600

Era	Period	Epoch	Millions of years ago	Development of Plants and Animals
Cenozoic	Quaternary	Holocene	0.01	Humans develop
		Pleistocene	2.6	
	Tertiary — Neogene	Pliocene	5.3	Large mammals flourish
		Miocene	23.0	
	Tertiary — Paleogene	Oligocene	33.9	
		Eocene	56.0	Extinction of dinosaurs and many other species
		Paleocene	66.0	
Mesozoic	Cretaceous			First known flowering plants
	Jurassic		145.0	First known birds
	Triassic		201.3	Dinosaurs flourish / First known mammals
			252.1	
Paleozoic	Permian		298.9	Extinction of trilobites and many other marine animals
	Carboniferous — Pennsylvanian		323.2	First known reptiles / Large coal swamps
	Carboniferous — Mississippian		358.9	Amphibians abundant
	Devonian		419.2	First insect fossils / Fishes dominant
	Silurian		443.8	First land plants
	Ordovician		485.4	First known fishes / Cephalopods abundant
	Cambrian		541.0	Trilobites abundant / First organisms with shells / First multicelled organisms
Precambrian			~4600	First one-celled organisms / Origin of Earth

* Hadean is the informal name for the span that begins at Earth's formation and ends with Earth's earliest-known rocks.